Agrobiodiversity Management for Food Security

A Critical Review

Dedication

For Dandylion

Agrobiodiversity Management for Food Security

A Critical Review

J.M. Lenné and D. Wood

North Oldmoss Croft, Fyvie, Turriff, Aberdeenshire, UK

www.cabi.org

CABI is a trading name of CAB International

CABI Head Office
Nosworthy Way
Wallingford
Oxfordshire, OX10 8DE
UK

Tel: +44 (0)1491 832111
Fax: +44 (0)1491 833508
E-mail: cabi@cabi.org
Website: www.cabi.org

CABI North American Office
875 Massachusetts Avenue
7th Floor
Cambridge, MA 02139
USA

Tel: +1 617 395 4056
Fax: +1 617 354 6875
E-mail: cabi-nao@cabi.org

A catalogue record for this book is available from the British Library, London, UK.

Library of Congress Cataloging-in-Publication Data

Agrobiodiversity management for food security : a critical review / J. M. Lenné and D. Wood [editors].
p. cm.
Includes bibliographical references and index.
ISBN 978-1-84593-761-4 (alk. paper)
1. Agrobiodiversity. 2. Food security. I. Lenné, Jillian M. II. Wood, D. (David), 1939-
S494.5.A43A475 2011
631.5′8--dc22

2010043978

ISBN-13: 978 1 84593 761 4

Commissioning Editor: Rachel Cutts
Editorial Assistant: Alexandra Lainsbury
Production Editor: Fiona Chippendale

Typeset by Columns Design XML Limited, Reading, Berkshire.
Printed and bound in the UK by Antony Rowe

Contents

Contributors

Jonathan Gressel, Plant Sciences, Weizmann Institute of Science, Rehovot 76100, Israel. E-mail: jonathan.gressel@weizmann.ac.il

Kenneth E. Giller, Plant Production Systems Group, Wageningen University, PO Box 430, 6700 AK Wageningen, the Netherlands. E-mail: ken.giller@wur.nl

K.D. Joshi, Advanced Research in International Agriculture Development (CARIAD), South Asia Office, c/o CIMMYT – South Asia, PO Box 5186, Kathmandu, Nepal. E-mail: kdjoshi@mos.com.np; k.joshi@bangor.ac.uk

Thomas W. Kuyper, Department of Soil Quality, Wageningen University, PO Box 47, 6700 AK Wageningen, the Netherlands. E-mail: thom.kuyper@wur.nl

J.M. Lenné, North Oldmoss Croft, Fyvie, Turriff, Aberdeenshire AB53 8NA, UK. E-mail: jillian.lenne@btopenworld.com

Rodomiro Ortiz, Martín Napanga 253, Apt. 101, Miraflores, Lima 18, Perú. E-mail: rodomiroortiz@gmail.com

B.R. Sthapit, Centre for Bioversity International, Office for South Asia, National Agricultural Science Centre, DPS Marg, Pusa Campus, New Delhi 110012, India. E-mail: b.sthapit@cgiar.org

D.S. Virk, Centre for Advanced Research in International Agriculture Development (CARIAD), Bangor University, Bangor, Gwynedd LL57 2UW, UK. E-mail: d.virk@bangor.ac.uk

J.R. Witcombe, Centre for Advanced Research in International Agriculture Development (CARIAD), Bangor University, Bangor, Gwynedd LL57 2UW, UK. E-mail: j.r.witcombe@bangor.ac.uk

D. Wood, North Oldmoss Croft, Fyvie, Turriff, Aberdeenshire AB53 8NA, UK. E-mail: agrobiodiversity@btinternet.com

Acknowledgements

We are especially grateful to the contributors to this book who kindly provided their expertise in key areas of client-oriented crop breeding (John Witcombe, Krishna Joshi, Daljit Virk and Bhuwon Sthapit), transgenic crop breeding (Jonny Gressel), soil biodiversity management (Thom Kuyper and Ken Giller) and agrobiodiversity management for climate change (Rodomiro Ortiz).

We would also like to thank CAB International for inviting us to write this book and the following people who kindly provided us with literature and illustrations: Mauricio Bellon, Tim Chancellor, Rachel Cutts, David Grzywacz, Eugene Hettel, Marcus Knapp, Zeyaur Khan, Kristin Mercer, Dagmar Mithöfer, Sam Mohanty, Peter Neuenschwander, Rodomiro Ortiz, Jörg Romeis, Fritz Schultess, Mark Tester and Henry Wainwright.

1 Agrobiodiversity Revisited

J.M. Lenné and D. Wood

the dynamism of agrodiversity, a constantly changing patchwork of relations between people, plants, and their environment, always coping with new problems, always finding new ways.

Brookfield (1998)

Introduction

Biodiversity refers to all living things and the interactions between them: a vast array of organisms with an almost infinite complexity of relationships. Agricultural biodiversity, that is, 'agrobiodiversity', is an exceptionally important subset of biodiversity. Agrobiodiversity has been defined by Qualset *et al.* (1995) as including all crops and livestock and their wild relatives, and all interacting species of pollinators, symbionts, pests, parasites, predators and competitors. This definition formed the foundation for our first book *Agrobiodiversity: Characterization, Utilization and Management* (Wood and Lenné, 1999) and remains an important basis for this second book, *Agrobiodiversity Management for Food Security: a Critical Review*.

Agrobiodiversity through agriculture, that is, the management of the interactions between crops and domestic animals and their associated biodiversity and the environment, provides most of our food with less than 5% coming from the wild (Prescott-Allen and Prescott-Allen, 1986 [for the USA]; Wood and Lenné, 1999). Most of our food is also derived directly or indirectly from plants. It has been estimated that more than 80% of our calories and edible dry weight comes from crop plants (Evans, 2003). Less than 20 species provide most of the world's food and three staple crops – rice, wheat and maize – account for about 60% of the calories and 56% of the protein that humans consume directly from plants. Wheat and rice alone contribute about 44% of edible dry weight directly; root crops less than 10%; sugar crops about 8%; vegetables and fruit about 7%; and pulses about 3%. Future global food security is therefore firmly anchored in improved productivity and appropriate management and use of crop plant agrobiodiversity, especially of rice, wheat and maize.

But agrobiodiversity includes far more than the husbandry of crops and farm animals. As Brookfield (1998) observed, 'the dynamism of agrodiversity, a constantly changing patchwork of relations between people, plants, and their environment, always coping with new problems, always finding new ways', the dynamic interactions of this food agrobiodiversity with other agrobiodiversity in agroecosystems – both beneficial and harmful and both above- and below-ground – are critical to determining if we harvest more or less food. The almost limitless combinations of more or less intensive management, the varied local biotic and abiotic environments, and the human ability to introduce crops and their pests and diseases from elsewhere, and then select within and between these varieties, resulted in a diversity of planned agrobiodiversity and a yet greater diversity of associated, unplanned and, even, unmanageable and unpredictable agrobiodiversity (Wood and Lenné, 1999). Agrobiodiversity is complex and dynamic, both in the crop and associated components. Too often, however, the term is used narrowly to denote only 'beneficial diversity' based on the common belief that biodiversity is a 'good thing' (see www.bioversityinternational.org). Consideration of harmful biodiversity, e.g. insect pests, pathogens and weeds in agroecosystems, is usually excluded in much of the current literature. Yet, as a key factor in substantially reducing food production, it must be addressed as an important component of agrobiodiversity management for food security.

In the introductory chapter to Wood and Lenné (1999), we discussed the importance and functional biotic components of agrobiodiversity based on a classification by Swift and Anderson (1994) in wild and agroecosystems. Their classification of productive, beneficial and destructive biota underpins the role of farmers and agricultural research practitioners in managing agrobiodiversity for increasing crop and animal productivity by encouraging beneficial biota and discouraging destructive biota. In Wood and Lenné (1999), we comprehensively covered most aspects of agrobiodiversity in agroecosystems. In this second book, we wish to sharply focus on and critically review current issues in agrobiodiversity management in key food cropping systems. Not only does our current food security depend on such systems but our future food security increasingly depends on innovative, science-based solutions to making such systems even more productive using less inputs and from the same land base.

Agrobiodiversity Within the International Biodiversity Agenda

For many years, agrobiodiversity was largely neglected in the international biodiversity debate in spite of its massive economic value (Wood and Lenné, 1999). Global initiatives, including the World Heritage Convention for protected sites and the UNESCO Man and the Biosphere Programme for biosphere reserves largely neglected agrobiodiversity. The 1992 Convention on Biological Diversity (CBD) was the first international initiative to specifically highlight the importance of agrobiodiversity (UNEP, 1992). The CBD recognizes domesticated or cultivated species, the need for scientific research on genetic resources, and *in situ* and *ex situ* conservation. In CBD Article 1, emphasis was given to *the sustainable use of biodiversity in meeting food needs* and Agenda 21 gives a comprehensive coverage of agriculture.

However, the main implementing mechanism for the CBD – the Global Environment Facility (GEF) of the World Bank, UNDP and UNEP – has only allocated a very small proportion of its funding to agrobiodiversity. As a result of this neglect, there has been substantial growth in nature reserves, often at the expense of agricultural land. However, the ethics of continued expansion of these reserves in the light of the urgent need to increase food production should now be critically examined.

The UN Food and Agriculture Organization (FAO) international conference on 'Plant Genetic Resources for Food and Agriculture' (PGRFA) held in Leipzig 1996, on one hand emphasized the importance of agrobiodiversity but on the other, conveniently used it as a synonym for PGRFA thus excluding crop-associated biodiversity. This conference produced a 'Global Plan of Action' (FAO, 1996) for the conservation and sustainable use of PGRFA, but new funding for the plan was notably lacking.

The most significant recent development for agrobiodiversity internationally has been the coming into force of the International Treaty for Plant Genetic Resources (ITPGR) in 2004 (see www.planttreaty.org). The International Treaty (IT) further reinforced the perceived synonymy between 'crop diversity' and agrobiodiversity, again ignoring the importance of crop-associated diversity. This was closely followed by the establishment of the Global Crop Diversity Trust in 2006, an independent international organisation, which endeavours to support the conservation of distinct and important crop diversity (see www.croptrust.org). In the past 2 years, the Trust has raised $100 million in contrast to the IT for which no significant new funding has emerged. The implications of these recent developments will be analysed in detail in Chapter 10, this volume.

Agrobiodiversity Within the International Environmental and Development Agenda

Wood and Lenné (1999) noted that the continued neglect of agrobiodiversity by the international environmental agenda is due to the long standing antipathy of environmentalists to agriculture, as noted by Borlaug '*environmental activists ... are a threat to progress on global food security*' (Bailey, 2009). At its most strident, the debate over 'feeding people versus saving nature' finds in favour of nature. In spite of the recent food crisis and growing concerns over the need to feed nine billion people by 2050, there are many who still argue in favour of nature (BirdLife International, 2008).

Although the environmentalists' arguments are based on the premise that intensive agriculture causes environmental degradation which leads to declines in the well-being of poor people who depend on ecosystem services, paradoxically the Millennium Ecosystem Assessment (MEA) found that human well-being has increased despite declines in some ecosystem services (MEA, 2005; Raudsepp-Hearne *et al*., 2010). Human well-being dramatically increases with access to more food, which is far more important than other ecosystem services (Everson and Gollin, 2003).

Encouragingly, the growing dilemma of how to produce more food from less land without damaging the resource base to feed future populations is now being given greater importance in global debates. Support for increasing food production through further agricultural intensification and livelihood diversification without converting more land into food production is growing (Evans, 1998, 2003; Lee and Barrett, 2000; Dixon and Gulliver, 2001; IRRI, 2008a, b). However, many still criticize modern/industrialized agriculture and argue that farming in developing countries can only be made more productive through extensive systems and organic agriculture rather than through agricultural intensification (Pretty *et al*., 2006; McIntyre *et al*., 2009; also see Chapter 11, this volume).

Fortunately, a less polarised and realistic view is beginning to emerge (World Bank, 2008; Royal Society, 2009; Spielman and Pandya-Lorch, 2009; IRRI, 2010; Nature, 2010), at least partly in response to a recent rise in global food prices. Primarily, this view acknowledges the significant contribution of modern agriculture through intensification in irrigated and high-potential farming systems, especially Asia, in continuing to meet rising food demand from burgeoning populations while, at the same time, reducing the rate of conversion of natural ecosystems into agricultural land and dealing with climate change. The Green Revolution alone is estimated to have saved over 80 million ha of land from being converted to agriculture from 1960 to 2000 and slowed the pace of global warming (Evans, 2003; Bergeron, 2010). About half of the world's present population would not have been sustained without this intensification. During 1970 to 1990, an estimated one billion people benefited from increased access to food and/or income (Spielman and Pandya-Lorch, 2009). At the same time, these analyses emphasize that future food increases will need to be achieved more equitably and sustainably through more efficient use of energy, fertilizer and water. In addition, a substantial boost to funding for public sector agricultural research in and for developing countries will be needed to feed the additional two to three billion people (Evans, 2003; Nature, 2010). Managing the linkages and synergies between agriculture, natural resource conservation, the environment and funding support must increasingly become an integral part of using agriculture for development to achieve more sustainable food production systems (Evans, 2003; World Bank, 2008; Nature, 2010). The productive and sustainable management of agrobiodiversity *sensu lato* is paramount to the success of this approach.

There also appears to be a growing consensus among agricultural and environmental scientists that they must work together to deal with climate change. Agriculture and climate change are linked in important ways (Nelson, 2009). Rising temperatures, altered rainfall patterns and more frequent extreme events will increasingly affect crop production and agriculture, but precisely where and how much is still uncertain. Agriculture can help mitigate climate change and poor farmers in developing countries will need help in adapting to climate change. In fact, the advances in modern agriculture achieved in the past 40 years have helped slow the pace of global warming by reducing the amount of biomass burned when land is cleared for farming (Bergeron, 2010). It has been estimated that emissions have been reduced by over 0.5 trillion t of carbon dioxide. For example, irrigated rice under multiple cropping sequesters considerable amounts of carbon (IRRI, 2010). Adaptation of staple food crops through plant breeding and mitigation through improved management will support climate change goals of enhancing the well-being of people who manage and depend on agriculture, especially in the developing world. The failure of the 15th Conference of Parties of the UN Framework Convention on Climate Change (UNFCCC) held in Copenhagen in December 2009 to reach a consensus and agree a global plan of action that includes agrobiodiversity management is therefore very disappointing. These issues will be considered in detail in Chapter 12.

Unfortunately, there are negative developments associated with the UN system. Despite the relative failure of the UNFCCC in Copenhagen there is more to come. The technical body advising the UNFCCC – the UN Intergovernmental Panel on Climate Change (IPCC) – is itself widely mistrusted: American scientists have charged the IPCC with the politicization of science and pointed out the need to 'bring the focus back to credible science, rather than invented hyperbole' (Anon, 2010). Remarkably, given this mistrust and criticism, there is a worrying new UN focus on 'biodiversity and ecosystem services', with advanced planning to set up a panel of scientists modelled on the IPCC. This will be called the 'intergovernmental science-policy platform on biodiversity and ecosystem services' (IPBES, 2010). The topic is highly contentious yet the IPBES will inevitably attempt to bring its collective wisdom to bear on global agriculture. Equally inevitably it will be subject to the input of environmentalist NGOs such as WWF (which provided grossly wrong information to the IPCC) and that have a track record of hostility to agriculture. We can, with certainty, predict that this new panel will exaggerate the value of the diversity of wild biodiversity for agriculture, dabble in critiques of vegetable oil production in developing countries, insist on yet more development-limiting conservation and ignore the most important ecosystem service of all: photosynthesis, directly, or indirectly through animals, driving agriculture and providing food for ourselves and all heterotrophs. The proponents of 'ecosystem services' are already there – suggesting the need to divert funding to 'agroecology', 'organic agriculture', and topics such as natural resource management (NRM) and ignoring ecosystem services such as photosynthesis, biological control, nitrogen fixation and pollination. We will deal with this unfortunate digression from food production in relevant chapters of this book.

The Importance of Agrobiodiversity for Food Security

The main objective of *Agrobiodiversity: Characterization, Utilization and Management* (Wood and Lenné, 1999) was to address the misconceptions, neglect and ignorance over agrobiodiversity, its potential and its managment. Part of the reason for neglect and ignorance was the poor presentation of agrobiodiversity in the international arena and the lack of synthesis of the vast agricultural knowledge base into an agrobiodiversity agenda. This contrasted strikingly with the success of the promotion of wild biodiversity within the international biodiversity agenda, with the subsequent rapid expansion of nature reserves. International donors and development policy makers have continually failed to give agrobiodiversity and food production the importance and funding it merits, as will be discussed in detail in Chapter 2. Unfortunately, lack of understanding of how effective agrobiodiversity management can substantially contribute to food security widely persists, 10 years on from Wood and Lenné (1999). And, attacks on modern agriculture have increased (Pretty *et al.*, 2006; McIntyre *et al.*, 2009; Herren and Ishii-Eiteman, 2010).

Agricultural scientists quite rightly continue to concentrate on science and the increasing need to develop improved technologies to meet the food needs of an ever-expanding global population. Hence opportunities continue to be lost to promote the importance of agrobiodiversity to food security internationally. Although they have limited time to contribute to international debates, scientists should try to seize appropriate opportunities to participate in policy debates to influence investment decisions on the science that underpins food production:

> By hesitating to enter the debate, we can only accede the field to the biologically naive and find ourselves able to serve only as peripherally significant technicians in the pursuit of the objectives of the uninformed.
>
> Namkoong (1991)

The 2008 food crisis, which pushed an additional 100 million people into hunger, is, however, beginning to focus international attention on the critical role of science in sustainable intensification of agriculture for ensuring global food supplies (World Bank, 2008; Royal Society, 2009; Nature, 2010). The time is therefore ripe to revisit, reassess and re-emphasize agrobiodiversity management as governments and policy makers begin to rediscover the need to be more concerned about current and future food security.

Agriculture is the largest global user of biodiversity (Wood and Lenné, 1999). Agriculture has selected and added value to wild biodiversity over more than 10,000 years of managing agrobiodiversity. Agriculture has conserved biodiversity on the hoof and as seed and planting materials over this long period. Agriculture extracts value from biodiversity at each harvest or cull, but nurtures the productive and renewable base. Indeed, it is certain that the most immediately valuable part of global biodiversity is the agrobiodiversity on which farming and, in turn, global food security, depends.

Wood and Lenné (1999) was premised on the fact that agrobiodiversity is irreplaceably important in its own right, for providing most of our food. The management of agrobiodiversity will determine our future, both in cities and the countryside. Agroecosystems – mediated through agrobiodiversity – have always provided the essential ecosystem service of food production, and can be designed to deliver a further range of ecosystem services as needs and knowledge change. Present knowledge extends from a greater appreciation of traditional agriculture and the needs of farmers, through classical agricultural research in animal husbandry, genetics, statistics, replicated experiments, plant breeding, agronomy, crop protection, rural sociology, information management and many more, through to biotechnology. Contributors to the first book reviewed the practical knowledge of agrobiodiversity and its management with the objective of giving it greater prominence in the global debate over biodiversity and sustainable development. We now briefly revisit the scope and main findings of Wood and Lenné (1999) as a basis for introducing the objectives and coverage of this book.

Agrobiodiversity: Characterization, Utilization and Management: a Brief Synopsis

Wood and Lenné (1999) provided a broad, technically sound, functional view of agrobiodiversity: what it is made up of; how it is managed; how it is conserved; and how it can best be utilized. This book covered the status of the concept and usage of the word agrobiodiversity and its relation to wild biodiversity; the components of agrobiodiversity and how they relate together functionally, how they impact on agricultural production, and how agrobiodiversity can best be managed for sustained food production; and whether this extensive knowledge of the management of agrobiodiversity can provide models and practices for the wider management of biodiversity. Emphasis was given to tropical agrobiodiversity as there is more of it and its management is more important for the food security for the poor. Most importantly, the first book highlighted and demonstrated the extensive knowledge base generated by ten thousand years of crop and animal production and the multitude of interacting organisms in a wide range of terrestrial environments. These reasons remain valid for this second book.

Chapter 1 of the first book discussed the importance of agrobiodiversity and highlighted the problematical relationships between biodiversity, agriculture and the environment. In this chapter, we have revisited and updated these relationships. Chapter 2 reviewed the historical dimensions of agrobiodiversity with particular emphasis on crop domestication. It considered the management of diversity before agriculture; the transition to agriculture; the domestication process; and the human impact on diversity. It stressed that the process of domestication focused on a limited range of species in few families in nuclear areas and occurred over a short time span. Subsequent to domestication, there was strong selection pressure by farmers for varietal purity. Even with the last 100 years of plant breeding, there have been few recent additions to the crop portfolio of early farmers. These issues are developed further in this second book.

The next five chapters of the first book examined the nature, role and function of important components of agrobiodiversity. First, Chapter 3 looked at genetic diversity among and within crops and ways in which crop diversity is distributed, assessed and organized into agroecosystems. It also highlighted that the species and varieties in any cropping system are largely determined by farmers and influenced by economic, social, cultural, natural and historical forces. In this second book, we place these key findings into a food security context. Chapter 4 reviewed the biodiversity of domesticated animals used for food, including its nature, extent, erosion, conservation and importance. Management of domestic animals for global food security remains very relevant today. Chapter 5 considered the regulation and functional significance of soil biodiversity. It critically analysed how agricultural practices such as intensification impact upon the biodiversity of the below-ground system in the context of crop productivity. It stressed that there is limited consistent support for the view that intensification has detrimental consequences for soil biodiversity. Soils can be strongly abused yet still continue to produce yields, indicating the robust nature of below-ground biodiversity. The extension and application of these key findings are addressed in this new book. Chapters 6 and 7 provided interesting contrasts between the harmful nature of pathogen diversity and the beneficial character of arthropod biodiversity in agroecosystems. Chapter 6 analysed the evolution of disease in plants, emphasized the mechanisms by which pathogen diversity arises, the functional diversity of pathogens in agroecosystems and the consequences of pathogen diversity for effective disease management in agroecosystems. Pathogen diversity is seen as harmful, to be managed or even eliminated rather than to be conserved to generate useful disease resistance. It concluded that trade-offs are needed between the two views for future crop improvement. Chapter 7 considered the function of beneficial arthropod biodiversity, optimizing insect biodiversity in agroecosystems, and critical issues in biological control and conservation biology. In this second book, we further explore the impact of above-ground crop-associated biodiversity with particular emphasis on using beneficial biodiversity to manage harmful biodiversity, i.e. biological control.

Chapter 8 introduced the third theme – the management of agrobiodiversity – through a consideration of the agroecosystem in the landscape. The differences between agroecosystems from an ecological context were examined through a comparison of a traditional agropastoral system in Spain and a paddy rice system in Thailand. This provided a basis for a discussion of the effects of intensification in agroecosystems. Traditional management of agrobiodiversity was reviewed in Chapter 9 through a series of case studies on the management of diversity by farmers in specific crops – cassava, maize, common bean and rice – and of domesticated animals. The remarkable parallels across crops, cultures and continents and the continuing need for farmer management of agrobiodiversity were highlighted. In this second book, various aspects of these key findings are further discussed. Chapter 10 critically reviewed the effects of plant breeding on genetic diversity in crops with emphasis on the role of farmer participation in the breeding process and the potential impact of modern plant breeding on agrobiodiversity. It noted that the expansion of modern cultivars has in many cases lead to an increase in diversity, particularly when participatory methods and more innovative plant breeding strategies are employed. This second book extends and updates these key findings through several case studies. The effects of pest management methods on biodiversity in agroecosystems were addressed in Chapter 11. It concluded that vegetational diversity in agroecosystems is unpredictable as the outcome is generally site-specific and may be either beneficial or detrimental to the crop. In this second book, we build on these key findings with emphasis on biological control. Chapter 12 looked at the effects of alternative methods of tillage on agroecosystem function through a comparative analysis of wheat with maize and rice systems. The relationships between seed management systems and genetic diversity were addressed in Chapter 13 through a comparison of traditional farmer-managed and modern commercial systems. The need to integrate the two seed supply systems for food production and sustainability was highlighted. The issues raised in both chapters remain very relevant today.

The next two chapters of the first book considered the conservation of agrobiodiversity. Chapter 14 addressed approaches and justifications for the conservation of agrobiodiversity with emphasis on crops. The policy and technology of conservation were then discussed followed by the complementarity of conservation methods in the context of the agroecosystem. Chapter 15 presented a conceptual framework for valuing crop genetic resources on-farm to support strategic decisions about which crop populations are suitable candidates for conservation. The importance of farmers' preferences and the opportunity costs of maintaining specific varieties were highlighted. In this second book, we review recent developments in *ex situ* and *in situ* conservation of crop biodiversity.

Chapter 16 looked at the effects of regulatory issues on agrobiodiversity. It identified the issues where input regulation can have a significant impact on agrobiodiversity and examined possible changes in common regulatory practices to more effectively promote or protect agrobiodiversity. Some negative aspects of regulation are further discussed in this second book. Chapter 17 looked at the parallels between natural ecosystems and agriculture and stressed that agriculture and agrobiodiversity can be linked conceptually and biologically with all nature, rather than, as hitherto, only the more complex parts of nature. It also noted that farmers have done as nature does – employed a range of separate and different systems to meet different conditions and requirements, e.g. the common combination of field and garden within a farming system. These concepts are developed further in this second book, in particular, the critical need to recognize that modern monocultures, essential to current and future food security, have evolved from natural monocultures of cereals under farmer management.

The final chapter of the first book looked at ways of optimizing agrobiodiversity for productive agricultural development. It emphasized that study, increased understanding and the sustainable management of agrobiodiversity may well be critical not just for agricultural production, but also to the future of biodiversity globally. In the concluding chapter of this second book, we build on these conclusions through a more detailed analysis of the policies required to ensure that sound management of agrobiodiversity will achieve global food security.

Objectives of Agrobiodiversity Management for Food Security

All of the concepts and much of the information presented in Wood and Lenné (1999) is just as relevant, important and useful today. Rather than produce a second edition of Wood and Lenné (1999), we believe that there is a need for a new book that emphasizes and justifies the central role of agrobiodiversity in the global effort to ensure food security for today and the future. The main objectives of this second book are therefore to build on and extend this wealth of information to show how agrobiodiversity can effectively and efficiently be managed for food security. We feel that this critical review is timely in the light of the serious challenges facing global food production during the next 20–30 years and the growing attacks on modern, intensive agriculture. In particular, we will refute the plethora of bogus claims and misinformation about the roles of agroecology, organic and subsistence agriculture and their proposed contributions to sustainable agriculture and food security (Pretty *et al.*, 2006; McIntyre *et al.*, 2009). Our analyses will be based on sound scientific principles, the wealth of agricultural research knowledge, and new and emerging biological advances available for achieving sustainable intensification of agriculture. A growing number of reports clearly show the continuing important role for science and technology and that research can have a decisive impact by enabling productive and sustainable agriculture (see World Bank, 2008; Royal Society, 2009; Spielman and Pandya-Lorch, 2009). Most importantly, we wish to bring to the attention of policy makers, especially those responsible for future national and international food security strategies, that knowledgeable, practical and realistic management of agrobiodiversity is the most important toolbox available for significantly and sustainably contributing to global food security. If possible, this book should be read as a companion volume to the first book.

Brief Outline of Chapters in this Volume

Chapter 1 updates the role of agrobiodiversity in the international biodiversity, environmental and development agendas and re-emphasizes the importance of agrobiodiversity management for food security. It provides a brief synopsis of Wood and Lenné (1999) and highlights the key linkages to this second book. Chapter 2 sets the context of the book by defining food security, food sovereignty and food self-sufficiency; looks at current and future food needs; places food security in the context of the international development agenda; and considers growing support for the sustainable intensification of agriculture for food security.

Chapters 3 to 12 critically review many of the past, current and emerging issues affecting agrobiodiversity management for future food security. Chapter 3 chronologically explores in some detail where agrobiodiversity came from. It considers the wild progenitors of crops; the ecological settings of wild crop relatives; pre-domestication management; the impact of the Pleistocene to the Holocene transition, including the important Younger Dryas period; and cropping analogues of the impacts of fire and flood on wild relatives. It concludes with some lessons for modern farming. Chapter 4 discusses the important role of crop introduction in agrobiodiversity management. It looks at the origin and distribution of crops; the Columbian Exchange – the most important period of crop introduction and exchange; systematic crop introduction, especially in the past century; co-evolved pests and diseases and local adaptation; re-encounter and new-encounter diseases; and the importance of plant quarantine. It concludes with some lessons for agrobiodiversity management. Chapter 5 examines the role of crop diversity for food security. It looks briefly at the origin, generation and utilization of crop diversity and considers why farmers need crop diversity and how they cultivate it both within fields and between fields. It then highlights some notable achievements from past investments in crop science for food security and concludes with a taste of future crop diversity technologies to achieve food security. Chapter 6 reviews the impact of modern varieties on crop diversity through three detailed case studies on rice: (i) cultivar replacement in high-altitude rice in Nepal; (ii) cultivar replacement in upland rice in eastern India; and (iii) client-oriented breeding in low-altitude areas of Nepal. It concludes with a discussion of the impact on varietal diversity of these successes in improving local food security. Chapter 7 provides a readable analysis of whether transgenics, often referred to as genetically modified (GM) crops, can enhance crop diversity. It considers the use of transgenics to breach the genetic glass ceiling or yield barriers in certain crops; whether the current use of transgenics is appropriate; regulatory impediments to enhancing agrobiodiversity; and new molecular methods that could assist enhancing crop diversity. It concludes that crop diversity can be enhanced by transgenic approaches and emphasizes the need for scientific-based and *not* emotional- and politically-based risk analysis.

The function and management of crop-associated biodiversity above- and below-ground are reviewed in Chapters 8 and 9. The main focus of Chapter 8 is the use of beneficial crop-associated biodiversity (CAB) above-ground as an ecosystem service to manage harmful CAB in the context of enhancing food security in an environmentally benign manner, thus extending issues raised in the first book. It briefly reviews the roles of some important components of beneficial CAB, successful examples of their application, and some advantages and limitations. It also considers the importance of pollinators; the effects of GM crops on non-target insects; and the role of associated vegetation in managing harmful CAB. It concludes with lessons learned from various strategies used. Chapter 9 reviews biodiversity and ecosystem functioning below-ground through a consideration of the effects of agricultural intensification on soil biodiversity, extending some of the arguments put forward in the first book. It critically analyses whether the relation between biodiversity and ecosystem function can be extended to the more specific association between soil biodiversity, agroecosystem functioning and sustainable food production. It concludes that evidence for a relation between soil biodiversity and sustainable agroecosystem functioning is at best anecdotal and scattered and the case for a causal link between soil biodiversity and ecosystem functioning has been overstated.

Chapter 10 briefly reviews the history of conservation of crop genetic resources and discusses recent developments in both *ex situ* and *in situ* conservation of crop biodiversity, especially in relation to policy. It considers the impacts of the Convention on Biological Diversity and the International Treaty for Plant Genetic Resources on *ex situ* conservation and food security. It also highlights the lack of progress in developing a sound scientific basis for both *in situ* conservation of crop wild relatives and on-farm conservation of landraces. Chapter 11 reviews the recent International Assessment of Agricultural Knowledge, Science and Technology for Development (IAASTD) (McIntyre *et al.*, 2009) process to develop a future roadmap to ensure future global food security. It suggests that a paradigm appears to have emerged from the IAASTD global synthesis report due to a series of highly challengeable assertions based on largely unfounded and blanket criticisms of many existing agricultural knowledge, science and technology approaches, assumptions of questionable technical merit and much incorrect or flawed evidence. The chapter critically examines the key elements of this paradigm, including criticisms of the Green Revolution and GM crops; agroecological approaches; and reliance on organic and small-scale agriculture, exposing the deficiencies in the assertions and evidence provided. It concludes by highlighting the deficiencies and dangers in the 'global assessment' approach.

Chapter 12 addresses the important issue of agrobiodiversity management for climate change. It considers climate change impacts on agrobiodiversity and food security; the neglect of agrobiodiversity by the Inter-Governmental Panel on Climate Change; coping with climate change through knowledge-based agricultural research; and on-going research to adapt and mitigate climate change impacts in major staple food crops such as wheat, rice and maize as well as other important food crops. It highlights the need for improved public awareness of the important role that agrobiodiversity can play in dealing with climate change.

Finally, Chapter 13 attempts to make policy makers and investors in agriculture more clearly aware of: (i) the value of supporting proven and promising i.e. 'good' approaches to increasing food production; and (ii) the serious pitfalls in supporting unproven, flawed and failed i.e. 'bad and ugly' approaches. It draws on the demonstrably successful approaches to agrobiodiversity management for feeding millions highlighted throughout this book (see Chapters 2, 3, 4, 5, 6, 7, 8, 9, 10 and 12, this volume) and warns policy makers against supporting the unproven, pseudo-science-based alternative approaches analysed in Chapter 11, this volume. Above all, it emphasizes that enhanced policy support and significant increases in government and international donor investment will be essential for future growth in agricultural productivity and global food security.

We hope that this second book will build on the process begun with the previous book to redress the past neglect of agrobiodiversity and demonstrate that the long and productive history of the human management of agrobiodiversity can provide a corpus of knowledge and practice, which is both of supreme value in its own right and also of the greatest value as a model for wider biodiversity conservation and utilization for global food security (Wood and Lenné, 1999). By critically reviewing both the positive and the negative developments of the past 10 years, especially in the context of the management of agrobiodiversity for food security, we feel that this book will be useful for agricultural practitioners, researchers and, especially, policy makers.

References

Anon. (2010) An Open Letter from Scientists in the United States on the Intergovernmental Panel on Climate Change and Errors Contained in the Fourth Assessment Report: Climate Change 2007. Available at: www.openletterfromscientists.com (accessed 18 August 2010).

Bailey, R. (2009) Norman Borlaug: the man who saved more human lives than any other has died. Available at: http://reason.com/blog/show/136043.html (accessed 7 March 2010).

Bergeron, L. (2010) High-yield agriculture slows the pace of global warming, say Stanford researchers. *Stanford Report*, 14 June 2010.

BirdLife International (2008) State of the world's birds: indicators for our changing world. Available at: www.bespacific.com/mt/archives/019483.html (accessed 12 December 2009).

Brookfield, H. (1998) Review of Zimmerer, K.S. (1996) Changing Fortunes, Biodiversity and Peasant Livelihood in the Peruvian Andes. *Annals of the American Association of Geographers* 88, 180–182.

Dixon, J. and Gulliver, A. (2001) *Farming Systems and Poverty 2001: Improving Farmers' Livelihoods in a Changing World.* FAO and the World Bank, Rome and Washington, DC.

Evans, L.T. (1998) *Feeding the Ten Billion.* Cambridge University Press, Cambridge.

Evans, L.T. (2003) Agricultural intensification and sustainability. *Outlook on Agriculture* 32, 83–89.

Evenson, R.E. and Gollin, D. (2003) Assessing the impact of the Green Revolution, 1960 to 2000. *Science* 300, 758–762.

FAO (1996) *Global Plan of Action for the Conservation and Sustainable Utilization of Plant Genetic Resources for Food and Agriculture.* FAO, Rome.

Herren, H. and Ishii-Eiteman, M. (2010) Genetically modified crops are not the answer. Available at: http://thehill.com/opinion/op-ed/93907-genetically-modified-crops-are-not-the-answer (accessed 18 August 2010).

IPBES (2010) See web link: http://ipbes.net (accessed 2 September 2010).

IRRI (2008a) *Background Paper: The rice crisis: What needs to be done?* International Rice Research Institute (IRRI), Los Baños, the Philippines. Available at: www.irri.org (accessed 24 November 2009).

IRRI (2008b) *Responding to the rice crisis: How IRRI can work with its partners.* International Rice Research Institute (IRRI), Los Baños, Philippines. Available at: www.irri.org (accessed 24 November 2009).

IRRI (2010) GRiSP International Rice Research Institute (IRRI), Los Baños, Philippines. Available at: www.irri.org (accessed 20 July 2010).

Lee, D.R. and Barrett, C.B. (2000) *Tradeoffs or Synergies? Agricultural Intensification, Economic Development and the Environment.* CAB International, Wallingford, UK.

McIntyre, B.D., Herren, H.R., Wakhungu, J. and Watson, R.T. (eds) (2009) *Agriculture at the Crossroads. The global report of the International Assessment of Agricultural Knowledge, Science and Technology.* Island Press, Washington, DC.

Millennium Ecosystem Assessment (MEA) (2005) *Ecosystems and Human Well-being: Biodiversity Synthesis.* World Resources Institute, Washington, DC.

Namkoong, G. (1991) Biodiversity issues in genetics, forestry and ethics. *The Forestry Chronicle* 68, 438–443.

Nature (2010) How to feed a hungry world. *Nature* 466, 531–532.

Nelson, G. (2009) Climate change impacts on agriculture. *Development and Cooperation* 50, 370–372.

Prescott-Allen, C. and Prescott-Allen, R. (1986) *The First Resource: Wild Species in the North American Economy.* Yale University Press, Newhaven, Connecticut.

Pretty J., Noble, A., Bossio, D., Dixon, J., Hine, R.E., Penning de Vries, P. and Morison, J.I.L. (2006) Resource conserving agriculture increases yields in developing countries. *Environmental Science and Technology* 40, 1114–1119.

Qualset, C.O., McGuire, P.E. and Warburton, M.L. (1995) 'Agrobiodiversity': key to agricultural productivity. *California Agriculture* 49, 45–49.

Raudsepp-Hearne, C., Peterson, G.D., Tengo, M., Bennett, E.M., Holland, T., Benessaiah, K., MacDonald, G.K. and Pfeifer, L. (2010) Untangling the environmentalist's paradox: why is human well-being increasing as ecosystem services degrade? *BioScience* 60, 576–589.

Royal Society (2009) *Reaping the Benefits: Science and the Sustainable Intensification of Global Agriculture.* RS Policy Document 11/09, Royal Society, London.

Spielman, D.J. and Pandya-Lorch, R. (2009) *Millions Fed: Proven Successes in Agricultural Development.* International Food Policy Research Institute, Washington, DC.

Swift, M.J. and Anderson, J.M. (1994) Biodiversity and ecosystem function in agricultural systems. In: Schulze, E.-D. and Mooney, H.A. (eds) *Biodiversity and Ecosystem Function.* Springer, Berlin, pp. 15–41.

UNEP (1992) *Convention on Biological Diversity.* UNEP, Geneva, Switzerland.

Wood, D. and Lenné, J.M. (eds) (1999) *Agrobiodiversity: Characterization, Utilization and Management.* CAB International, Wallingford, UK.

World Bank (2008) *Meeting Growing Demand for Agriculture through Innovations in Science and Technology.* World Development Report 2008, World Bank, Washington, DC.

2 Food Security and Agrobiodiversity Management

J.M. Lenné

The Green Revolution demonstrated that human well-being dramatically increases with access to more food, which is far more important to well-being than any other ecosystem service.

Raudsepp-Hearne *et al.* (2010)

Food Security Defined

Sufficient, quality food is essential for people to lead healthy and productive lives. Food production is more crucial than other ecosystem services for human well-being, and trends in the Human Development Index are clearly correlated with food provisioning services (Raudsepp-Hearne *et al.*, 2010). In past decades, advances in food crop productivity, food processing and trade have substantially increased and improved food availability, stability, access and utilization. This fundamental role of agriculture and the diversity it contains has long been recognized. Yet, at the beginning of the 21st century, achieving global food security is looking increasingly uncertain productively, economically and politically.

The concept of a *Right to Food* as a human right is a binding obligation defined as 'the right of every man, woman and child, alone or in community with others, to have physical and economic access at all times to adequate food or means for its procurement in ways consistent with human dignity'. Although well established under international law, and recognized in the Universal Declaration on Human Rights and the International Covenant on Economic, Social and Cultural Rights, the obligation is not currently enforced. There is a need for states to be proactively engaged in strengthening people's access to, and their utilization of, resources as well as the means to ensure their livelihood and thereby food security in the longer term (UK APPG, 2010).

Food security exists when all people, at all times, have physical, social and economic access to sufficient, safe and nutritious food that meets their dietary needs and food preferences for an active and healthy life. This definition was agreed at the World Food Summit in 1996 (FAO, 1996, 2009). Household food security is the application of this concept to individuals within a household. Food insecurity exists when people do not have adequate access to food as defined above. Undernourishment occurs when the calorific intake is below the minimum dietary energy requirement, which varies among countries and years depending on the gender and age structure of the population.

Food security, as defined by FAO, is a valuable concept and provides a useful goal towards which the world should strive (Pinstrup-Anderson, 2009). At both the national and global level, food security tends to focus mainly on food supply. But availability does not ensure access. At household level, access must be an integral part of food security. Furthermore, food safety and food quality are increasingly being discussed as part of future global food security.

Food Sovereignty

'Food sovereignty' is a policy framework and discourse proposed by Via Campesina, an international peasant farming movement, as a response to the inclusion of agriculture within the world trading system through the Agreement on Agriculture (Lee, 2007). Food sovereignty was defined as the right of each nation to maintain and develop its own capacity to produce basic foods respecting cultural and productive diversity (Via Campesina, 1996). It is based on seven principles: food as a basic human right; agrarian reform; protecting natural resources; reorganizing food trade; ending hunger; social peace; and democratic control.

There is no international definition of food sovereignty and there also does not appear to be a universally agreed concept. For example, McIntyre *et al.* (2009) define food sovereignty as the right of peoples and sovereign states to democratically determine their own agricultural and food policies. According to Windfuhr and Jonsen (2005), food sovereignty is essentially a political concept.

No one would argue with Via Campesina's support for countries to develop and maintain their own capacity for staple food production as this should be under the control of national governments. To meet future national food needs, it is paramount that developing countries stimulate the recovery of their national food producing capacity (Rosset, 2008). The unfortunate reality is that in many developing countries in the past 20–30 years, there has been a substantial reduction in national investment in agricultural research and development (discussed later in this chapter). Unless developing-country governments begin to put significant financial support to agricultural research, agricultural infrastructure including rural roads, transport, markets and processing, and appropriate policies (see Hazell *et al.*, 2007), practical food sovereignty will remain an illusive concept.

Although Via Campesina argues that food sovereignty is a precondition to genuine food security, we suggest the contrary – that food security is a precondition for food sovereignty. If nations develop a sound and sustainable system to ensure national food security, food sovereignty then becomes a realistic concept. At the same time, the impact of climate change will create future uncertainties for both national food security and food sovereignty. Furthermore, even with improvements in small-scale farming, the ability of many developing countries to produce enough food to feed their growing urban populations only from small-scale farms has been seriously questioned (Wiggins, 2009). Many developing countries will continue to rely on imported food.

As the only principle of food sovereignty that directly impacts on agrobiodiversity management is protecting natural resources through the universal adoption of 'agroecological production methods', we do not feel that the concept merits further discussion here. However, the role of agroecological approaches is discussed in Chapter 11, this volume.

Population Growth, Food Needs and Food Prices: Implications

The goal of the 1996 World Food Summit was to reduce the number of undernourished people by half between 1990 and 2015, that is, from 840 to 420 million people globally (FAO, 1996). Even before the recent food price rises and the economic crisis, the number of undernourished people in the world had been steadily increasing for more than a decade (FAO, 2009; Table 2.1). Thus, no progress had been made towards the World Food Summit target even before these events further exacerbated food insecurity.

In the second half of 2009, some 105 million additional people were forced into chronic hunger and malnutrition. Worsening of the situation in 2010 is likely as the world remains firmly in the grip of the economic crisis. It is estimated that 1.02 billion people will be undernourished in 2009, one-sixth of all of humanity (FAO, 2009). Unsurprisingly,

Table 2.1. Number of undernourished people in selected regions 1990 to 2008 (Adapted from FAO, 2009).

Region	Period	No. undernourished millions
Asia & Pacific	1990–1992	680
	1995–1997	530
	2000–2002	550
	2004–2006	560
	2008	670
Sub-Saharan Africa	1990–1992	160
	1995–1997	190
	2000–2002	205
	2004–2006	210
	2008	240
Latin America & Caribbean	2008	50
Near East & North Africa	2008	40

the vast majority of these people are in Asia and sub-Saharan Africa (Table 2.1). There are now more hungry people than at any time since 1970, although, as a result of rising population, the proportion of hungry people globally has declined. Food, the most basic of all human needs, is no longer affordable to the poor. Furthermore, the fact that hunger was increasing even before recent crises suggests that current investment in actions to reduce hunger is woefully inadequate.

At the current rate of increase, the world's population is predicted to rise from 6 billion to over 9 billion by 2050, rising at a rate of 6 million a month (Evans, 1998; FAO, 2009). Africa's population alone is projected to nearly double from 1 billion to 2 billion. Estimates suggest that to meet the most basic of needs for this increased global population, food production will need to increase substantially. At the same time, 2009 marked the first point in human history where urban populations outnumbered rural ones, a trend set to continue as urban populations swell to approximately 5 billion by 2030 (UK APPG, 2010). This continued growth in urbanization will result in additional and larger cities, which will need to be serviced with food, water and energy from a reduced national food production capacity due to rural urban migration. Furthermore, rural areas in developing countries are home to many millions of poor people who will also continue to need support to produce and market food. Feeding the 9 billion people expected to inhabit our planet by 2050 will be an unprecedented challenge (Ash *et al.*, 2010).

What does this mean for global food security? Simply put, the world must produce 50% more food, on less land, with less fresh water, using less energy, fertilizer and pesticide – by 2030 – a daunting challenge that *must* be met (Beddington, 2010; UK APPG, 2010). Let us look at what this means for rice, the world's staple for 2.4 billion people. The International Rice Research Institute (IRRI) estimates that by 2015, the world will need an additional 50 million t of rice annually (IRRI, 2008a,b, 2010). Consumption is rising by 1.5% each year as the population of rice consumers increases. However, the rate of yield increase has been slowing for major cereals such as rice as well as wheat (Fischer and Edmeades, 2010). Despite this, significant yield gains could still be made by narrowing the gap between potential yield and yield on farm, especially in developing countries (Evans, 1998; Fischer and Edmeades, 2010). IRRI (2010) stresses that a combination of improved rice technology and better farm management could lift rice output in countries such as India, the Philippines and Thailand (currently less than 4 t/ha) to the levels attained in China of

6.5 t/ha. Furthermore, Tester and Langridge (2010) recently noted that increasing yield by 1 t/ha or more in low-yielding areas will deliver a much higher relative increase than would the same increase in high-yielding environments by virtue of the much larger areas of low-yielding land globally. Lower-yielding environments offer a great opportunity for substantial increases in global food production by tackling key yield limitations, for example, pests and diseases, salinity, heat and drought. IRRI, after 50 years, is putting even more effort into distributing further improved, high-yielding rice varieties with multiple resistances to pests and diseases as well as tolerance to heat, flooding, salinity and drought in the face of changing climate through the development of a Global Rice Science Partnership (IRRI, 2010). Clearly, more support and investment will be needed to ensure that the global rice science community can continue to develop and promote yield-enhancing technologies even more effectively and rapidly to meet the predicted increased demand for rice. This initiative provides a model for other major global food crops such as wheat, maize, soybean and potato.

Many poor people spend 30–50% of their income on staple food. Price increases in staple cereals can therefore significantly impact on the food security of such people as a greater proportion of income must be spent on basic food needs. Although the general trend in relative food prices has been downward since the early 1970s, there have been quite remarkable increases in grain prices in recent years. International market prices for major cereals, especially rice, surged during the second half of 2007 and the first half of 2008, before falling again later in 2008 (Dorosch, 2009; Fig. 2.1). Rice, wheat and maize prices were 100%, 127% and 106% above their 1998–2007 average. Production shocks played a major role, especially for wheat. The increase in demand for biofuels is also blamed but this was only a major influence on maize, contributing to a 54% increase in price during 2006–2007. Such steep rises in food prices, global food safety scares and continued volatility in agricultural commodities resulted in food riots and social unrest in over 20 countries over four continents (UK APPG, 2010). Rapid economic growth and urbanization in Asia leading to increased demand for meat raised on livestock feed derived from maize and soybean also contributed to the food shortages (Hubert *et al.*, 2010). Such conflicting demands for food, livestock feed and biofuels as well as increases in the frequency of unfavourable climatic events (e.g. droughts, floods etc.) are likely to continue well into this century, with continuing impact on food security and food prices. Clearly, increased food production and security will be essential to combat food price volatility.

Stimulated by the food price crisis and predicted changing climate, a number of science-based reports and papers have recently proposed solutions and action plans to address future food security. These have included NRC (2008), World Bank (2008), Dorosh (2009), Evans (2009), FAO (2009), Royal Society (2009), Spielmann and Pandya-Lorch (2009), Von Braun (2009), Conway *et al.* (2010), Godfray *et al.* (2010b), Hubert *et al.* (2010) and UK APPG (2010) among others. All agree that achieving future food security will require actions on many fronts and across different time scales. In general, there is an emerging consensus on the actions urgently needed to: (i) address the current and near-term needs for food security; and (ii) build a stronger food system that can respond to future challenges. Different assessments place different emphasis on the following critical interventions:

- Significantly increasing investment in agricultural research and development, especially in the developing world;
- Expanding social safety net interventions (food aid, nutritional programmes and humanitarian assistance) to food-insecure poor households;
- Strengthening markets and facilitating fair global and regional trade; and
- Supporting policy development for science and technology, social and trade interventions.

Von Braun (2009) and FAO (2009) also highlighted the importance of improving global governance to address fundamental

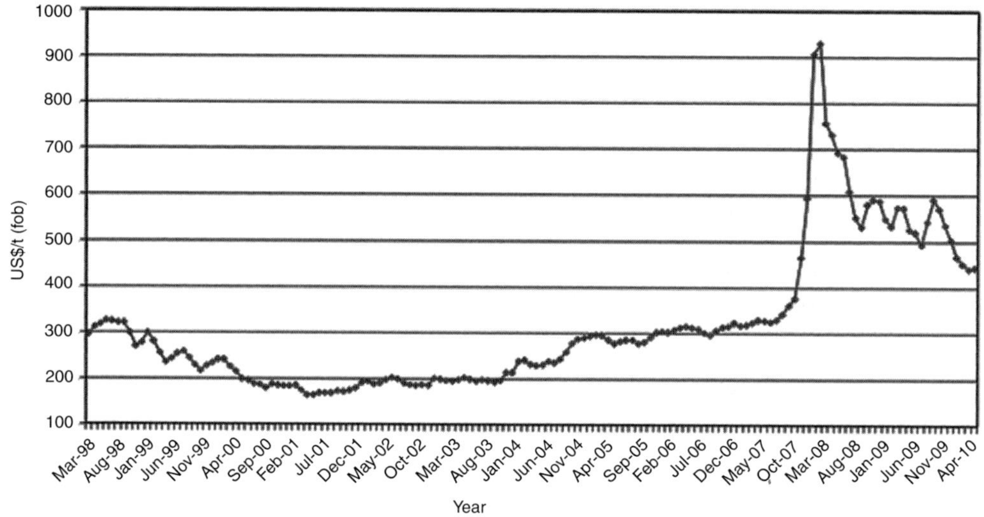

Fig. 2.1. Thai 5% broken rice price (March 1998 to July 2010). (Source of raw data: The Pinksheet, World Bank, courtesy of Sam Mohanty, IRRI.)

weaknesses in systems governing food, nutrition and agriculture. Stable and effective policies, regulatory and institutional mechanisms and functioning market infrastructures that promote investment in the agriculture sector are paramount. Hubert *et al.* (2010) further highlighted the importance of food safety and quality as part of food security and an increasing future role for technologies to safely preserve and process food. Godfray *et al.* (2010a) noted that we have perhaps 40 years to radically transform agriculture, work out how to grow more food without exacerbating environmental problems and simultaneously cope with climate change. Furthermore, Byerlee *et al.* (2009) have stressed that globalization, integrated value chains, rapid technological and institutional innovations, and environmental constraints have changed the context of agriculture's role in global development. They argue for a new paradigm that recognizes agriculture's multiple functions in development in this emerging context, including providing food security, triggering economic growth, reducing poverty, narrowing income disparities and delivering environmental services. There is an intrinsic link between the challenge we face to ensure food security through the 21st century and other global issues, most notably climate change, population growth and the need to sustainably manage the world's rapidly growing demand for energy and water (UK APPG, 2010).

At the same time, it is often forgotten that there is a long time lag of from 15 to 25 years before the products from agricultural research filter through to farmers and realize impact on food production (Normile, 2008; Pardey and Pingali, 2010). To this must be added the initial time spent in conducting basic and applied research to develop the technology, which could add another 10 to 15 years. The Green Revolution was built on decades of agricultural research generated by the science systems of the USA, UK, Germany, France and Japan among others (UK APPG, 2010). In this context, the outputs from research in progress today may not be delivering food to hungry people until 2030 to 2040. Furthermore, much of today's and probably tomorrow's investment in agricultural research is targeted at 'maintaining' yields and productivity, that is, to prevent yield decline from diseases, pests and environmental pressures (drought, declining fertility etc.), rather than directly for additional productivity increases (Alston *et al.*, 2009; Peng *et al.*, 2010). It is therefore critical that investments in agricultural research target the key interventions that will *both* prevent yield decline and increase food production.

Food Security and the International Development Agenda

Until the late 1980s, the international development agenda strongly supported agricultural research for increasing food production to reduce hunger (Pardey and Beintema, 2001). International donors such as the World Bank, the US government, the European Community (EC) and UK Department for International Development (DFID) provided significant support to agricultural research and training through national research institutes, the Consultative Group for International Agricultural Development (CGIAR) and bilaterally. As a result, during the 1970s and 1980s, growth in gross world food production outpaced population growth from almost the same land base and good progress was made in reducing chronic hunger (FAO, 2009). In addition, agricultural productivity gains helped lift millions out of poverty. This past investment in agricultural research was critical to utilizing agriculture as a driver of poverty relief (UK APPG, 2010).

However, over the past 30 years, support for agricultural research and development (R&D) has been significantly eroded (Pardey and Beintema, 2001; Pardey *et al.*, 2006). Increasingly, donor support to developing countries has been redirected away from agriculture towards poverty reduction, health and education. Globally, Official Development Assistance (ODA) spent on agriculture fell dramatically from US$6.2 billion in 1980 to US$2.3 billion in 2002. Yet at the very same time, global ODA levels have increased massively by 65%, meaning that not only is international assistance for agriculture lower now than it was in 1980 in real terms, but that as a share of total ODA it has fallen even more devastatingly from 17% in 1982 to just 3.7% in 2002 (UK APPG, 2010). Just as seriously, the OECD countries' agricultural subsidies are now almost US$1 billion per day, ten times as much as the total cost of global agricultural R&D (Evans, 2003).

National support for agricultural R&D in developing countries has shown a similar decline from about 20% in the 1980s to 10% today (Pardey and Pingali, 2010). More worryingly, 83% of the developing world's total agricultural R&D spending was in three BRIC countries only in 2006 – notably China, India and Brazil. As a result, these emerging economies are enjoying a food production boom, with agricultural outputs growing three times as fast as in the USA and Europe (OECD, 2010). The priority given to agriculture in most developing country national plans is far too low and must increase, to avoid reliance on subsidized imported food and food aid. Without such further investment, any call for food sovereignty will remain unanswered.

Let us look at some specific examples. During 1987–1998, EC support for agriculture declined from 12% of the aid budget to 4% in 1998. The World Bank's lending for agricultural projects was reduced from 26% to 10% of total lending from the 1980s to 2000 (Pardey and Beintema, 2001). In the 1970s, the USA provided significant support for the development of the Indian agricultural university system as well as contributing to agricultural universities in Africa, Latin America and elsewhere in Asia (Federoff, 2009). Such programmes are now a pale reflection of what they once were: investments in human capital development, science and long-term institutional building have nearly disappeared. DFID's spending on agriculture halved from 1995 to 2005 and there was a significant drop in the number of advisory staff with the requisite technical skills (Heath, 2007). This seriously challenges DFID's capacity to sponsor a strategy of agriculture-led growth in future. It is likely that a shared complacency throughout the international donor community of adequate global food production and availability associated with demands from other sectors, e.g. health and education, resulted in this steady erosion of investments in agricultural research for development.

There is little doubt that decades of under-investment in the agricultural research which underpins yield growth by major donors have been a contributing factor to cereal yield decline (Pardey and Pingali, 2010). Global yields of rice, wheat and maize grew rapidly from 1961 to 2007 by 2.2, 2.6 and 2.6 times, respectively (Alston *et al.*, 2009). However, for all crops, rates of yield growth were slower from 1990 to 2007 than during 1961 to 1990 (Pardey and Pingali, 2010), strongly correlated to the erosion in investment. Average yield growth has fallen from 6% to 1.5% in developing countries. The Royal Society (2009) highlights the slowing of increases in productivity as a driver of chronic food insecurity.

Ironically, the unwillingness of major donors to support agriculture flies in the face of its demonstrable investment returns. Many studies have clearly shown that investment in agricultural research and development achieves high returns (Evenson *et al.*, 1979; Alston *et al.*, 2000; Hossain *et al.*, 2003; Raitzer, 2003; Raitzer and Kelley, 2008; Evans, 2009; Renkow and Byerlee, 2010; Table 2.2). Investment in agricultural science has paid off handsomely, with an average rate of return of 43% in 700 projects evaluated in developing countries (World Bank, 2008). A 10-year evaluation of research conducted by the Consultative Group on International Agricultural Research (CGIAR) showed that for every dollar spent on high-quality international agricultural research, US$9 were returned in benefits to poor communities with rates of return of between 40% and 80% (CGIAR, 2009). This should have been more than enough to justify an increased rate of growth in funding for agricultural research (Alston *et al.*, 2009). And the investment continues to pay off, year after year, unlike one-off food aid.

As well as complacency about food security, an additional contributing factor to the erosion of investment in agricultural research for development was the wide adoption of the Millennium Development Goals (MDGs) by UN member states in 2001 (UN, 2001). Notwithstanding the success of the MDGs in improving social and economic conditions in some of the world's poorest countries (DFID, 2009), no progress has been made to address MDG 1c, the *only goal* referring to hunger (see Box 2.1). Worryingly, no explicit reference is made to increasing agricultural productivity or food security as the main means to not only reducing hunger but also providing a demonstrable route out of poverty (NRC, 2008; Byerlee *et al.*, 2009). Clearly, the international development agenda did not give enough attention to the linkages between hunger and social conditions. As a result, gains made in reducing poverty have been lost as recent increases in the cost of staple foods have pushed more people back into poverty (FAO, 2009). Currently the world is letting MDG 1c slip through its fingers (UK APPG, 2010). It has been recently estimated that US$30 billion of additional funds will be needed annually to meet MDG 1 by 2015 (FAO, 2009).

Moreover, the MDG silos are ill-suited to address complex development challenges (Conway *et al.*, 2010). This exclusive focus on specific MDGs has ignored the complex interrelationships and linkages between poverty, hunger, health and education – these linkages need urgent attention. The reality is that success in one MDG is predicated on success in others. A much more inclusive response to the MDGs is urgently needed.

Table 2.2. Returns from publicly-funded agricultural research and extension (Examples from Echeverria (1989) adapted from Evans (2009)).

Country	Crop/s	Years	Rate of return %
Bangladesh	Wheat and rice	1961–1977	30–35
Brazil	Soybean	1955–1983	46–69
Brazil	Irrigated rice	1959–1978	83–119
Colombia	Rice	1957–1964	75–96
Mexico	Wheat	1943–1963	90
Pakistan	Wheat	1967–1981	58
Philippines	Rice	1966–1975	75

Box 2.1. The Millennium Development Goals

Goal 1: Eradicate extreme poverty and hunger
Target 1a: Reduce by half the proportion of people living on less than a dollar a day
Target 1b: Achieve full and productive employment and decent work for all, including women and young people
Target 1c: Reduce by half the proportion of people who suffer from hunger

Goal 2: Achieve universal primary education
Target 2a: Ensure that all boys and girls complete a full course of primary schooling

Goal 3: Promote gender equality and empower women
Target 3a: Eliminate gender disparity in primary and secondary education preferably by 2005, and at all levels by 2015

Goal 4: Reduce child mortality
Target 4a: Reduce by two-thirds the mortality rate among children under 5

Goal 5: Improve maternal health
Target 5a: Reduce by three-quarters the maternal mortality ratio
Target 5b: Achieve, by 2015, universal access to reproductive health

Goal 6: Combat HIV/AIDS, malaria and other diseases
Target 6a: Halt and begin to reverse the spread of HIV/AIDS
Target 6b: Achieve, by 2010, universal access to treatment for HIV/AIDS for all those who need it
Target 6c: Halt and begin to reverse the incidence of malaria and other major diseases

Goal 7: Ensure environmental sustainability
Target 7a: Integrate the principles of sustainable development into country policies and programmes; reverse loss of environmental resources
Target 7b: Reduce biodiversity loss, achieving, by 2010, a significant reduction in the rate of loss
Target 7c: Reduce by half the proportion of people without sustainable access to safe drinking water and basic sanitation
Target 7d: Achieve significant improvement in lives of at least 100 million slum dwellers, by 2020

Goal 8: Develop a global partnership for development
Target 8a: Develop further an open, rule-based, predictable, non-discriminatory trading and financial system
Target 8b: Address the special needs of the least developed countries
Target 8c: Address the special needs of landlocked developing countries and small island developing States (through the Programme of Action for the Sustainable Development of Small Island Developing States and the outcome of the 22nd special session of the General Assembly)
Target 8d: Deal comprehensively with the debt problems of developing countries through national and international measures in order to make debt sustainable in the long term

Breaking down the MDG silos will allow the international development agenda to address the inter-connectedness of the MDGs and, hopefully, reduce hunger in a sustainable manner.

The recent food crisis propelled agriculture and food security back on to the front pages of newspapers and to the top of policy makers' agendas (FAO, 2009). The *Joint Statement on Global Food Security (L'Aquila Food Security Initiative)*, produced by the G8 and partner governments, agencies and institutions in July 2009, gave a strong, renewed commitment by the global community to a coordinated, comprehensive strategy for sustainable agriculture development through mobilizing US$20 billion over 3 years. However, the technological challenges facing food production in the 21st century are more daunting than those of previous decades (World Bank, 2008). One must ponder: is the Global Food Security Initiative – *too little, too late?* (Pardey *et al.*, 2006). Not only will a significant amount of 'catch-up' funding be needed, but substantial commitment and investment by developing country governments will also be essential if the outcomes of technical advances are to reach the poor.

'Sustainable' Intensification of Agriculture

Agriculture has always had elements of a Faustian bargain in its trade-offs between productivity and sustainability (Evans, 1998). Although intensification is often viewed as the enemy of sustainability, about half the world's current population could not have been sustained without intensification (Evans, 2003). Furthermore, intensification has not, so far, reduced the efficiency of food production in terms of total input energy.

Key options for increased crop production for food security are well-known:

- Increase the area of land under cultivation;
- Increase the yield per hectare per crop;
- Increase the number of crops per hectare per year;
- Replace lower-yielding by higher-yielding crops;
- Reduce losses caused by pests, diseases and weeds; and
- Reduce postharvest losses.

With the exception of increasing the area of land under cultivation, all the above interventions have proven successful and should be even more actively pursued for agricultural intensification to produce more food. However we accomplish increased global food supply to feed growing populations, the imperative of ongoing yield increases remains a priority, in spite of recent slowing of the rates of yield growth.

While acknowledging the significant contribution of intensification technologies to meeting global food needs in the past 50 years, both the World Bank (2008) and the Royal Society (2009) highlight the need for a future strategy of 'sustainable' intensification of global agriculture in which yields are increased without adverse environmental impact and without the cultivation of more land. In order to achieve this, increased production of food and agricultural products to meet global needs will have to come from gains in productivity per unit of land and resource inputs, as noted above. Major attributes of a sustainable, intensive production system include:

- Utilization of crop varieties with higher productivity per external input;
- Efficient utilization of external inputs, especially water and fertilizer;
- Efficient exploitation of nutrient cycling, biological nitrogen fixation, allelopathy, predation and parasitism;
- Minimization of the use of technologies that have adverse impacts on the environment and human health, e.g. pesticides;
- Productive use of human knowledge and capacity to adapt and innovate and to resolve common landscape-scale problems; and
- Minimization of the impacts of system management for food production on greenhouse gas emissions, clean water availability, carbon sequestration, conservation of biodiversity and dispersal of pests, pathogens and weeds.

At the same time, we should retain a realistic perspective on the effects of agricultural intensification on ecosystem services. A recent study has shown that the benefits of food production currently outweigh the costs of declines in other ecosystem services at the global scale (Raudsepp-Hearne *et al.*, 2010).

Above all, sustainable intensification will require sustainability of funding for such research (Evans, 1998; Pardey *et al.*, 2006). It is notable that, in recent years, ongoing calls for the former have not been matched by the supply of the latter. It is now imperative that funding is substantially increased to support sustainable agricultural intensification for future food security.

Agrobiodiversity Management for Food Security

Agrobiodiversity management for food security includes crop introduction, genetic manipulation, crop breeding, genetic resources conservation, agronomy, soil management and crop protection as well as delivering appropriate technologies and knowledge to farmers (Wood and Lenné, 1999). Sound agrobiodiversity management therefore provides the main building blocks for appropriate and practical sustainable intensification of agricultural production for food security.

In the next decades, agricultural research will be seriously challenged to produce the science-based knowledge and technologies needed to: (i) sustainably increase the productivity of improved crop varieties; (ii) manage the associated biodiversity above- and below-ground to further enhance productivity and reduce losses; and (iii) improve the capacities of farmers and supporting institutions to access and use resources wisely to manage their systems sustainably. Recently, there have been a number of useful and comprehensive assessments of the kinds of agricultural research knowledge, methodologies and technologies that will be needed in future to ensure food security and feed 9 billion people by 2050 (World Bank, 2008; Royal Society, 2009; Spielman and Pandya-Lorch, 2009; IRRI, 2010). Table 2.3 lists some major strategies and interventions for achieving increased productivity and food security based on these assessments. Many are based on proven approaches to the scientific characterization, utilization and management of agrobiodiversity. Throughout this book, we will consider many of these interventions in more detail.

It should be noted that the interventions include: ongoing current and new applications of existing knowledge and technologies; the generation of new knowledge and technologies; as well as radical and innovative approaches. Clearly, there will be a need to balance investment in radical new approaches that may result in major increases in productivity, e.g. conversion of C3 crops such as rice to C4 photosynthesis (Sheehy *et al.*, 2007), with investment in approaches which deliver modest but incremental improvements to more poor people with shorter research lags, e.g. ongoing staple food crop improvements (Royal Society, 2009). If the interventions and approaches are underpinned by sound agricultural science, they should not lead to any deterioration of biodiversity or the environment.

Meeting global food security through science-based agrobiodiversity management will require, in many cases, simultaneous attention to several interventions. For example, expanded use of cereal hybrids and higher yielding varieties with a range of abiotic stress tolerances may need to be accompanied by improvements in local and national seed systems, integrated pest and weed management, as well as reduced postharvest losses and strengthened market chains. In addition, meeting the challenges imposed by changing climate is likely to require simultaneous improvements in food crop resistances to heat, drought and diseases, accompanied by improved approaches to integrated pest and weed management. Such multiple problems will demand a diversity of approaches, specific to cropping systems, cultures and environments (Royal Society, 2009; Spielman and Pandya-Lorch, 2009; Beddington, 2010). Such diversity demands that the breadth of relevant scientific enquiry is equally diverse, and that science needs to be combined with social, economic and political considerations, as will be discussed throughout this book.

Future global food security can be achieved with the concerted application of current and pipeline technologies, given sufficient political will (Godfray *et al.*, 2010b). But to do so sustainably in the face of climate change, equitably in the face of social and regional inequalities, and successfully in an uncertain investment environment, remains one of humanity's greatest challenges (Evans, 1998). Investment in research sooner rather than later is essential to enable the food system to cope with both known and unknown challenges in the coming decades.

Table 2.3. Agrobiodiversity management interventions for food security.

Strategies	Interventions
Increasing investment in agricultural research and development	
Improved genetic resource management and utilization	Safe and secure management and conservation of plant genetic resources
	Informed and targeted crop and varietal introduction e.g. fruits and vegetables
	Exploring crop genomes including wild relatives for useful traits
Increased staple food crop production	Revitalising yield growth in rice and wheat in high potential systems
	Expanded use of crop hybrids and higher-yielding varieties
	Fostering local and national quality seed systems
	Expanded use of genomics, MAS, genetic modifications, high-throughput systems
	Increasing photosynthetic efficiency: C3 plants e.g. rice to C4
Reduced losses caused by biotic stresses	Improved deployment of existing and novel pest and disease resistances
	Innovative management of crop-associated biodiversity for diseases
	Integrated pest management through crop-associated biodiversity
	Enhanced use of biological control
	Integrated weed management through crop-associated biodiversity
Reduced losses caused by abiotic stresses	Enhanced deployment of resistances to heat, cold, drought, submergence, salinity, and infertile and toxic (e.g. Al, Fe) soils
Improved soil fertility and conserving soil	Wider use of green manure crops
	Innovative management of soil-based crop-associated biodiversity
	Zero-tillage, crop rotation, intercropping, mulching, biochar etc.
Increased efficiency of water and fertilizer use	Enhanced deployment of more efficient crop varieties e.g. aerobic rice
	Improving existing nitrogen fixation processes
	Developing nitrogen fixation systems in other crops
	Enhancing crop phosphorus-uptake through improved mycorrhizal processes

Improve the nutritional quality of food	Enhanced deployment and promotion of more nutritional foods Development and deployment of crop varieties with higher protein, mineral and vitamin contents Establishing processing facilities and promoting processing to preserve nutritional quality
Improved food safety	Elimination of potentially toxic compounds from foods Reducing microbial toxins in foods that impact on human health
Relevant knowledge and technologies delivered to small farmers	Most of the above where appropriate, practical and affordable
Expanding social and safety net interventions	
Appropriate and practical food crop diversity promoted	Home gardens, especially for vegetables and fruits Community-based food production options in urban areas
Strengthening markets and facilitating trade	
Food crop market chains improved	Reducing post-harvest losses and inefficiencies in the market chain Extending storage-life or delaying ripening Connecting small farmers equitably into market chains Improving infrastructure and transport linkages
International and regional trade fostered and increased	Building capacity of small farmers to meet quality and regulatory standards Connecting small farmers to high value and export value chains Facilitating equitable and rule-based international and regional trade
Improving policy support for food security	Developing and expanding policies based on proven scientific approaches Involvement of scientists in informing policy makers Avoiding policies based on unproven and flawed approaches

References

Alston, J.M., Marra, M.C., Pardey, P.G. and Wyatt, T.J. (2000) *A Meta-Analysis of Rates of Return to Agricultural R&D: Ex Pede Herculem?* International Food Policy Research Institute, Washington, DC.
Alston, J.M., Beddow, J.M. and Pardey, P.G. (2009) Agricultural research, productivity and food prices in the long run. *Science* 325, 1209–1210.
Ash, C., Jasny, B.R., Malakoff, D.A. and Sugden, A.M. (2010) Feeding the future. *Science* 327, 797.
Beddington, J. (2010) Global food and farming future. *Philosophical Transactions of the Royal Society B* 365, 267.
Byerlee, D., de Janvry, A. and Sadoulet, E. (2009) Agriculture for development: toward a new paradigm. *Annual Review of Resource Economics* 1, 15–31.
CGIAR (2009) *Annual Report 2008*. Consultative Group for International Agricultural Research, Washington, DC.
Conway, G., Waage, J. and Delaney, S. (2010) *Science and Innovation for Development*. UK Collaborative on Development Sciences, London.
DFID (2009) *Eliminating World Poverty: Building our Common Future*. White Paper, Department for International Development, TSO (The Stationery Office), Norwich.
Dorosh, P.A. (2009) Price stabilization, international trade and national cereal stocks: world price shocks and policy response in South Asia. *Food Security* 1, 137–149.
Evans, A. (2009) *The Feeding of the Nine Billion. Global Food Security for the 21st Century*. A Chatham House Report, London. Available at: www.chathamhouse.org.uk (accessed 3 November 2009).
Evans, L.T. (1998) *Feeding the Ten Billion*. Cambridge University Press, Cambridge.
Evans, L.T. (2003) Agricultural intensification and sustainability. *Outlook on Agriculture* 32, 83–89.
Evenson, R.E., Waggoner, P.E. and Ruttan, V.W. (1979) Economic benefits from research: an example from agriculture. *Science* 205, 1101–1107.
FAO (1996) *Rome Declaration and World Food Summit Plan of Action*. FAO, Rome.
FAO (2009) *The State of Food Insecurity in the World. Economic crisis: impacts and lessons learned*. FAO, Rome.
Federoff, N.V. (2009) Science diplomacy in the 21st century. *Cell* 136, 9–11.
Fischer, R.A. and Edmeades, G.O. (2010) Breeding and cereal yield progress. *Crop Science* 50, S85–S98.
Godfray, H.C.J., Beddington, J.R., Crute, I.R., Haddad, L., Lawrence, D., Muir, J.R., Pretty, J., Robinson, S., Thomas, S.M. and Toulmin, C. (2010a) Food security: the challenge of feeding 9 billion people. *Science* 327, 812–818.
Godfray, H.C.J., Crute, I.R., Haddad, L., Lawrence, D., Muir, J.F., Nisbett, N., Pretty, J., Robinson, S., Toulmin, C. and Whiteley, R. (2010b) The future of the global food system. *Philosophical Transactions of the Royal Society B* 365, 2769–2777.
Hazell, P.B.R., Poulton, C., Wiggins, S. and Dorward, A (2007) The future of small farms for poverty reduction and growth. *International Food Policy Research Institute 2020 Vision Discussion Paper* 42.
Heath, J. (2007) *DFID's 2005 Agriculture Policy: An Interim Evaluation*. DFID Evaluation Report EV672, DFID, London.
Hossain, M., Gollin, D., Cabanilla, V., Cabrera, E., Johnson, N., Khush, G.S. and McLaren, G. (2003) International Research and Genetic Improvement in Rice: Evidence from Asia and Latin America. In: Evenson, R.E. and Gollin, D. (eds) *Crop Variety Improvement and its Effect on Productivity: The Impact of International Agricultural Research*. CAB International, Wallingford, UK.
Hubert, B., Rosengrant, M., van Boekel, M.A.J.S. and Ortiz, R. (2010) The future of food: scenarios for 2050. *Crop Science* 50, S1–S18.
IRRI (2008a) *Background Paper: The rice crisis: What needs to be done?* International Rice Research Institute (IRRI), Los Baños, the Philippines. Available at: www.irri.org (accessed 24 November 2009).
IRRI (2008b) *Responding to the Rice Crisis: How IRRI can work with its partners*. International Rice Research Institute (IRRI), Los Baños, the Philippines. Available at: www.irri.org (accessed 24 November 2009).
IRRI (2010) GRiSP International Rice Research Institute (IRRI), Los Baños, Philippines. Available at: www.irri.org (accessed 20 July 2010).
Lee, R. (2007) Food security and food sovereignty. *Centre for Rural Economy Discussion paper Series No. 11*. University of Newcastle Upon Tyne, UK.
McIntyre, B.D., Herren, H.R., Wakhungu, J. and Watson, R.T. (eds) (2009) *Agriculture at the Crossroads. The global report of the International Assessment of Agricultural Knowledge, Science and Technology*. Island Press, Washington, DC.

Normile, D. (2008) Reinventing rice to feed the world. *Science* 321, 330–333.
NRC (2008) *Emerging Technologies to Benefit Farmers in Sub-Saharan Africa and South Asia.* National Academy of Sciences, National Academies Press, Washington, DC.
OECD (2010) *Agricultural Outlook Report.* United Nations OECD.
Pardey, P.G. and Beintema, N.M. (2001) *Slow Magic: Agricultural R&D a Century after Mendel.* International Food Policy Research Institute, Washington, DC.
Pardey, P.G. and Pingali, P.L. (2010) *Reassessing International Agricultural Research for Food and Agriculture.* Global Conference on Agricultural Research for Development 2010, Background Paper. Available at: www.gcard2010.net (accessed 30 March 2010).
Pardey, P.G., Alston, J.M. and Piggott, R.R. (eds) (2006) *Agricultural R&D in the Developing World: Too Little, Too Late?* International Food Policy Research Institute, Washington, DC.
Peng, S., Huang, J., Cassman, K.G., Laza, R.C., Visperas, R.M. and Khush, G.S. (2010) The importance of maintenance breeding: A case study of the first miracle rice variety-IR8. *Field Crops Research* 119, 342–347.
Pinstrup-Anderson, P. (2009) Food security: definition and measurement. *Food Security* 1, 5–7.
Raitzer, D.A. (2003) *Benefit Cost Meta-Analysis of Investment in the International Agricultural Research Centres of the CGIAR.* Science Council Secretariat, FAO, Rome.
Raitzer, D.A. and Kelley, T.G. (2008) Benefit–cost meta-analysis of investment in the International Agricultural Research Centers of the CGIAR. *Agricultural Systems* 96, 108–123.
Raudsepp-Hearne, C., Peterson, G.D., Tengo, M., Bennett, E.M., Holland, T., Benessaiah, K., MacDonald, G.K. and Pfeifer, L. (2010) Untangling the environmentalist's paradox: why is human well-being increasing as ecosystems degrade? *BioScience* 60, 576–589.
Renkow, M. and Byerlee, D. (2010) The impacts of CGIAR research: a recent review of evidence. *Food Policy* 35, 391–402.
Rosset, P. (2008) Food sovereignty and the contemporary food crisis. *Development* 51, 460–463.
Royal Society (2009) *Reaping the Benefits: Science and the Sustainable Intensification of Global Agriculture.* RS Policy Document 11/09, Royal Society, London.
Sheehy, J.E., Mitchell, P.L. and Hardy, B. (eds) (2007) *Charting new Pathways to C4 Rice.* International Rice Research Institute, Los Baños, the Philippines.
Spielman, D.J. and Pandya-Lorch, R. (2009) *Millions Fed: Proven Successes in Agricultural Development.* International Food Policy Research Institute, Washington, DC.
Tester, M. and Langridge, P. (2010) Breeding technologies to increase crop production in a changing world. *Science* 327, 818–822.
UK APPG (2010) Why No Thought for Food? A UK Parliamentary Inquiry into Global Food Security. The All Party Parliamentary Group on Agriculture and Food for Development, London. Available at: www.agricultureandfoodfordevelopment.org/inquiry (accessed 16 February 2010).
UN (2001) Millennium Development Goals. Available at: www.un.org/millenniumgoals (accessed 30 March 2010).
Via Campesina (1996) *Food Sovereignty: A Future without Hunger.* Position Statement, 2nd International Conference of Via Campesina, Tlaxacala, Mexico, 1996.
von Braun, J. (2009) Addressing the food crisis: governance, market functioning, and investment in public goods. *Food Policy* 1, 9–15.
Wiggins, S. (2009) Can the smallholder model deliver poverty reduction and food security for a rapidly growing population in Africa? *Paper prepared for the FAO Expert Meeting 'How to Feed the World in 2050'*, FAO, Rome, 24–26 June, 2009.
Windfuhr, M. and Jonsen, J. (2005) *Food Sovereignty: Towards Democracy in Localised Food Systems.* ITDG Publishing, Rugby, UK.
Wood, D. and Lenné, J.M. (eds) (1999) *Agrobiodiversity: Characterization, Utilization and Management.* CAB International, Wallingford, UK.
World Bank (2008) *Meeting Growing Demand for Agriculture through Innovations in Science and Technology.* World Development Report 2008, World Bank, Washington, DC.

3 Agrobiodiversity Management and the Origins of Agriculture

D. Wood

... it is fair to state that ecological research is the weak link in agricultural origins studies.... In comparison, both archaeological and genetic investigations are going great guns...

Blumler (1998)

Introduction

Achieving food security through productive agriculture depends on the sustainable management of agroecosystems. However, there remains considerable uncertainty as to just how to manage the wide range and diversity of farming systems that produce most of our food. One approach is to search in wild ecosystems for appropriate models for agriculture. The obvious departure point in this search is the identification of vegetation containing the wild relatives of plants in regions where crop agriculture first began. By our definition, these wild relatives are included in 'agrobiodiversity': they gave rise to crops and are of high importance for plant breeders.

The prime candidate in the search for relevant wild ecosystems is the 'Near Eastern' centre of crop origins – the arc from Palestine, Jordan and Israel, through Syria, southern Turkey, Iraq and south-western Iran. As the source of important cereal and pulse crops (wheats, barley, pea, lentil, faba bean and others) this region has been the focus of extensive botanical, genetic and, to a lesser extent, ecological research, which has resulted in a multiplicity of theories on the origins of plant domestication.

While these theories may one day be resolved into practical suggestions for managing crops or domesticating new crops, surprisingly little attention has been given to identifying models for the ecological management of crops. Yet for many crops – and especially cereals – each wild relative is little different from the crop it developed into not much more than ten thousand years ago.

As ecology and 'agroecology' have emerged as a banner for some forms of agriculture, we thought it useful to bring together information on some of the environmental determinants of the ecology of wild relatives. May (1999) has pointed out an ecological relationship between low species diversity and high productivity and suggested that this is a key issue for sustainable cereal cropping. We agree and we were further intrigued by the statement that many fields of present-day wild cereals were: 'as productive as are varieties of durum and barley planted in ground prepared by a wooden plough' (Hassan, 1977). How could this be after ten thousand years of domestication? The implication is that the ecology of wild cereals was at least as important to traditional farming as the 'domestication syndrome' (the genetic changes in crops) that hitherto has been the almost exclusive focus of debate. Can we learn lessons from the ecological underpinnings of the productivity of the wild relatives of crops that can be more generally applied to food security?

Where Did Agrobiodiversity Come From?

Before the advent of agriculture there was the long-standing environmental knowledge of hunter-gatherers, living in scattered or seasonal encampments, and much of this knowledge was an essential precursor to the management of agrobiodiversity in a more settled agricultural way of life. For example, crops were invariably domesticated from known wild food plants. The knowledge already existed on where and under what conditions wild food plants would grow, the season of harvest, how to harvest most efficiently, how to store grain securely, and how to prepare palatable food from the harvested grain.

The transition from hunter-gathering to animal herding and the growing of crops first happened more than ten thousand years ago. This was a major stage in the human exploitation of the environment. This transition placed the production of food under human control, and, with the greater control of food supply, allowed the growth of human settlements and civilization coupled with the gradual marginalization of hunter-gathering as a way of life.

But this transition did not happen once, in one place, with one crop or one domestic animal: it happened many times, independently, in many regions, and with many crops. In addition, at least some crops had multiple domestication events separately from the same or similar wild relatives. Examples include Asian rice, sorghum and common bean (Burger *et al.*, 2008). This diversity and multiplicity of origins is a key feature of domestication. It removes the possibility that domestication was a chance, even arbitrary, discovery, perhaps flawed, or was a discovery that, with hindsight, could be bettered and improved. It reinforces the idea that domestication was a process of knowledge and skill, taking the best biodiversity available and changing it to agrobiodiversity, much of which is still a vital part of our food supply many thousand years later. We should note that very few new crops or domesticated animals have been produced by recent breeding: early farmers got it very right indeed.

We believe that a greater appreciation of the obvious success of the independent and multiple domestication of crops is a valuable resource for the future management and sustainability of agriculture. But the very richness in number of domesticated species and the many locations of domestication can be confusing. There is, unfortunately, still considerable unresolved debate over the process of domestication. For example, a commentary on a major recent 'conversation' on agricultural origins noted that:

> because of the rapid and still accelerating accumulation of relevant new information over the past quarter-century, many of the extant universalist explanations for agricultural origins now represent more a distraction than an advance in understanding of what is increasingly recognised as a set of long, complex and independent developmental trajectories in different regions of the world.
>
> (Zeder and Smith, 2009)

Another comment on the 'conversation' noted that contributors: 'remain locked into conceptual frameworks and interpretive positions that arose 20 to 30 years ago,' and that researchers needed 'an openness to new directions in conceptualizing and investigating early agriculture' (Denham, 2009).

Baldly stated, explanations of agricultural origins are now in a state of flux, even impasse. This admission is important, because if there is no general agreement over the process of domestication and agricultural origins, then no general recommendations can be made based on presumed patterns of early agriculture, its relations with 'nature' and its subsequent management. For example, there are now numerous attempts to claim that modern agriculture – to ensure sustainability – must mimic nature (reviewed by Wood and Lenné, 1999). But any mimic of nature must know what form 'nature' took at the time of domestication. We extend the theme of whether agriculture should mimic nature in Chapter 11: here we are more concerned with the ecology of crop relatives as a contribution to the management of agrobiodiversity.

Zeder and Smith's (2009) subtitle 'Talking Past Each Other in a Crowded Room' and their mention of a complex process presents us with a 'Gordian knot': an intellectual impasse around the origins of agriculture. This is compounded by the increasing evidence of multiple domestication events for many crops: 'a consensus on the number of origins has not been reached for some of our most important crop species. This is particularly true for several cereal crops of the narrowly delimited Fertile Crescent' (Burger *et al.*, 2008).

If we are to understand fully the process of domestication, this Gordian knot needs to be cut. But does the future of agrobiodiversity management and our food security depend on a fine-tuned understanding of the process of domestication? We think not. It is worth noting that some of the past debate on agricultural origins is certainly not relevant to future agricultural management decisions, driven by the imperative to feed 9 billion people by 2050. Certainly, with our increasing present human population, it makes no sense to further debate 'The Food Crisis in Pre-History' (Cohen, 1977) or just what were the socio-economic drivers that forced the adoption of agriculture, as we now can never return to hunter-gathering for all our food. Also, we feel that a continual emphasis on the 'domestication syndrome' – the genetically based characters that distinguish domesticates from wild species – has consumed too much time and effort with little practical reward for agrobiodiversity management. For example, one domestication feature is larger seed: yet for early cereal domesticates larger seed is not a defining character of domestication. Another criterion of domestication is non-shattering seed heads – that is, seed is maintained on the plant. Yet this can be a result simply of harvesting technique: sweeping into a basket favours shattering even for crop species otherwise fully domesticated. There are further problems of definition: Spriggs (1993) draws attention to what he calls: 'the fetishisation of domestication as an important threshold between cultivation and agriculture' and goes on to argue that a sharp break between cultivation and 'wild plant-food production' is somewhat arbitrary.

We wish to refocus the debate to other facets of agricultural origins perhaps of greater importance for future agrobiodiversity management and our food security than current academic controversies over the origins of agriculture. We will look at the biological and, in particular, the *ecological* continuities between the wild and the crop, rather than the differences (such as the syndrome of domesticated characters). We will search for evidence that the ecology of wild relatives was and still is a realistic model for farming. We have two priorities: first, to understand why certain types of wild plant species rather than others came to be domesticated; secondly – and more importantly – to relate the traditional management of *crops* to the ecological settings of *wild relatives*.

Our approach is chronological over a time span of more than 10,000 years – at each stage highlighting features of the ecology of crop relatives that could throw light on the present and future management of agrobiodiversity. Also, given the rapid and severe climate swings around the time of domestication, knowledge of the ecology and management of crop lineages could perhaps help present and future farming better to secure our food in the face of possible climate change. We consider:

- The wild progenitors of crops;
- Ecological settings of wild relatives;
- Pre-domestication management;
- The impact of the Pleistocene to Holocene transition;
- Cropping analogues of the impacts of fire and flood on wild relatives; and
- Conclusions: lessons for farming.

The Wild Progenitors of Crops

Identifying the wild progenitors of domesticates is one of the least contentious areas of agricultural origins: in most cases, ancestral wild species of a crop can readily be identified.

Although the least contentious, knowledge of the characteristics of immediate wild relatives can be of great advantage to crop management.

Notwithstanding the advantages of knowing more of wild relatives, there is a wide range of types of crops, each with its corresponding wild relatives. The range of wild relatives cannot usefully be lumped into one category. There are several different biological methods of categorizing wild relatives of crops that could help us understand why they were domesticated, and further, to better manage crops once they have been domesticated. For example, Frankel *et al.* (1995) have argued for an ecological and evolutionary pedigree for crops in their relations to natural vegetation: this pedigree would be of survival value – a result of millions of years of adaptation as a wild relative, versus only thousands of years as a crop.

Food type: why were some wild plants suitable for domestication and not others?

The most outstanding feature of most food crops is that we are eating food reserves that the plant has laid down in its tissues for a specific purpose: successful reproduction. If we harvest that food, then in some way we are damaging the survival of that species by undermining its long-standing strategy of high food storage to allow enhanced competitive ability. Compensating the plant in some way can ensure the survival of that plant locally and, of course, our continued access to food. For example, tilling and weeding restore the competitive ability of food-producing species which had been reduced by our harvesting. But this management can occur long before full domestication and it is the management, rather than domestication syndromes, that is more important for our future management of agrobiodiversity.

There are four main classes of reproduction-related food stores in wild relatives. First, for annual species, reserve food is stored in seeds and the plant itself dies, avoiding the inimical dry or cold season. The larger the seed, the better the seedling can compete the following season, the more chance that the wild relative can exclude competitors, and the more attractive the dense stand of a large-seeded species will be to pre-agricultural seed gatherers. This syndrome seems to have been the situation in the heartlands of crop domestication, the Near East. Several different species of large-seeded grasses and legumes occurred in dense stands, were widely exploited pre-domestication and were later domesticated (Zohary, 1969, for cereals). A great advantage of this type of food is that, once harvested, it can be stored for long periods (Willcox, 1998, p. 30).

Second, for perennial species, food reserves in the root or tuber allow the plant a competitive advantage early in the next growing season of a seasonal climate. The aerial parts of the plant die down, and the food reserve allows rapid growth and either subsequent seeding (for example, biennial crops including onions, leeks, carrot) or seeding and also clonal multiplication by roots and tubers (for example, potato). In more equable climates, food may be stored by the plant for several or many years to allow a flush of flowering and seeding (for example, sago palm and, for an alcoholic drink, *Agave*).

Third, wild relatives can store food of various kinds in fruit and seeds to attract animals that are able to disperse fruits and seeds. This mechanism is predominantly found with fruits and nuts ranging from strongly seasonal climates (temperate fruits) to more equable climates (a range of tropical fruits, including bananas, papaya, pineapple). Many of these fruit species are distinctly unpalatable with latex (papaya, jak fruit) or sourness (*Citrus*, apple) or, indeed, are toxic (*Blighia*) before the seed is ripe. A *variorum* of this pattern is the ultra-large seeds that need mechanical protection to avoid being eaten before they are ready for dispersal (Brazil nut). Water-dispersed fruits, while not adapted to attract animal dispersal, need food reserves for the germinating seed – with notably large reserves in coconut.

In all three cases described above, harvesting the reproductive tissue of plants for food reduces the plant's reproductive ability. If harvested to excess, this will exterminate the plant population *unless we compensate the plant in other ways*. This compensation – and the intensity of it – may be crucial to the domestication of some species. For annuals such as wild cereals the yearly removal of all seed will exterminate the plant locally (but see below for Anderson, 1998 who argued that this would not happen). For perennial plants for which our food source is their reproductive tissue (corms, tubers, bulbs), harvesting these will compromise the plants' reproduction. For both of the above, food preparation – grinding and cooking food – will kill the reproductive/propagative tissue and thus will prevent dispersal of the species through discarded kitchen debris. At the other extreme, for long-term perennials, use by humans of their fruit – still normally eaten raw – is a positive advantage to the plant, as rejected (or ingested) seed will allow the harvested species to colonize new areas. There is even a functional difference between type of fruit: the 'bite and swallow' for strawberry-type fruits with small seeds; and the 'suck and spit' for stone fruits such as plums and often tropical species in, for example, the plant families Sapotaceae and Annonaceae. On a scale of intensity of domestication, the annual food species need the greatest compensation to survive, are therefore most dependent on human intervention and reach a greater level of domestication. In contrast, many fruits and nuts are barely domesticated and can readily survive in the wild. For example, it is common to find tropical fruit trees from one continent surviving around long-abandoned homesteads in another continent. Mangoes, originally from India, wherever they are found in Africa and the Americas, are commonly thought to be native and can certainly survive apparently as wild species.

Vegetables from leaves represent a fourth type of human food: here we simply act as a grazing animal – taking leaf and stem tissue the plant needs for its own immediate survival. But, as with other food types taken from plants, if we damage the survival and persistence of the plant population by harvesting our food, then we must compensate the plant in some way to achieve a sustainable food supply for ourselves. This compensation can be done through agriculture: tilling to reduce competition; sowing to give the seed an advantage; dispersing seed and fruits; and generally protecting plants from other animals eating the plant food reserves. But the nurturing of food plants by modifying their ecology certainly long pre-dated agriculture.

Ecological Settings of Wild Relatives

We argue that a greater knowledge and appreciation of the ecological settings of wild crop relatives is an underutilized resource for present-day farming. This resource reflects the long-term co-evolution of wild relatives with their biotic and abiotic environments, long pre-dating the relatively recent evolution of domesticates.

Succession and climax vegetation

A key concept of wild ecology is the idea of plant succession. Simply put, bare ground will be colonized by smaller, annual plants with easily dispersed seed. Over time, these pioneer plant species will be replaced in a process of succession by larger, perennial plants and, eventually, if conditions permit by woodland with large, perennial species – that is, a squeezing out of smaller species by increased competition for light, water and nutrients. But, as with many ideas in ecology, concepts of succession have changed over time. Previous ideas of an orderly, indeed inevitable succession of plant communities to a climax of forest determined by regional climate were successfully challenged in a classic paper by Tansley (1935).

Tansley distinguished between: 'autogenic succession, in which the successive changes are brought about by the action of the plants themselves on the habitat, and allogenic succession in which the changes are brought about by external factors'. In contrast to previous ideas (Clements, 1916; Phillips, 1934), which thought that succession was always progressive (from simpler to more complex communities) and always driven by biotic interactions, Tansley argued that succession could be retrogressive, moving from more luxuriant to poorer vegetation. Tansley then suggested alternatives to 'climatic climaxes': *'edaphic'* climaxes determined by soil; *'physiographic'* climaxes determined by land-relief; *'fire'* climax determined by burning; and a *'mowing'* climax determined by periodic cutting. There is even a *'plagioclimax'* (bent climax) where vegetation comes into equilibrium with any factor deflecting vegetation away from the climatic climax.

Note that Tansley (1935) considered the various types of climaxes as 'relatively stable' in a dynamic equilibrium with controlling factors. Here it is important to accept, with Tansley, that the 'controlling factors' of stable climaxes of the types he lists and that occur naturally (soil, fire and flooding) can also be anthropogenic. Indeed, in order of ease of management by farmers, fire, flooding and soil disturbance characterize important types of agriculture, including shifting cultivation, paddy rice cultivation and most arable farming. More importantly, Tansley's view on allogenic succession – brought about by external factors – and on alternative states for climax vegetation – for example, fire climax and plagioclimax – can be highly relevant to the sustainable management of agrobiodiversity in agriculture.

It is even possible that annual species can form climax vegetation. Whyte (1968: 98) thought that under certain conditions of surface and slope, annual vegetation is the ultimate stage of evolution or climax. This is highly significant, as many of our most important food crops are annuals.

It is evident that wild relatives of the three of the four broad classes of food sources we have described above – 'seeds', 'roots and tubers' and 'fruit and nuts' – have quite different roles to play at different stages in succession to climax vegetation. First, the highly important cereals are wind-pollinated grasses of annual habit which will not be found in climax forest: their natural habitat in the wild is grassland, a 'plagioclimax' maintained by fire, grazing, flooding or other form of disturbance. Second, perennial or biennial roots and tubers have the food reserves to persist in more closed and less disturbed vegetation, but they are not the dominant species of climax forest or woodland. Third, fruits and nuts are produced by a range of plant types from herbaceous perennials to large trees. The role of fruits and nuts is to obtain seed dispersal: seed dispersal can be valuable in any stage of succession. For example, coconut is the largest seed of any crop, with reserves of oil and water. Wild types of coconut, dispersed by sea, can establish on the landward side of beach sand and form the basis for subsequent succession. However, it is uncommon for wild relatives of crops to be a dominant species of climax forest: to dominate they need to be big and long-lived. This precludes a fast rate of domestication. Indeed, some of our plantation tree crops were only fully domesticated under formal breeding programmes (for example, rubber and oil palm). Others, including mango, tea and coffee, have a longer history of domestication but are scarcely dominants in the wild (although, in the wild or in abandoned plantations, tea and coffee can become substantial trees). For the fourth type of crop, leafy vegetables, these species tend to be herbaceous, either as part of an early succession or as a plagioclimax free of trees. For example, wild cabbages (*Brassica*) and wild beets (*Beta*) in Europe are found on shingle beaches maintained by wave wash. Some trees do produce edible leaves, but less so than herbaceous plants, as the leaves of many tree species are protected by anti-grazing toxins.

Despite the forceful arguments of Tansley (1935) that relatively stable climaxes could be maintained in simple vegetation, discussion on the origin of agriculture remains dogged by older ideas of climatic climaxes and early succession species (Hawkes, 1969). There are value judgements in favour of climatic climaxes: 'The ecological climax, as opposed to earlier stages of succession, is a conserver of energy, wasting very little, rather it builds up a store of wealth within the ecosystem' (Fraser Darling, 1956). Fraser Darling designates 'Man the climax breaker' but omits to point out that this 'wealth' of climate climaxes is mainly in the form of wood, which we cannot eat, rather than in the stored plant food which we can eat, characteristic of crop relatives found in other types of climax vegetation.

In particular, much ink has been expended to show that crop ancestors were weedy – that is, early succession species that would inevitably be replaced by other, more successful species, over time. This idea was linked to the suggestion that early cultivation – that is, disturbance by humans – was thought to be essential to maintain this weedy vegetation of our early crops. For example, Hawkes (1969, p. 21) is explicit: crop ancestors were 'ecological weeds' with large food reserves to resist drying out and which 'naturally colonized the bare ground and rubbish heaps provided by man.'

In contrast to the views of authors who emphasize the weedy habit for crop ancestors (Hawkes, 1969; De Wet and Harlan, 1975; Jackson, 1980), we take a directly contrasting view that focuses on two linked features: the climax status of vegetation of crop relatives (discussed above) and also the ecological role of large seeds. In this we are following the insights of Blumler (1996), who sees the domestication of large-seeded annual cereals in seasonally dry climates almost to be inevitable: excluding wild rice (*Zizania*) 'the only region with large-seeded annual grasses that was not a centre of agricultural origin is Morocco/southern Spain.'

Large seeds and monodominant vegetation

Large seeds are not characteristic of weeds: rather, they are an adaptation to one or both of two situations. First, they are characteristic of monospecific stands of trees forming mature climax vegetation. The largest monocotyledonous one-seeded fruit is the coco-de-mer palm (*Lodoicea maldivica),* a very large palm growing in dense stands on two islands in the Seychelles. The largest dicotyledonous one-seed fruit is the mora tree (*Mora excelsa*) growing in monospecific stands on the Caribbean coast of South America. In each case large seed size allows the germinating seed of dominant species of climax vegetation to out-compete invading species for light and nutrients. These two immense woody plants are the very least 'weedy' of all species of plants. Second, and more relevant to our argument for the management of agrobiodiversity, large seed size, while allowing species dominance of vegetation to persist, also is an adaptation to the deep burying of seed needed to escape annual fires. The larger the seed, the deeper in the ground the seed can successfully germinate and emerge. Species, such as annual cereal crop relatives, with large seeds, can survive a fire regime that destroys smaller-seeded competitors. They can then form monodominant stands of annual vegetation. Notably, their local dominance and large seeds would be an attraction to pre-agricultural human food gatherers.

Similar arguments can be used for other plant species with large food reserves: the food reserve allows competitive dominance and monodominant stands.

So, it seems that rather than pinning the ecologically-loaded term 'early-succession species' – that is, weeds of unstable plant communities – on crop ancestors, it makes more ecological sense to look at things Tansley's way. We suggest that the massive stands of cereal crop relatives well documented in the Near East by the likes of Harlan and Zohary (and often related to the origin of agriculture) are in fact species of climax vegetation (and not early stages of a succession depending on human disturbance). A key criterion of the climax nature of vegetation formed by these close relatives of crops is their large seed size. In contrast to the large seed size of cereal crop relatives, weeds, as invaders of bare or highly disturbed ground, characteristically have small, easily dispersed seed. This small seed size of weeds would reduce their value to pre-agricultural food gatherers. For example, the extremely large plant family, the Compositae – often with small seeds with a pappus allowing wind dispersal – are strongly weedy, but very under-represented as crops (small-seeded crops would be too labour-demanding to harvest). In contrast, the large seed size of relatives of our cereals – *coupled with their massive stands of vegetation* – would make such species attractive to food gatherers.

A small seed size is no general bar to qualifying a plant species as a food crop: but small-seeded plants tend to have another plant structure as the human food, rather than the seed. For example, in the sugarcane, a grass with wind dispersed plumose fruits containing the seed, the sugary stem, rather than the tiny seeds, provides human food; in fleshy fruits such as strawberry the swollen fruit, rather than the seed, is our food source.

With the range of survival strategies and the different roles in vegetational succession of wild relatives, it is very unlikely that crops directly derived from disparate types of wild relatives can be shoehorned into one type of agricultural management. Yet this has been attempted. The recent International Assessment of Agricultural Knowledge, Science and Technology for Development (IAASTD) promotes 'agroecology' and notes in a glossary that: 'Agroecological functions are generally maximized when there is high species diversity/perennial forest-like habitats' (McIntyre, 2009). In the light of the information we are reviewing about wild relatives, this assertion of the IAASTD is highly questionable. Wild relatives are not noted for their 'forest-like habitats' – rather the reverse. We return in Chapter 11 to question the validity of the current fashion for agroecology.

An example of the ecology of crop ancestors: cereals in the Near East

We have suggested the ecological link between large seeds and some form of climax status for vegetation of large-seeded crop relatives and also hinted at one ecological determinant (fire) of this vegetation. Where is the evidence for these 'massive pure stands' of crop ancestors and what, in addition to fire, are the environmental factors that could maintain such vegetation? Note that in attempting to step back more than ten thousand years we assume that the controlling influences on present-day vegetation were operational then.

The high importance of the Near Eastern region of crop domestication has produced a wealth of evidence on the presumed ecology of crop relatives. Very notable are reports of the occurrence of wild relatives in dense stands over large areas. In an excellent review of the ecology of wild cereals, Zohary (1969) talks of extensive masses and 'wild fields' characteristic of two wild wheats *Triticum boeoticum* (wild einkorn) and *T. dicoccoides* (wild emmer). Zohary shows that such wild wheats are found in primary habitats – that is, undisturbed by man. Similarly, for wild barley (*Hordeum spontaneum*), Zohary writes that it is 'massively and continuously spread over primary habitats.' In contrast to their 'massive stands' in primary habitats in the heart of their distribution ranges, all three species, in the periphery of their ranges, are found in disturbed, weedy, habitats. Zohary writes of extensive 'natural fields' of wild cereals which are frequent dominant annuals. Also: 'In their total growth and mass, these wild fields of wheat, barley and oats are not inferior to their cultivated counterparts.' Note that Zohary is describing the present-day ecology of wild cereals: at a time before agriculture they were probably more intensively exploited by man.

Botanists draw parallels with cultivated fields:

- 'Over many thousands of hectares it would be possible to harvest wild wheat today from natural stands almost as dense as a cultivated wheat field' (Harlan and Zohary, 1966, of south-eastern Turkey).
- 'On uncultivated slopes, natural fields of these wild cereals extend over many kilometres. In their growth and total mass, these wild fields of wheat, barley and oats are not inferior to their cultivated counterparts' (Zohary, 1969: 55–56).
- 'Even now, stands of wild cereals develop as dense as sown cultivated fields when protected from livestock' (Harlan, 1981).
- 'It is therefore possible to envisage a vast expanse of wild einkorn expanding across the erstwhile steppe, and resembling a seemingly limitless, if patchy, field' (Hillman, 1996: 189).

Other examples of monodominant vegetation of crop relatives

Wild relatives of crops originating beyond the near East have been reported in monodominant stands (reviewed by Wood and Lenné, 1999, 2001). Prain (1903: 357) described the ecology of *Oryza coarctata*, a wild relative of the world's most important food crop, rice. It was the most common and most plentiful grass species in the Sundarabans mangrove swamps of Bengal and: 'the first species to establish itself on the compensation banks of alluvium that are formed on the opposite bank of a river whenever the "set" of the current produced erosion. Such banks vary in size from a few square yards to several acres; whenever they occur they are closely and uniformly covered by a sheet of *Oryza coarctata*.'

For sorghum, Harlan (1989b: 336) identified the *verticilliflorum* race of *Sorghum bicolor* as the progenitor of cultivated sorghums, and noted that it was found as the chief dominant, in enormous quantities, of the extensive tall-grass savannah of Sudan and Chad. Harlan (1992) also noted for Africa: 'Massive stands of truly wild races of sorghum can be found widely distributed over the savanna zones.'

Other monodominant vegetation of grass species

While these examples of dense natural stands are of large-seeded relatives of cereals, there are very many examples of monodominant vegetation of perennial grass species. Any field botanist or ecologist will know of numerous examples of monodominant grass vegetation. Very commonly these monodominant species are found in habitats somehow marginal for plant growth, with obvious abiotic limiting factors. But low diversity need not be correlated with low environmental productivity: for example, salt marshes and estuaries are examples of low species diversity in productive environments (May, 1999). Salt marshes on the margin between land and sea in Europe and North America are often dominated by species of the grass genus *Spartina*. Net annual primary productivity of *Spartina alterniflora* marshes has been reported as up to 6000 g/m^2, a figure close to the highest dry matter yields of intensively managed arable crops (Long and Woolhouse, 1979: 338). Significantly, these authors report that many of these *Spartina* marshes 'consist of extensive monotypic stands of no greater complexity as ecosystems than a field of an arable crop.' For fresh water, monodominant reed beds of *Phragmites australis* grow at the margins of lakes in Europe (stands of *Phragmites* can have an age in excess of 1000 years, Rodwell, 1995: 147).

Other examples include Harlan (1976), who recorded that for the indigenous blue grama grass (*Bouteloua gracilis*) a 'vast expanse on continuous stands covers many thousands of square kilometers of the high plains of the Central United States.' Blue grama grass is favoured by fire. European cheatgrass (*Bromus tectorum*) is found in 'competitive monocultures' on 5 million ha of Idaho and Utah (Pimental *et al.*, 2000). The US Forest Service database (US Forest Service, 2010) (www.fs.fed.us/database/feis/plants/graminoid/brotec/all.html) records that cheatgrass (an introduced species) may remain the *de facto* climax dominant, regardless of site potential and can maintain dominance for many years on sites where native vegetation has been eliminated or severely reduced by grazing, cultivation or fire.

Why is monodominant vegetation so successful?

Although there is a wide knowledge of the occurrence of monodominant vegetation, how do we explain its success? Blumler argues that dense stands of wild relatives are a 'paradox and a puzzle' (Blumler, 1996). The title of May's paper (May, 1999) is 'Unanswered questions in ecology.' Yet answering the questions and resolving the puzzle could be of far greater benefit to agrobiodiversity management in farming than the continual – and we feel erroneous and unsubstantiated – 'tyranny of diversity' in recommendations for agriculture based on an outmoded or irrelevant 'Clementsian' view of climax vegetation. By this we mean that if stable and climax monodominant vegetation can be demonstrated to have a long ecological and evolutionary pedigree, then generic recommendations, from Jackson (1980) with his call for 'increasing the productivity of complex (as opposed to monoculture) farming systems', through Johnson (1998), that monocultures should be replaced by polycultures, to McIntyre (2009), are wrong and dangerous to the future of agriculture. But in the face of these false prescriptions for farming, there is certainly a great and urgent need for field ecologists to discover just how monodominant vegetation – and, in particular, dominant stands of wild cereals – remain stable and productive despite repeated claims that our cereal monocultures are unnatural and fragile. For example, 'It is not nature's way to allow large expanses of land to be planted to a single crop' (Fowler and Mooney, 1990: 42) and 'monocultures in order to function must be predominantly subsidized by chemical inputs' (Altieri, 1998: 69) seem to be perverse statements. In the light of the evidence we have presented above, these views on the nature and management of monocultures are both wrong and dangerous to food production and our future food security.

Pre-domestication Management

Once we establish that there are dense stands of wild cereals with large seeds in the Near Eastern area of crop origins we are part of the way to understanding the ecological management of our cereals fields. These dense stands are apparently stable climax vegetation of annuals. But they are not climatic climax vegetation, which in the region is open oak woodland. Some other ecological factor or factors must be responsible for the persistence of dense stands of annual grasses in the Near East in the face of competition from woodland. For the Near East fire is a probable ecological driver of dense stands of large-seeded annual crop relatives.

But is the Near East unique? Apparently not, as we have given examples for *Sorghum* in Africa and *Oryza coarctata* in coastal India and Bangladesh. For *Sorghum bicolor* seasonal fire is probably the ecological factor favouring dense stands, for *Oryza coarctata* regular saltwater flooding. What is the range of ecological factors that can favour dense stands of plants that provide food for gatherers? More broadly, for other regions and other food plants, are there features of the natural environment or human management of that environment that could cause concentrations of food plants of value to gatherers? Knowledge of these factors could provide an ecological heritage for crops and also an ecological guidance for field management regimes. There could be many candidates for ecological management of wild food species, many of which are linked to overall climate and seasonality – including fire, edaphic features such as gravel fans caused by river erosion, flooding of seasonal rivers, grazing by wild herbivores and even climatic change itself.

Throughout this section we will attempt to understand first, which factors of the natural environment determined the presence and abundance of food plants and, second, if these factors were understood and even manipulated by gatherers to enhance human food supply, could this management act as a model for crop cultivation. The question throughout this chapter is: would the gatherers of wild food before crop domestication have been aware of such ecological relationships? We suggest that they certainly would have been.

Climate and the annual habit

Harlan (1981: 17) noted that the annual floras from which so many major crops are derived mainly evolved under the constraints of long dry seasons and that these long dry seasons appeared to be necessary for the most productive ecosystems for plant domestication. This long dry season was a major feature of the environment of the Near Eastern wild relatives of cereals which grow in a strongly seasonal climate, with a wet season in the mild winter and a hot dry summer. Annual grasses may also dominate under some conditions of soil, slope and ecoclimate in summer rainfall regions of what Whyte (1968) has called 'the grasslands of the Monsoon', stretching from Africa through southern Asia to the more tropical regions of Australia. Harlan (1989a) argued that annual species tend to yield more and be more dependable than perennials. With specific reference to wild rices in West Africa, Harlan noted that the annual *Oryza barthii* seeds abundantly, but *Oryza longistaminata,* a perennial, is relatively shy seeded.

The yield advantage of annuals has been explained by the need of annuals to partition more of their growth into seed (as their sole survival strategy). Spears and Rowe (1991: 140), in a review of tree-based farming systems, note the disadvantages of tree-based mixed species systems included the fact that the yield of annual crops is usually much higher than that of perennials. Further, annuals have a high net primary productivity, much of which is allocated to the reproductive or storage organs that are harvested for food or other purposes, whereas only a small proportion of the total biomass of perennial crops is harvested, except perhaps in the case of species grown specifically to provide fuelwood.

The annual habit – as seen in wild relatives of cereals – is one adaptive response to strong seasonal variation (McCorriston and Hole, 1991). After a relatively rapid growth, flowering and seeding, spikelets carrying the seeds fall and, aided by the awns, may become buried in cracks in the drying soil. The ability to bury the seeds has a multiple adaptive value to the dry season: to escape predation by birds and small mammals; to escape dry-season fires (see below); and, a feature of the large seeds of wild cereals, to permit emergence from deeper in the soil than competing plants with smaller seeds. Indeed, the ability of fallen seed to survive always allowed spontaneous re-seeding from spikelets that had escaped the harvest even under pressure of experimental harvesting of wild cereals stands (Anderson, 1998). This re-seeding was so effective that Anderson argued that the actual cultivation of wild cereals would be unnecessary, as they could always be harvested from wild stands without sowing.

The broad scale of climate and weather was, and still is, impossible to manipulate on a large scale (the best we can do now is glasshouse crops). The association of settlements and agriculture (with the ability to store cereals) was a mechanism for surviving through adverse seasons (cold or dry) without having to relocate to warmer or wetter locations. For most of human history, particularly in strongly seasonal climates, transhumance was the pattern: following and then gathering food sources determined by seasons. The most widespread and best known recent examples of this are the seasonal herding of cattle to graze mountain pastures and also the tracking of rainfall patterns by pastoralists seeking fresh grazing for their herds. Prehistoric examples of hunting depended on the same feature: following the movement of game animals to seasonal pasture. Several areas of crop origins are characterized by pronounced topography, with mountains and valleys, allowing many microclimates and harvesting in seasonally productive environments.

This topography was useful when harvesting annual grasses. Zohary (1969) sheds light on this. There are two features: one is the rapid maturation of annual cereals which grow in masses and are readily harvested. But in 1 or 2 weeks nothing is left but 'barren dry stalks'. The second feature is the differing maturation time of wild cereals in differing localities and, particularly, at different altitudes. Zohary gives details for wild emmer. This ripens at sea level around the end of April, at 700–800 m in mid-May, and at 1400–1600 m in early June. As this progressive seed ripening is a feature of the wider region – Turkey, Iraq and Iran – gatherers can start their harvest on lower slopes and extend the harvest period by progressively moving to the higher slopes. Such was the mass of wild cereals that 500–800 kg/ha could be harvested at an efficiency of 1–4.5 kg/h. Harlan (1989a) estimates that in a 3.5 h period gatherers could harvest enough grain for a 10-day supply. Combining both Zohary's and Harlan's figures, a difference in altitude spreading the harvest over 6 weeks would permit sedentary human populations to harvest the mountain slopes above their settlements and to return home with well over a year's supply of grain. This links the steep topography of areas of origin of crops with the sedentary life (with grain storage and milling facilities), which is almost certainly a prerequisite for farming.

Fire

Most of the warm-temperate and monsoonal grasslands of the world are subject to strong dry seasons – in particular, the Mediterranean climate of the Near-Eastern region of origin of many crops. Under seasonal drought, the ecological filter of fire is responsible for some of the world's most dramatic monodominant grassland – albeit of a perennial species – those of *Imperata cylindrica* in monsoon climates of South-east Asia. Geertz (1963: 25) commented on the 'notorious imperata savannah grass which has turned so much of Southeast Asia into a green desert.' Merrill (1946: 65) noted that *Imperata* was persistent, dominant and occupied vast areas.

While perennial grasses can survive burning from their underground rhizomes, annual grasses of seasonally burned grassland may be highly adapted to burying their seeds in the soil. One mechanism is the geniculate awn, attached to the fallen spikelet with one or more seeds, and capable of hygroscopic twisting with changes in humidity.

Wild cereals have the ability to bury seeds. Zohary (1969) describes this in detail: 'As annuals under Mediterranean summer-dry conditions, these wild cereals are heavily dependent on efficient mechanisms to disperse and plant their seed, protect them in the long dry summer, and facilitate germination when rains start in the subsequent season.' In addition, annual grasses are more flammable than perennial grasses (Moritz and Svihra, 1998). Zohary notes that the seed-bearing structures in wild wheats and barley are: 'specialized arrow-shaped dissemination units which very effectively insert the mature fruiting units into the soil.' In summer these wild cereal 'fields' – Zohary's word – appear as dry barren stalks and big awns protruding from the ground. Zohary concludes that the large size of the seed is both a necessary adaptation to the condition of germination (deep in the soil) and, very significantly, a 'pre-adaptation to domestication', providing 'large quantities of big, easy-to-collect-and-store kernels.' The insertion of seeds into the soil protects the seed from dry-season fires.

Interestingly, awns – often suggested as a mechanism for burying seed (Murbach, 1900; Garnier and Dajoz, 2001; Elbaum *et al*., 2007; Kulić *et al*., 2009) – are often lost early in the stages of full domestication (Fuller and Allarby, 2009). At this time the burying of seed by awns would be redundant, as the agricultural process of sowing and tilling would both remove seed from harm by fire or predation and also place it at the appropriate depth in the soil for successful germination and establishment.

What is apparent with fire – and there are further examples below for flooding – is that quite natural processes occurring independently of humans can have profound effects on vegetation. These effects would be well known to hunter-gatherers who could and did adopt fire as an aid in food procurement. For example, Galinat (1995) suggested that an accidental fire may have set the stage for farming when it burned trees and brush and, thereby, opened the land to the growth of annuals and herbaceous perennials that sprouted from seeds and rhizomes. He thought that the obvious lesson was that fire could serve as a management practice to bring forth the growth of food plants.

One feature of fire is that it is an evolutionary force for adaptation: species that in some way adapt to fire can survive and out-compete other species that have not adapted. If annual grasses can escape dry-season burning which kills trees and shrubs, then succeeding generations of the grass will not have to compete with woody plants for light and water. There is another reason for the need to control competition from trees. Grasses are wind pollinated; a tree canopy reduces the effectiveness of wind pollination (wind-pollinated trees themselves tend to be deciduous and flower before the leaves unfold).

Their resistance to fire helps to explain why grasses are so important as cereal crops. There are other ecologically robust features of grasses that make them ideal crops. Grasses have the ability to resist disturbance, indeed to thrive under seasonally disturbed conditions. Clayton and Renvoize (1986: 16–17) suggest that:

- Grasses are physiologically adaptable to saline, alkaline and seasonally water-

logged soils, forming edaphic grasslands in such environments;
- Grasses benefit from a fire regime that is lethal to many other plants, and, having co-evolved with herbivores, can sustain a level of predation sufficient to cripple many competitors; and
- Grasses have evolved a versatile life style adapted to unstable or fluctuating environments, particularly those associated with strongly seasonal rainfall regimes or the early stages of succession following disturbance.

Flood

In a perceptive passage the Chinese historian Ssu-ma Ch'ien wrote of the Yangtse Valley in 148 BC: 'where the land is tilled by fire and hoed by water' (reported in Grist, 1975: 4). This relates to one of the impacts of both fire and flood on vegetation – physical disturbance clearing existing perennial vegetation and providing a substrate free of competition for crops. This, indeed, is what farmers worldwide spend effort to achieve. There is another common and positive effect of both fire and flood: to provide nutrients for the growing plants.

Flooding, like fire, has a similar impact on vegetation. It offers a selective and often seasonal stress on vegetation that under some conditions favours dense stands of harvestable food plants. As with fire, there is a synergistic bonus for gatherers: under the ecological 'stress' of flooding, large seeds or tubers allow for massive pure stands; the massive stands and the edible seeds or tubers can be efficiently harvested by gatherers. However, while it is easy for gatherers to mimic nature by setting dry-season fires, it is difficult or impossible for gatherers to mimic the effect of most types of natural flooding – the scale is too large.

But all flooding of value to gatherers is natural and seasonal. Floodwaters rise in the wet season and recede in the dry season. For major rivers the geographic scale of this can be enormous and depend on rainfall several thousand kilometres away. The Niger River arises in high rainfall tropical rainforest vegetation, flows north-east through an inland delta in Mali (NASA, n.d.), passes through full desert at Timbuktu then flows south into the Gulf of Guinea through a mangrove delta. The monsoon rains falling on the Himalayas provide seasonal flooding to some of the world's greatest rivers, including the Indus, Ganges, Brahmaputra, Irrawaddy, Mekong, Yangtze, Salween, and the Red and Yellow rivers (the home of 2.5 billion people). During floods, dry season vegetation is swept away (or eaten by fish) and banks are eroded and silt banks deposited. As floods retreat, previously flooded areas and silt banks become re-vegetated. In the example we give above from Prain's (1903) observation of *O. coarctata* growing in dense stands in the rivers passing through the Sundaraban mangroves there are two sets of ecological factors at work. One is the destructive force of the river eroding (and then re-depositing) banks of silt; the other is the increasing salinity. These two forces prevent the establishment and growth of all but the most closely adapted species and these are then able to grow in pure stands without competition. This is an example of May's (1999) argument that species diversity was not correlated with environmental productivity, for example, salt marshes and estuaries are examples of low species diversity in productive environments. This is exactly the environment of *O. coarctata*, which is now used in rice breeding as a source of salt tolerance.

One of us (DW) has walked over pure stands of another grass – *Sporobolus spicatus* – in shallow saline lagoons of Indian Ocean atolls. This is an ecologically tough species, widespread in Africa, where it is eaten as a famine food, and is said to be the most alkali-tolerant grass in Kenya. In Tanzania (at Lake Rukwa) it grows as a pure stand over soda-impregnated soil (Michelmore, 1939). In this paper, Michelmore noted generically that floodplains liable to frequent flooding are nearly treeless and dominated by grasses: we will argue that this 'nearly treeless' observation is important in understanding crop domestication and cereal farming.

Other forms of flooding are less dramatic, with floodplains where the river rises above its dry season course, and wetlands, swamps and marshes, often surrounding seasonally enlarged lakes – termed 'aquatic prairie' (Wigham *et al.*, 1993: 54, derived from Chevalier). As with permanent lakes, vegetation around and in water is usually zoned, with characteristic species often in dense stands. Examples include: dense stands of *Echinochloa stagnina* in tropical West Africa; 'prairies' of the grass *Vossia cuspidate* around Lake Chad, and inundated zones of the Massenya floodplain covered by *Hyparrhenia rufa*, with the wild rice *Oryza barthii* in the more marshy areas (Wigham *et al.*, 1993).

The presence of African wild rice is often noted in descriptions of wetland vegetation. For example, in vegetation described as edaphic grassland or swamp savannah, Thompson (1985: 81) notes wild rice as part of the hydrosere (zonation from wetter to drier areas) of *Vossia–Oryza–Echinochloa–Hyparrhenia rufa*, with the deep-flooded *Oryza* dominating on the Kafue Flats, Zambia. Vesey-Fitzgerald (1970: 72) noted semi-floating mats of *Oryza perennis* in the seasonally-wet valley grasslands of Eastern Africa. Vesey-Fitzgerald regarded such grasslands as entirely natural, in contrast to the fire-maintained, and probably anthropogenic, savannah grassland. Harlan (1989a: 88–91 and Figs. 5.2–5.3) describes and illustrates harvests from dense stands of wild rice in Africa (*O. barthii*, progenitor of the African cultivated rice, *Oryza glaberrima*). *Oryza barthii* was harvested wild on a massive scale and was a local staple across Africa from the southern Sudan to the Atlantic. Evans (1998: 34) reports that the grain yields of wild rice stands in Africa and Asia can exceed 0.6 t/ha – an indication of the stand density of wild rice.

For Asia, Merrill (1946: 65) reported wild sugarcane (*Saccharum spontaneum*) forming dense stands 6 to 15 feet high in alluvial valleys. Yadava (1991: 42–43) reports different species of *Saccharum* as characteristic of grassland communities on recent alluvium of the Ganga and Brahmaputra valleys of north-eastern India, particularly in low-lying, ill-drained topography (the more xerophytic habitats with *Saccharum* are more properly savannah).

Exploitation of edible plants of flooded areas is well reported. For example, Harlan (1989a) reviewed the species of the swamp lands of West Africa, including the wild rices *Oryza longistaminata* of more permanent water and *O. barthii* of seasonally dry waterholes and shallow lakes. Wild rice in Asia (*Oryza rufipogon*, including *O. nivara*) is also harvested. Zong *et al.* (2007) report that Mesolithic foragers gathered aquatic perennial wild rice (*Oryza rufipogon*) in the middle Yangtze basin and the lower Yangtze region from the beginning of the Holocene. The dominant grass of the seasonally flooded inland delta of the Niger River and the shores of Lake Chad is usually *Echinochloa stagnina*, which may occur in 'massive, nearly pure stands' and is gathered as a wild cereal (Harlan, 1989a). This species and a few others were estimated to cover more than 250,000 ha on the bend of the Niger. Wild rice in Thailand is reported (Whyte, 1989) to grow in dense, nearly monospecific stands 'on land that during the course of the year is dampened, becomes slowly inundated, and then dries up.' Whyte then argues that, in this context, paddies – that is, fields of cultivated rice – may be considered as a close ecological analogy to the natural lake-edge environment.

The Impact of the Pleistocene to Holocene Transition

Throughout this chapter we have avoided becoming enmeshed in the unresolved disputes surrounding domestication and the origin of agriculture. Our argument has been that far more of value to the management of crop agrobiodiversity can be gained from knowledge of the ecology of the ancestors of crops prior to their domestication and the roles of fire and flood. In the following pages we will try to relate these two factors to the early management of agrobiodiversity to provide both a model and also a validation for present day arable farming.

However, first we look at climate. Climate is now, and certainly has been since the last glacial period, a major feature of plant ecology, determining for plants their adaptive strategies for survival and reproduction and their geographic extent. Was there a climate shift around the time of agricultural origins?

More importantly, did this have any bearing on the adaptive strategies of people and vegetation in regions of crop origin? If so, can this help with our present-day management of agrobiodiversity, especially with regard to adapting agriculture to the predicted future climate change?

Climate change: the Younger Dryas

The 'Younger Dryas' climatic reversal is named after the appearance in sediments in Northern Europe of the pollen of *Dryas octopetala*, a widely distributed tundra species that is not found in sediments laid down under warmer climate. The Younger Dryas marked the boundary from the last stages of the glacial Pleistocene to the warmer Holocene. The Younger Dryas was a colder, drier period around 11,000 years ago, lasting perhaps 800 years. Both the onset and departure were abrupt, with a rate of temperature change of 7°C in 50 years: note that this rate of warming at the end of the Younger Dryas is double the worst case scenario of the Intergovernmental Panel on Climate Change. Evidence for the Younger Dryas is frequently detected in a diverse array of climate proxies from all latitudes in the northern hemisphere. However, evidence is much weaker in the southern hemisphere, where proxy data often do not show a cold Younger Dryas period (Bettinger *et al.*, 2007).

It seems certain that the Younger Dryas spanned a major and rapid change in climate around the time of crop origins, with dramatic impacts on vegetation. For example, Bettinger *et al.* (2007) suggest that late glacial natural communities must have always been in the process of chaotic reorganization, as the climate varied too rapidly for communities to reach equilibrium. Piperno *et al.* (2007), studying the archaeology of maize domestication in Mexico, suggested that vegetation was re-sorting at the time of the Pleistocene to Holocene transition. Pollen records from the Mediterranean and California showed how much more dynamic plant communities were during the last glacial than in the Holocene. During the millennium or so of the Younger Dryas, foragers in western China were being 'whipsawed' from one climatic extreme to another, often within periods of less than a decade (Marsden and Elston, 2007). Marsden and Elston report that the Younger Dryas was unusual, in that it was a climatic episode that was even more sharply bounded and more dramatic than other climatic change and was the most volatile period of the last 14,500 years. Another paper in the volume argued that shorter term, but intense, centennial- to millennial-scale shifts in the monsoon climate of China appear to have acted as 'triggers' driving dramatic sociological and technological changes (Marsden *et al.*, 2007).

Zeder and Smith (2009) specifically link what they call an 'ice age flashback' that occurred during the Younger Dryas to agricultural origins in both the Near East and in China. For North America, Price (2009) noted that 'an eerie synchronicity in the timing of the first domesticates around the end of the Pleistocene is emerging.' For Mexico, the 'flora and fauna experienced dramatic changes as the Pleistocene was drawing to a close, and as we and others have argued, these ecological cascades created new selective pressures on human populations and their subsistence pursuits, leading to novel and ultimately successful strategies that included the cultivation and domestication of plants' (Piperno *et al.*, 2007).

It is widely known that climatic changes associated with the Younger Dryas around the time of agricultural origins changed the type and distribution of vegetation in the well-studied region of the Near East: for example, the suggestion that the Younger Dryas had a significant impact on climate, vegetation and human economy in southwest Asia (Moore and Hillman, 1992). This impact came after a period when wild cereals and pulses would have become more and more abundant and more important for human subsistence during the late glacial climatic amelioration (Willcox, 1998).

Indeed, the impact of the Younger Dryas on vegetation could have been magnified in the Near East. It is difficult for natural vegetation at the edge of its ecotone (and therefore its survival range) to cope with rapid climatic change (Dansgaard *et al.*, 1989). Large-scale vegetation zones and the boundaries between them (ecotones) are very closely spaced in parts of the Near East. Along the local ecotone between Mediterranean and Irano-Turanian vegetation zones the Younger Dryas, both at its onset and departure, could have caused a displacement of the forest–grassland boundary and had a major impact on vegetation and the distribution of large-seeded grasses. As argued by McCorriston and Hole (1991): 'Situated at the shifting juncture of mediterranean, continental, and monsoonal climates, the Southern Levant provided a unique and changing series of regional ecotones.'

How did the Younger Dryas drive cultivation and the origin of agriculture?

But just what was the impact of the Younger Dryas? The argument has been used that the rapid *onset* of the colder and drier Younger Dryas threatened this food supply by reducing natural stands of cereals and necessitated the change to cultivation (Salamini *et al.*, 2002) and subsequent domestication. That is, the worsening conditions for wild cereals caused by the onset of the Younger Dryas forced their cultivation (Harris, 2003; see also Fuller, 2007).

There is a problem with this argument: if agriculture was forced by food shortages, then there should be some evidence. But, as reported by Munro (2003), that although: 'it is tempting to assign the Younger Dryas a causal role in the adoption of agricultural economies, support for this hypothesis (in the form of food stress and resource intensification) does not currently exist.'

Significantly, as admitted by Harris (2003), the archaeobotanical evidence of domesticated crops at PPNA (Pre-Pottery Neolithic A) sites is extremely meagre. As in the Levant the period of the PPNA is within the Younger Dryas event; if the Younger Dryas had caused domestication, we would expect evidence. As reported by Harris (2003), it was only in the succeeding two millennia of the Pre-Pottery Neolithic B (PPNB *c.* 9500–7500 BP) that agriculture and pastoralism became the main system of food production supporting most of the human population of South-west Asia. It was only when climate conditions improved around 10,000 years ago (that is, on the departure of the Younger Dryas) that the peoples of the southern Levant immediately adopted cereal agriculture (Jones, 2004). This is reinforced by Munro (2003), who claims that the subsequent re-expansion of the Mediterranean forest and the return to warmer and wetter conditions coincides with the appearance of the first agricultural settlements in the Jordan Valley where rich alluvial soils provided a suitable setting for early agriculture.

Arguments that the end of the Younger Dryas provided a driver for cultivation

In the absence of clear evidence for domestication associated with the *onset* of the Younger Dryas, we suggest another possibility: that the Younger Dryas did indeed have an impact on the origin of agriculture, but this was an indirect impact mediated through the spread of trees. We know that the Younger Dryas forestalled the Holocene spread of trees and, at least temporarily, prevented tree vegetation from replacing the grasslands that contained the wild relatives of crops. This effect depends on the differential impact of climate on trees and grasses, with grasses being the most resistant to the effects of the harsh climate (Butler, 1998). Open areas encouraging the persistence of wild grasses would appear from another impact of the Younger Dryas. The sudden onset of the cold conditions of the Younger Dryas caused the death of many trees, with resulting fires (van der Hammen and van Geel, 2008). However, at *the end of* the Younger Dryas, as warmer and wetter conditions returned, trees spread greatly, increasing competition for wild cereals. This dramatic change is evident in Fig. 2 of Allen *et al.* (2000), which is a pollen diagram from a lake in southern Italy. There is the greatest loss of grasses for 80,000 years and an unprecedented increase in woody species at the time of the end of the Younger Dryas (dated 12,800 BP, Table 2 in Allen *et al.*, 2000).

In a classic paper, Hillman (1996) described the movement of vegetation eastwards from the eastern Mediterranean during the late Pleistocene. Grasses advanced in front of trees, with wild cereals probably a conspicuous and possibly dominant component in even the first 'bow-wave' of invaders. Hillman noted 'In the absence of dense tree cover, wild einkorn in particular tends to form dense stands, and its yields per square metre often match those of cultivated wheats under traditional management. ... Similarly, huge dense stands can be produced by wild barley and wild annual rye.'

Our suggestion is that the rapid spread of trees in the early Holocene, as the Younger Dryas abruptly ended and climate became warmer and wetter, threatened the wild cereal subsistence base of foragers that had provided reliable food for thousands of years through to the Pleistocene to Holocene transition. The 'ecological imperative' for cereals – adapted to grassland – would have been then, as now, to escape the domination of trees, which can out-compete grasses and also interfere with wind pollination of all our cereals. A key decision for early cereal farmers would be to identify the ecological conditions under which their new crops would be safe from the Holocene spread of trees.

Was domestication an escape from trees: the 'fight or flight' hypothesis?

The human response to the threat of climate change and subsequent tree spread could have taken either of two pathways: a 'fight' or a 'flight' response.

Fight

One 'fight' approach for humans would be to control the spread of trees through dry-season burning (when the seeds of the annual wild cereals would be below ground). This use of fire was effective. As reported by Turner *et al.* (2009), South-west Asia's grasslands reached their greatest extent during the early Holocene. Grasslands were maintained by dry-season burning that helped to delay the spread of woodland by up to 3000 years, at the same time as Neolithic settlement became established across this grass parkland landscape. Interestingly, a major review of the role of fire in domestication (Lewis, 1972) neglects the impact of fire in the suppression of competition from trees in early agriculture.

The use of fire for tree control would have to have been skilled – things can go badly wrong. The vast green deserts of the perennial rhizomatous grass *Imperata cylindrica* of South-east Asia are a result of the misuse of agricultural fire. The management of domestic grazing animals could also maintain vegetation tree-free, but again, can go wrong and result in invasion by unpalatable, toxic and spiny scrub.

Africa has natural and vast grassland plains, naturally maintained by dry-season fires and grazing – not least by elephants and giraffe, which can tackle most trees, even large ones. The swathe of grassy plains south of the Sahara has given rise to important drought-resistant cereals, including sorghum and pearl millet. In these two species, domestication seems to have proceeded in contact with wild relatives ('sympatric domestication'), with no suggestion that, as perhaps in the Near East, wild cereals had to be moved to enable domestication ('allopatric domestication').

There is an additional dependence on naturally tree-free areas in Africa: for both sorghum and pearl millet early cultivators developed varieties able to grow under *décrue* cultivation in the tree-free seasonally flooded inland delta of the River Niger (Harlan, 1989b). This passive 'fight' mechanism – with nature doing the fighting by flood – could initially use seasonally flooded tree-free river and lake margins for cropping (and later to move to control of irrigation and then terracing). The first records of agriculture in the western Loess Plateau of China at Dadiwan are characterized by deposits resulting from short-term flooding that would have produced disturbed mud flat microhabitats highly suited to agriculture. Dadiwan is thought to be the location of domestication of broomcorn millet (*Panicum miliaceum*) (Bettinger *et al.*, 2007).

Flooding was thought to be important for agricultural origins by Allan (1965): 'Systematic agriculture ... may have begun in the flood plains of the great rivers – first by utilising the natural floods and then by controlled flooding or irrigation, for the step from one to the other is natural and not very difficult.... This is not at all surprising, for these soils are the most persistently fertile in the world: they have an almost inexhaustible supply of available plant nutrients brought down from the upper lands drained by the rivers.' We can reinforce this suggestion by linking the fertility of floodplains noted by Allan to the many examples we have given above of the presence of dense stands of crop relatives in flooded areas of wild ecosystems. Allan's suggestion is further reinforced by multiple examples in the literature of flood-related early agriculture.

In a paper specifically intended to establish the ecological conditions under which the early management of wild rice and the subsequent transition took place, Zong *et al.* (2007) sampled a more than 2000 year sequence of deposits at Kuahuqiao, in the Lower Yangtze coastal region of eastern China. They established that by 7700 calibrated years before present Neolithic people selected lowland swamps for their rice cultivation and settlement, using fire to clear alder-dominated wetland scrub and prepare the site for occupation, then to maintain wet grassland vegetation of paddy type. Regular flooding by slightly brackish water was probably controlled by 'bunding' to maintain crop yields (that is, a 'landesque' improvement). They report very high-intensity clearance and management of the coastal marsh vegetation by fire. Of interest to our argument is that there was evidence that abundant *Typha* stands at the site, encouraged by human management activity, had provided another highly productive food from its starchy roots. They concluded that it was: 'likely that floodwater input to the cultivated areas was also controlled by humans, as the proportion of tidal brackish water influence is maintained at a consistently low level throughout the later cultural phases. The earliest system of rice cultivation in China may well have been a form of "receding-flood" water regulation, with artificial bunding used to retain some nutrient-rich seasonal floodwater, prevent major inundation and provide rice with the consistent water regime it requires.' We have reported this at some length as it provides evidence for both the use of fire and flood over a period of transition to early rice agriculture, in addition to the management of a wild food crop, *Typha*. This is an environment we are encountering often in association with the origins of agriculture: tree-free and of high fertility and productivity.

Flight

The 'flight' approach would be to move crops to a tree-free environment. Earlier in this chapter when dealing with wild relatives of crops, we have a section on 'flood'. The natural flooding of lake shores and rivers can result in moist, fertile and tree-free soils ideal for agriculture. Several suggestions have been made that, at a time of agricultural origins, incipient crops were moved from their areas of natural distribution. For example, Hillman (1996) mentions moisture-enhanced soils on small terraces, and the lower reaches of major *wadi* systems (one of us has worked in Yemen, and seen exactly these environments still being used for agriculture). Flannery (1965) talks of hard-grained grasses (wild cereals) being transported far from the 'biotopes' or niches in which they had been at home and transplanted to new environments. Willcox *et al.* (2008) specifically use the presence of wild einkorn, emmer and wild rye in charred plant remains from the sites in northern Syria and dated to the 10th and 9th millennia calibrated BC beyond their natural ecological range as evidence of cultivation before domestication – because under cultivation 'these cereals would have been able to thrive in adverse climatic and edaphic conditions, because fields would have been situated in favourable microhabitats where competition had been removed.' This is what any modern gardener does on bringing wild plants into gardens and controlling competition by soil preparation and weeding.

Maize: fight or flight?

For another highly important crop globally – maize – the evidence for fight or flight from trees was less clear than in the Old World but research is now providing a similar picture of an impact of the Pleistocene to Holocene transition. What is certain is that fire has played an important role.

Piperno (2006) identified 'environmental perturbations' that occurred during the transition from the Pleistocene to the Holocene related to human occupation of the lowland tropical forest and also the geography and chronology of agricultural origins. Fire was employed by hunters and gatherers and farmers alike during the past 11,000 years as a primary tool of forest modification. This had profound effects on the ancient pre-Columbian development of plant food production and, later, on slash and burn agriculture between *c.* 10,000 BP and 4000 BP in lowland forests from Mexico to the Amazon Basin. Flood impact was also important for maize. For example, palaeoecological data from the Balsas River valley in Mexico suggested that maize and squash were being planted in the productive soils near lake edges that were exposed during the dry season as lake levels fell and that this would have provided attractive yields for minimal effort (Ranere *et al.*, 2009).

In the same Balsas River watershed, during the late glacial period (14,000–10,000 BP), lake beds were dry, the climate was cooler and drier, and open vegetational communities were more widespread than after the Pleistocene ended (Piperno *et al.*, 2007). Records of tree pollen showed the late Pleistocene climate had cooled and then warmed in the early Holocene (exactly as we saw in the Near East), grasses were common and there were other indications that the late Pleistocene climate was drier and cooler than that of today. The same authors suggested that as evidence continues to mount on maize domestication, researchers should consider that at the end of the Pleistocene, probably continuing into the early Holocene, maize was taken under cultivation and domesticated. Significantly, the direct ancestor of maize – Balsas teosinte (*Zea mays* subsp. *parviglumis*) – may have been common in lower elevation areas where it does not now occur.

Of relevance to our thesis of the role of fire and flood, the relative impact of these two agencies on early maize changed. The end of the Pleistocene brought significant shifts in climate and vegetation around the Balsas River watershed. With what appear to have been substantial increases in temperature and precipitation between 11,000 and 9000 BP (13,000–10,000 cal BP), Piperno *et al.* (2007) report that lowland tropical forest expanded on the landscape, and once-dry lake beds filled with water. That is – our suggestion – the open, tree-free and lake-shore environment needed for the cultivation and domestication of maize at the end of the Younger Dryas was no more: maize presumably moved into shifting cultivation, where competition from trees was strictly limited by controlled firing.

Impact of 'fight or flight' on introgression with wild types

The 'fight or flight' hypothesis can be directly related to two types of domestication. As we note above, when domestication proceeds in contact with wild relatives it is 'sympatric domestication'; when wild cereals had to be moved to enable domestication it is 'allopatric domestication'. The 'fight' hypothesis favours sympatric speciation – the crop remaining in contact with wild relatives. In contrast, the 'flight' hypothesis favours allopatric speciation. This contrast has an important consequence for the management of crop agrobiodiversity. For example, with fire management for both sorghum and pearl millet continuing introgression with the wild types occurs. While this can result in continuing enrichment of the gene pool of the crops (Cox and Wood, 1999), it can also cause problems for farmers, as the wild state of natural shattering of the inflorescence causes loss of harvest and subsequent weed problems. For pearl millet there are hybrid swarms between the wild type and the crop, and between cultivated and the weed (*shibra*). In sorghum, hybrids between wild and crop are known as 'shattercanes' and are serious weeds well beyond the region of origin of sorghum.

Interestingly, another pair of crop species that escape competition from trees by 'fight' against tree cover (albeit indirectly by favouring seasonally flooded areas) also show hybrid swarms. These are the highly important Asian rice (*O. sativa*), which forms hybrids with the wild relative *O. rufipogon*; and the African rice (*O. glaberrima*), which has hybrid swarms with the wild *O. barthii*. It seems that natural environments that allow wild species and crops to escape tree competition, also allow introgression between the crop and the wild type. To our knowledge, sorghum, pearl millet and Asian rice show extraordinary morphological variation – evidence of strong human selective pressure that could perhaps counter the genetic fusing tendency of continual contact with wild relatives.

Cropping Analogues of the Impacts of Fire and Flood on Wild Relatives

The independent origins of farming in geographically separate regions of the world precludes a single intellectual revolution at the time of the origin of farming. Far more likely there was a gradual transition to agriculture based on age-old and widespread knowledge of the use of wild relatives of crops as gathered food. This knowledge would certainly include the ecological settings of wild relatives. Is it reasonable that the 'tilling of fields' (that is, the defining character of agriculture) was based on pre-agricultural knowledge of the environmental determinants of dense stands of wild relatives? Our suggestion is that field management was and still is a mimic of natural factors determining the ecological success of wild relatives – in particular, natural factors restricting competition from trees.

This is not a novel suggestion. Grime (1979: 124) has drawn a parallel between the ecology of natural dense stands and agriculture, with the example of *Impatiens glandulifera*, a large summer annual which in Europe colonizes extensive areas where the margins of watercourses have been disturbed by erosion, flooding and silt deposition and which attains the status of a dominant. In a significant reference to agriculture, Grime noted that:

> It is interesting to note that the objective of many forms of arable farming, especially cereal cultivation, is to achieve weed control by creating conditions in which the crop plant attains the status of dominant. As in the example of *I. glandulifera*, dominance by a cereal crop depends primarily upon the synchronous germination of a high density of large seeds followed by the rapid development of a dense vegetation cover composed of a large number of plants of comparable age and maturity.

Whyte (1989) described wild rice on lake margins in Thailand and suggested that, in this context, paddies – that is, fields of cultivated rice – may be considered as a close ecological analogy to the natural lake-edge environment.

Fire

There is no archaeological evidence – nor should we expect any – that pre-agricultural forest burning was deliberately associated with the management of gathered resources. But one of the effects of the burning of forest land during shifting cultivation would certainly have been evident to pre-agriculturalists. Repeated burning of once forested land removes trees and can lead to an extensive monodominant cover of grass, the most notorious of which is *Imperata cylindrica*, a coarse grass which is now widespread in tropical burnt-over areas.

There is ample but scattered evidence that fire is used in field management for several purposes. A major use is in preparing the seedbed from competition from other plants, for example, *chitemene* in Zambia and *hariq* in Sudan (Bartlett, 1956). Jones (1960), for Hampshire in England, reports burning for land preparation. In the Pacific Northwest (US EPA, 2000) burning is considered by some people to be an essential tool because it removes crop residue, makes seeding easier, helps maintain crop yields and reduces the use of chemicals by combating plant diseases, harmful insects and weeds. Burning can also reduce the need for tilling, which makes soil less susceptible to erosion and reduces water quality and windblown dust problems. Burning has been reviewed as a method of plant disease control (Hardison, 1976). For Ghana, farmers believe burning reduces labour cost, suppresses the growth of weeds, allows for more planting to be done per unit area of land, improves yield and is mandatory for successful cultivation of some crops (maize, pepper, tomatoes and yam) (Amissah, 2009).

However, the most notable and widespread use of fire in farming today is for shifting cultivation in tropical seasonally dry forested regions. It is still found throughout the global tropics and subtropics and was formerly used in temperate areas. Lianas are cut, underbrush slashed and trees ring-barked and all is burned at the end of the dry season. Annual and perennial crops are sown. There are several advantages: (i) to remove tree competition for growing crops; (ii) a weed-free seedbed; and (iii) the supply of nutrients from ash. After 2 or more years weed competition builds up, nutrients become exhausted, crop yields drop and the plot is abandoned for many years to allow trees to grow back as a fallow to suppress weeds. As the cropping phase moves into fallow, competition with herbaceous, and later, woody weeds leads to the early suppression of herbaceous crops. Competition from trees completes this: herbaceous crops – including cereals – are completely absent from woody stages of the unmanaged fallow.

At this stage an entirely different trajectory is possible for forests. Fallows are enriched with woody crops and intensively managed. This leads to forest gardening on permanent plots – also pan-tropical – with roots and tubers, fruit and nut-producing trees, and palms (but few or no cereals). We are not reviewing this rich and interesting crop production system as there is very little evidence of the archaeology of domestication for forest gardens (but see Piperno, 2006, cited above, for accumulating evidence that fire was employed as a primary tool for forest modification).

While shifting cultivation is widespread, and a very effective method of controlling the competition from trees that has dogged farmers since agriculture began, its use is now contentious. It is so effective at tree control that it is thought to destroy forests. However, there is wide evidence that most tropical forests have been submitted to numerous cycles of management and abandonment by human societies since remote times (Gómez-Pompa and Burley, 1991). This has had so little impact on forest structure that many are now mistakenly considered to be 'pristine' (Wood, 1993).

Flood

As we have reviewed above for both crop relatives and other wild plants, the occurrence of dense stands of plants in areas of natural flooding would have provided a ready model for farming: as Allan (1965) noted above: 'the step from one to the other is natural and not very difficult.'

As with fire, the earliest examples of the use of flooded fields in farming would not be apparent from field archaeology as there was no disturbance of natural conditions to obtain the flood. The term 'floodwater farming' (Bryan, 1926) refers to floodplains planted into crops after flood waters have receded or where fields are watered by flash floods on alluvial fans – no regular diversion of water is maintained. In describing floodwater farming, Denevan (1995) noted that such 'natural irrigation' was undoubtedly earlier than canal irrigation but left little archaeological evidence. One important example of this 'natural irrigation' is the method of *décrue* farming practised along the Niger and Senegal River floodplains (Harlan, 1989b). Lake Titicaca on the Peru–Bolivia border also demonstrates the same feature of *décrue* agriculture as in Africa: the planting of crops as water levels fall. Over a series of dry years, a 1 m fall in lake level can expose 200,000 ha of previously exposed lake bed with deep, organic-rich soil that is highly prized by local farming communities (Erickson, 1999). For what is undoubtedly the most important crop in the world – rice – van Liere (1985, see also Glover and Higham, 1996) suggested that the earliest form of rice cultivation would have been receding-flood agriculture around the many 'reservoir' lakes and oxbows in the middle courses of the great rivers of South-east Asia and South China.

However, in early farming, the advantages of water control could lead to mechanical methods of modifying water (or land) levels. This is one example of 'landesque' agriculture: 'any investment in land with an anticipated life well beyond that of the present crop, or crop cycle' (Blaikie and Brookfield, 1987: 9). We will pursue this only a short distance, as it leads to terracing and canal irrigation, beyond our topic of agrobiodiversity and agricultural origins.

The transition to landesque management is shown in the earliest records of agriculture in Papua New Guinea, at Kuk swamp in the highlands, showing a progressive development of wetland margin agriculture with taro and bananas. First records are of wetland margin cultivation around 10,000 cal BP, then mounding cultivation around 6800 cal BP, then ditched cultivation around 4000 cal BP (Denham *et al.*, 2003). Mounding, ditching and raised fields as methods of swamp management for agriculture are very widespread. The intention may have been to gain access to rich swamp soils and a controlled supply of water (determined by the height of the mound). The prehistoric Maya practised intensive cultivation of raised and drained field systems in wetland environments (Fedick and Ford, 1990).

There remains a problem with our suggestion that wetland farming mimics the dense natural stands of wetland vegetation. In fact, most wetland vegetation is perennial. While this could provide a model for perennial wetland crops such as taro (*Colocasia esculenta*), it would not do so for important annual cereals such as rice. However, for each of both the African and Asian cultivated rice species there is an annual wild species that grows in, and is harvested from, wetlands. For Asian rice (*O. sativa*) there may have been an extensive transition period, with the cultivation of pre-domesticated rice, before the slow transformation of cultivated rice (Kovak *et al.*, 2007).

Raised field agriculture has been documented in many areas of the Americas, and appears to have provided an important economic base for New World civilizations (Farrington, 1985; Erickson and Candler, 1989). For example, prehistoric raised fields of the margin of Lake Titicaca in Peru once covered at least 82,000 ha. This is an extreme farming environment, at over 3200 m above sea level. In Mesoamerica, Sluyter (1994) showed that intensive wetland agriculture in Mesoamerica was a productive and sustainable agroecosystem that could support dense populations. Whitmore and Turner (1992) report on perhaps the most famous raised field system – that of the 'chinampas' of the basin of Mexico around the Aztec capital of Tenochtitlán – noting that 'Few production systems in the world could match their sustained level of productivity.' Greenland (1997: ix) noted the sustained production of rice in the broad river deltas of Asia, dependent on the nutrients and fertile sediments carried with the seasonal floodwaters. Indeed, most of the world's rice grows on alluvium and annual silt deposits from the Himalayas.

It should be noted that either under natural flooding of lake shores and riverbanks, or bunded or terraced paddy, rice production is not only tree-free, it is predominantly free of competition from any other crop. Flooding is an effective method of providing our all-important rice with a biological '*tabula rasa*' for sustainable production quite free of damaging competition from trees.

Conclusions: Lessons for Farming

In this chapter we have argued and conjectured at some length about the origins and early development of crop production. We feel that some important issues have emerged around what to grow and how to grow it and, notably, these issues are ecologically validated. First, there is a sound reason for growing annual cereals that invest their entire reproductive future in their seed, seed that happens to be highly edible and easily storable by farmers and traders. Second, if the role of the Younger Dryas (a period of exceedingly rapid cooling and then warming) in halting the Holocene spread of trees was a factor in cereal domestication, the efforts of farmers worldwide in controlling trees by the disturbance of fire, flood and tillage is ecologically correct (and, in contrast, recommendations for agroecological farming with crops in tree cover are ecologically suspect). Third, a case can be made for the convenience and productivity of monocultures, as these seem to mimic monospecific stands of wild relatives that can be ecological dominants or even climax vegetation (although regrettably little is known of the present-day ecology of wild relatives). Following our earlier paper on this subject, Grime (2002) noted that 'Wood & Lenné (2001) have argued persuasively that the origins of arable farming and perhaps also its future are to be found as adaptations of naturally-occurring, productive ecosystems dominated by few species.'

References

Allan, W. (1965) *The African Husbandman.* Oliver & Boyd, Edinburgh, UK.

Allen, J.R.M., Watts, W.A. and Huntley, B. (2000) Weichselian Palyostratigraphy, Palaeovegetation and Palaeoenvironment: The Record from Lago Grande di Monticchio, Southern Italy. *Quaternary International* 73/74, 91–110.

Altieri, M.A. (1998) Biodiversity, ecosystem function, and insect pest management in agricultural systems. In: Collins, W.W. and Qualset, C.O. (eds) *Biodiversity in Agroecosystems.* CRC Press, Boca Raton, Florida, pp. 69–84.

Amissah, L. (2009) Indigenous fire management practices in Ghana. In: *Traditional Forest-Related Knowledge and Sustainable Forest Management in Africa.* IUFRO World Series Volume 23, 131–135.

Anderson, P.C. (1998) History of harvesting and threshing techniques for cereals in the prehistoric near East. In: Damania, A.B., Valkoun, J., Willcox, G. and Qualset, C.O. (eds) *The Origins of Agriculture and Crop Domestication.* ICARDA, IPGRI, FAO and UC/GRCP, Aleppo, Syria, pp. 145–159.

Bartlett, H.H. (1956) Fire, primitive agriculture, and grazing in the tropics. In: Thomas, W.L. (ed.) *Man's Role in Changing the Face of the Earth,* Vol. 2. University of Chicago Press, Chicago, Illinois, pp. 692–720.

Bettinger, R.L., Barton, L., Richerson, P.J., Boyd, R., Hui Wang and Won Choi (2007) The transition to agriculture in northwestern China. In: Marsden, D.B., Chen, F.H. and Gao, X. (eds) *Developments in Quaternary Science, 9, Late Quaternary Climate Change and Human Adaptation in Arid China.* Elsevier, Amsterdam, the Netherlands, pp. 83–101.

Blaikie, P. and Brookfield, H.C. (1987) *Land Degradation and Society.* Methuen, London.

Blumler, M.A. (1996) Ecology, evolutionary theory and agricultural origins. In: Harris, D.R. (ed.) *The Origin and Spread of Agriculture and Pastoralism in Eurasia.* University College Press, London, pp. 25–50.

Blumler, M.A. (1998) Introgression of durum wheat into wild emmer and the agricultural origin question. In: Damania, A.B., Valkoun, J., Willcox, G. and Qualset, C.O. (eds) *The Origins of Agriculture and Crop Domestication.* ICARDA, IPGRI, FAO and UC/GRCP, Aleppo, Syria, pp. 252–268.

Bryan, K. (1926) Flood-water farming. *Geographical Review* 19, 444–456.

Burger, J.C., Chapman, M.A. and Burke, J.M. (2008) Molecular insights into the evolution of crop plants. *American Journal of Botany* 95, 113–122.

Butler, A. (1998) Grain legumes: evidence of these important ancient food resources from early pre-agrarian and agrarian sites in Southwest Asia. In: Damania, A.B., Valkoun, J., Willcox, G. and Qualset, C.O. (eds) *The Origins of Agriculture and Crop Domestication.* ICARDA, IPGRI, FAO and UC/GRCP, Aleppo, Syria, pp. 102–117.

Clayton, W.D. and Renvoize, S.A. (1986) *Genera Graminum: Grasses of the World.* HMSO, London.

Clements, F.E. (1916) *Plant Succession.* Carnegie Institute Washington Publication 242. Carnegie Institute, Washington, DC.

Cohen, M.N. (1977) *The Food Crisis in Prehistory: Overpopulation and the Origins of Agriculture.* Yale University Press, New Haven, Connecticut.

Cox, T.S. and Wood, D. (1999) The nature and role of crop diversity. In: Wood, D. and Lenné, J.M. (eds) *Agrobiodiversity: Characterization, Utilization and Management.* CAB International, Wallingford, UK, pp. 35–57.

Dansgaard, W., White, J.W.C. and Johnsen, S.J. (1989) The abrupt termination of the Younger Dryas climate event. *Nature* 339, 532–534.

Denevan, W.M. (1995) Prehistoric agricultural methods as models for sustainability. *Advances in Plant Pathology* 11, 21–43.

Denham, T. (2009) A practice-centered method for charting the emergence and transformation of agriculture. *Current Anthropology* 50, 661–667.

Denham, T.P., Haberle, S.G., Lentfer, C., Fullagar, R., Field, J., Therin, M., Porch, N. and Winsborough, B. (2003) Origins of agriculture at Kuk Swamp in the Highlands of New Guinea. *Science* 301, 189–193.

De Wet, J.M.J. and Harlan, J.R. (1975) Weeds and domesticates: evolution in the man-made habitat. *Economic Botany* 29, 99–108.

Elbaum, R., Zaltzman, L., Burgert, I. and Fratzl, P. (2007) The role of wheat awns in the seed dispersal unit. *Science* 316, 884–886.

Erickson, C.L. (1999) Neo-environmental determinism and agrarian 'collapse' in Andean prehistory. *Antiquity* 73, 634–642.

Erickson, C.L. and Candler, K.L. (1989) Raised fields and sustainable agriculture in the Lake Titicaca Basin of Peru. In: Browder, J.O. (ed.) *Fragile Lands of Latin America: Strategies for Sustainable Development.* Westview Press, Boulder, Colorado, pp. 231–248.

Evans, L.T. (1998) *Feeding the Ten Billion: Plants and Population Growth.* Cambridge University Press, Cambridge.

Farrington, I. (ed.) (1985) *Prehistoric Intensive Agriculture in the Tropics.* British Archaeological Series, International Series, no. 232, Oxford.

Fedick, S.L. and Ford, A. (1990) The prehistoric agricultural landscape of the Central Maya Lowlands: an examination of local variability in a regional context. *World Archaeology* 22, 18–33.

Flannery, K.V. (1965) The ecology of early food production in Mesopotamia. *Science* 147, 1247–1256.

Fowler, C. and Mooney, P. (1990) *Shattering: Food, Politics, and the Loss of Genetic Diversity.* University of Arizona Press, Tucson, Arizona.

Frankel, O.H., Brown, A.D.H. and Burdon, J.J. (1995) *The Conservation of Plant Biodiversity.* Cambridge University Press, Cambridge.

Fraser Darling, F. (1956) Man's ecological dominance through domesticated animals on wild lands. In: Thomas, W.L. (ed.) *Man's Role in Changing the Face of the Earth.* University of Chicago Press, Chicago, Illinois, pp. 778–787.

Fuller, D.Q. (2007) Contrasting patterns in crop domestication and domestication rates: recent archaeobotanical insights from the Old World. *Annals of Botany* 100, 903–924.

Fuller, D.Q. and Allarby, R. (2009) Seed dispersal and crop domestication: shattering, germination and seasonality in evolution under domestication. *Annual Plant Reviews* 38, 238–295.

Galinat, W.C. (1995) The origin of maize: grain of humanity. *Economic Botany* 49, 3–12.

Garnier, L.K.M. and Dajoz, I. (2001) Evolutionary significance of awn length variation in a clonal grass of fire-prone savannas. *Ecology* 82, 1720–1733.

Geertz, C. (1963) *Agricultural Involution: the Processes of Ecological Change in Indonesia.* University of California Press, Berkeley, California.

Glover, I.C. and Higham, C.F.W. (1996) New evidence for early rice cultivation in South, Southeast, and East Asia. In: Harris, D.R. (ed.) *The Origins and Spread of Agriculture and Pastoralism in Eurasia.* Smithsonian Institution Press, Washington, DC, pp. 413–441.

Gómez-Pompa, A. and Burley, F.W. (1991) The management of natural tropical forests. In: Gómez-Pompa, A., Whitmore, T.C. and Hadley, M. (eds) *Rain Forest Regeneration and Management, Man and the Biosphere,* Vol. 6. Unesco, Paris, pp. 3–18.

Greenland, D.J. (1997) *The Sustainability of Rice Farming.* CAB International, Wallingford, UK.

Grime, J.P. (1979) *Plant Strategies and Vegetation Processes.* Wiley, Chichester, UK.

Grime J.P. (2002) Declining plant diversity: empty niches or functional shifts? *Journal of Vegetation Science* 13, 457–460.

Grist, D.H. (1975) *Rice.* Longmans, London.

Hardison, J.R. (1976) Fire and flame for plant disease control. *Annual Review of Phytopathology* 14, 355–379.

Harlan, J.R. (1976) Disease as a factor in plant evolution. *Annual Review of Phytopathology* 14, 31–51.

Harlan, J.R. (1981) Ecological settings for the emergence of agriculture. In: Thresh, J.M. (ed.) *Pests, Pathogens and Vegetation.* Pitman, London, pp. 3–22.

Harlan, J.R. (1989a) Wild-grass harvesting in the Sahara and Sub-Sahara of Africa. In: Harris, D.R. and

Hillman, G.C. (eds) *Foraging and Farming: the Evolution of Plant Exploitation.* Unwin Hyman, London, pp. 79–98.
Harlan, J.R. (1989b) The tropical African cereals. In: Harris, D.R. and Hillman, G.C. (eds) *Foraging and Farming: the Evolution of Plant Exploitation.* Unwin Hyman, London, pp. 335–343.
Harlan, J.R. (1992) *Crops and Man,* 2nd edn. American Society of Agronomy, Madison, Wisconsin.
Harlan, J.R. and Zohary, D. (1966) Distribution of wild wheats and cereals. *Science* 153, 1074–1080.
Harris, D.R. (2003) Climatic change and the beginnings of agriculture: the case of the Younger Dryas. In: Rothschild, L. and Lister. A.M. (eds) *Evolution on Planet Earth: Impact of the Physical Environment.* Academic Press, London, pp. 379–394.
Hassan, F. (1977) The dynamics of agricultural origins in Palestine: a theoretical model. In: Reed, C.A. (ed.) *Origins of Agriculture.* Mouton, The Hague, the Netherlands, pp. 589–609.
Hawkes, J.G. (1969) The ecological background of plant domestication. In: Ucko, P.J. and Dimbleby, G.W. (eds) *The Domestication and Exploitation of Plants and Animals.* Duckworth, London, pp. 17–29.
Hillman, G. (1996) Late Pleistocene changes in wild food plants available to hunter-gatherers of the northern Fertile Crescent: possible preludes to cereal cultivation. In: Harris, D.R. (ed.) *The Origin and Spread of Agriculture and Pastoralism in Eurasia.* University College Press, London, pp. 159–203.
Jackson W. (1980) *New Roots for Agriculture.* Friends of the Earth, San Francisco, California.
Johnson, I. (1998) Letter from the new Vice President Environmentally and Socially Sustainable Development. *Environment Matters Annual Review* The World Bank, Autumn 1998, p. 3.
Jones, E.L. (1960) Eighteenth century changes in Hampshire chalkland farming. *The Agricultural History Review* 8, 5–19.
Jones, E.L. (2004) Comments. *Current Anthropology* 45 (Suppl.), S25–S26.
Kovak, M.K., Sweeney, M.T. and McCouch, S.R. (2007) New insights into the history of rice domestication. *Trends in Genetics* 23, 578–587.
Kulić, I.M., Mani, M., Mohrbach, H., Thaokar, R. and Mahadevan, L. (2009) Botanical ratchets. *Proceedings of the Royal Society, London, B* 276, 2243–2247.
Lewis, H.T. (1972) The role of fire in the domestication of plants and animals in Southwest Asia: a hypothesis. *Man* N.S. 7, 195–222.
Long, S.P. and Woolhouse, H.W. (1979) Primary productivity in *Spartina* marshes. In: Jefferies, R.L. and Davy, A.J. (eds) *Ecological Processes in Coastal Environments.* Blackwell Scientific, Oxford, pp. 333–352.
Marsden, D.B. and Elston, R.G. (2007) Variation in Late Quaternary central Asian climates and the nature of human response. *Developments in Quaternary Sciences* 9, 69–82.
Marsden, D.B., Chen, F.H. and Gao, X. (2007) Changing views of Late Quaternary human adaptation in arid China. In: Marsden, D.B., Chen, F.H. and Gao, X. (eds) *Developments in Quaternary Science, 9, Late Quaternary Climate Change and Human Adaptation in Arid China.* Elsevier, Amsterdam, the Netherlands, pp. 227–232.
May, R.M. (1999) Unanswered questions in ecology. *Philosophical Transactions of the Royal Society B* 354, 1951–1959.
McCorriston, J. and Hole, F. (1991) The ecology of seasonal stress and the origins of agriculture in the Near East. *American Anthropologist (N.S.)* 93, 46–69.
McIntyre, B.D. (2009) International Assessment of Agricultural Knowledge, Science and Technology for Development. (2009) *Latin America and the Caribbean (LAC) report.* Washington, DC.
Merrill, E.D. (1946) *Plant Life of the Pacific World.* Macmillan, New York.
Michelmore, A.P.G. (1939) Observations on Tropical African grasslands. *Journal of Ecology* 27, 282–312.
Moore, A.M.T. and Hillman, G.C. (1992) The Pleistocene to Holocene transition and human economy in Southwest Asia: the impact of the Younger Dryas. *American Antiquity* 57, 482–494.
Moritz, R. and Svihra, S. (1998) *Pyrophytic vs fire resistant plants.* University of California Cooperative Extension, Oakland, California.
Munro, N.D. (2003) Small game, the Younger Dryas, and the transition to agriculture in the Southern Levant. *Mitteilungen der Gesellschaft für Urgeschichte* 12, 47–71.
Murbach, L. (1900) Note on the Mechanics of the Seed-Burying Awns of *Stipa avenacea. Botanical Gazette* 30, 113.
NASA (n.d.) Inland Delta of the Niger River. Available at: http://veimages.gsfc.nasa.gov/2243/Mali.A2001291.1045.250m.jpg (accessed 15 September 2010).
Phillips, J. (1934) Succession, development, the climax and the complex organism: an analysis of concepts. I. *Journal of Ecology* 22, 554–571.

Pimental, D., Lach, L., Zuniga, R. and Mossison, D. (2000) Environmental and economic costs of nonindigenous species in the United States. *BioScience* 50, 53–65.

Piperno, D.R. (2006) Quaternary environmental history and agricultural impact on vegetation in Central America. *Annals of the Missouri Botanical Garden* 93, 274–296.

Piperno, D.R, Moreno, J.E, Iriarte, J., Holst, I., Lachniet, M., Jones, J.G., Ranere, A.J. and Castanzo, R. (2007) Late Pleistocene and Holocene environmental history of the Iguala Valley, Central Balsas Watershed of Mexico. *Proceedings of the National Academy of Sciences* 104, 11874–11881.

Prain, D. (1903) Flora of the Sundribuns. *Records of the Botanical Survey of India* 1903, 231–370.

Price, T.D. (2009) Ancient farming in eastern North America. *Proceedings of the National Academy of Sciences* 106, 6427–6428.

Ranere, A.J., Piperno, D.R., Holst, I., Dickau, R. and Iriarte, J. (2009) The cultural and chronological context of early Holocene maize and squash domestication in the Central Balsas River Valley, Mexico. *Proceedings of the National Academy of Sciences* 106, 5014–5018.

Rodwell, J.S. (ed.) (1995) *British Plant Communities,* Vol. 4*: Aquatic Communities, Swamps and Tall-Herb Fens.* Cambridge University Press, Cambridge.

Salamini, F., Hakan Özkan, H., Brandolini, A., Schäfer-Pregl, R. and Martin, W. (2002) Genetics and geography of wild cereal domestication in the Near East. *Nature Reviews Genetics* 3, 429–441.

Sluyter, A. (1994) Intensive wetland agriculture in Mesoamerica: space, time, and form. *Annals of the Association of American Geographers* 84, 557–584.

Spears, J. and Rowe, R. (1991) Tree-based farming systems. In: Garbus, L., Pritchard, A. and Knudsen, O. (eds) *Agricultural Issues in the 1990s: Proceedings of the Eleventh Agriculture Sector Symposium.* World Bank, Washington, DC, pp. 129–154.

Spriggs, M. (1993) Pleistocene agriculture in the Pacific: why not? In: Smith, M.A., Spriggs, M. and Frankhauser, B. Sahul in review: Pleistocene archaeology in Australia, New Guinea and Island Melanesia. *Occasional Papers in Prehistory* 24, 137–143. Department of Prehistory, Research School of Pacific Studies, Australian National University, Canberra, Australia.

Tansley, A.G. (1935) The use and abuse of vegetational concepts and terms. *Ecology* 16, 284–307.

Thompson, K. (1985) Emergent plants of permanent and seasonally-flooded wetlands. In: Denny, P. (ed.) *The Ecology and Management of African Wetland Vegetation: a Botanical Account of African Swamps and Shallow Waterbodies.* Dr W. Junk, Dordrecht, the Netherlands, pp. 43–107.

Turner, R., Roberts, N., Eastwood, W.J., Jenkins, E. and Rosen, A. (2009) Fire, climate and the origins of agriculture: micro-charcoal records of biomass burning during the last glacial-interglacial transition in Southwest Asia. *Journal of Quaternary Science* 25, 371–386.

US EPA (Environmental Protection Agency) (2000) *Agricultural Burning.* Seattle, Washington.

US Forest Sevice (2010) *Bromus tectorum.* Available at: www.fs.fed.us/database/feis/plants/graminoid/brotec/all.html (accessed 15 September 2010).

van der Hammen, T. and van Geel, B. (2008) Charcoal in soils of the Allerød-Younger Dryas transition were the result of natural fires and not necessarily the effect of an extra terrestrial impact. *Netherlands Journal of Geosciences* 87, 359–361.

van Liere, W. (1985) Early agriculture and intensification in mainland Southeast Asia. In: Farrington, I. (ed.) *Prehistoric Intensive Agriculture in the Tropics.* British Archaeological Series, International Series, no. 232, Oxford.

Vesey-Fitzgerald, D.F. (1970) The origin and distribution of valley grasslands in East Africa. *Journal of Ecology* 58, 51–75.

Whitmore, T.M. and Turner II, B.L. (1992) Landscapes of cultivation in Mesoamerica on the eve of the Conquest. *Annals of the Association of American Geographers* 82, 402–425.

Whyte, J.J. (1989) Ethnoecological observations on wild and cultivated rice in northeastern Thailand. In: Harris, D.R. and Hillman, G.C. (eds) *Foraging and Farming: the Evolution of Plant Exploitation.* Unwin Hyman, London, pp. 152–158.

Whyte, R.O. (1968) *Grasslands of the Monsoon.* Faber and Faber, London, 325 pp.

Wigham, D.F., Dykyjová, D. and Hejný, S. (1993) *Wetlands of the World I: Inventory, Ecology, and Management.* Kluwer, Dordrecht, the Netherlands.

Willcox, G. (1998) Archaeobotanical evidence for the beginnings of agriculture in Southwest Asia. In: Damania, A.B., Valkoun, J., Willcox, G. and Qualset, C.O. (eds) *The Origins of Agriculture and Crop Domestication.* ICARDA, IPGRI, FAO and UC/GRCP, Aleppo, Syria, pp. 25–38.

Willcox, G., Fornite, S. and Herveux, L. (2008) Early Holocene cultivation before domestication in northern Syria. *Vegetation History and Archaeobotany* 17, 313–325.

Wood, D. (1993) Forests to fields: restoring tropical lands to agriculture. *Land Use Policy* 10, 91–107.
Wood, D. and Lenné, J.M. (1999) Agrobiodiversity and natural biodiversity: some parallels. In: Wood, D. and Lenné, J.M. (eds) *Agrobiodiversity: Characterization, Utilization and Management.* CAB International, Wallingford, UK, pp. 425–445.
Wood, D. and Lenné, J. (2001) Nature's Fields: a neglected model for increasing food production. *Outlook on Agriculture* 30, 165–174.
Yadava, P.S. (1991) Savannas of north-east India. In: Werner, P.A. (ed.) *Savanna Ecology and Management: Australian Perspectives and Intercontinental Comparisons.* Blackwell, Oxford, pp. 41–50.
Zeder, M.A. and Smith, B.D. (2009) A conversation on agricultural origins: talking past each other in a crowded room. *Current Anthropology* 50, 681–690.
Zohary, D. (1969) The progenitors of wheat and barley in relation to domestication and agricultural dispersal in the Old World. In: Ucko, P.J. and Dimbleby, G.W. (eds) *The Domestication and Exploitation of Plants and Animals.* Duckworth, London, pp. 47–66.
Zong, Y., Chen, Z., Innes, J.B., Chen, C., Wang, Z. and Wang, H. (2007) Fire and flood management of coastal swamp enabled first rice paddy cultivation in east China. *Nature* 449, 459–462.

4 Crop Introduction and Agrobiodiversity Management

D. Wood

If there was one thing that had been clearly shown by the experience of the nineteenth century, it was the potential value of crop introductions from one country to another. By 1900 this had become almost an article of faith rather than of policy, and this activity was the main preoccupation of many of the new Departments of Agriculture...

Masefield (1972, p. 63)

... the crops that now dominate the agricultural economies of the advanced industrial nations are not, for the most part, indigenous species. They have been introduced from elsewhere, principally from what is now the Third World. ... If the United States now has a food weapon, as former Secretary of Agriculture Earl Butz so bluntly put it, it is because nations such as Nicaragua, Ethiopia, Iran, and China have supplied, respectively, the corn, wheat, alfalfa, and soybean for its arsenal.

Kloppenburg (1988, p. 49)

Nonetheless, many exotic species provide important ecosystem services to humanity (e.g. many food plants and animals)...

Vellend *et al.* (2007)

Origin and Distribution of Crops

Crops and domestic animals originated from their wild relatives though single, or at the most, few events of domestication in limited regions. There has been dispute over how wide or narrow these regions were but no dispute over the fact that, on the larger scale, different continents had different suites of domesticates; some regions such as North America had few, and Australia and Southern Africa had none.

While wild animals simply move or migrate to where conditions suit them, it is more complicated for static plants. To colonize new areas, plants need to have natural dispersal mechanisms. These are of a wide variety of types, including wind, water and animals. But for crops, the main dispersal mechanism is humans, so much so that wild-type dispersal mechanisms may be lost by evolution – as with maize, where the seeds are enveloped by bracts. But remaining mechanisms – especially those involving animals – may disperse crop seed widely, as when a bird eats a soft-fruit crop and carries seed in its gut.

There is an additional mechanism for crops beyond the accidental dispersal of seed to new areas. People, since the dawn of agriculture, have moved seed deliberately, to extend the areas that can be farmed. This is true also for domestic animals, which can be herded hundreds of kilometres. This could be a gradual process of diffusion, as settlements were established away from the homelands of crops. But it could also be the long distance and deliberate transfer of crops along early trade routes. So very soon after domestication – and possibly as a direct result of the increased human population allowed by domestication – domesticates spread by deliberate introduction. By the early second millennium BC, there was significant wheat and barley production in China (Flad *et al.*, 2010) from introductions along what became known as the silk route, which could be easily accessed from the West Asian region of origin of wheat and barley. This trade route from one side of Asia to the other probably accounted for the eastwards introduction of sheep and cattle from West Asia, and for the westwards movement of *Panicum* and *Setaria* millets from China to West Asia and onwards to Africa. This early movement of domesticates to and from China is an example of the advantages of domesticates staying within a similar latitudinal band. Here day-length remains similar, avoiding problems of photoperiodism, and climate zones may be similar (homoclimes or analogous climates), with no greater extremes of winter and summer.

There were two other early examples of east to west movement of crops from Asia to Africa/Madagascar across the Indian Ocean, one for bananas and plantains (*Musa* spp.) of an uncertain date at least 1000 years ago, the other for rice to Madagascar, dating around AD 600. There are three features of interest to these introductions: (i) introduction was almost certainly by sea, as neither crop will grow in the dry conditions around the northern Indian Ocean; (ii) in each case the crop became the staple foodstuff, with rice in highland Madagascar and banana in the East African highlands; and (iii) also in each case – after what was a probable genetic bottleneck associated with long-distance introduction – a multitude of new varieties of each crop were selected. This was remarkable for bananas, as they are sterile triploids not producing seed. All variation must have been somaclonal. Rice in Madagascar also produced remarkable variants of a type not known in Asia, from crosses between introduced *indica* and *japonica* types (Ahmedi *et al.*, 1991). There was a similar prehistoric introduction of sweet potato from South America by the Polynesian navigators crossing the Pacific. As a result, sweet potato is widespread around the Pacific with notable introductions to New Zealand, where cold tolerant varieties evolved, and to the Highlands of Papua New Guinea.

These are examples of long-distance trans-oceanic introduction. Movement by land is also possible. Many of the early introductions from Africa to India and vice versa were dryland species (of crops and animals) that could readily have moved in stages (examples include sorghum, pigeon pea, pearl millet, mango and zebu cattle). Zebu cattle of Indian origin are thought to have been introduced to Africa around 4000 years ago but only started to become widespread in East Africa about AD 700 with the Arabic migrations into North and East Africa (MacHugh *et al.*, 1997).

The Columbian Exchange

The most important series of deliberate plant introductions started with European discovery of the Americas and continues to this day (known as the 'Columbian Exchange': Crosby, 1972). In the following centuries there were hundreds of thousands of trans-Atlantic movements of seed of thousands of varieties of hundreds of crops and dozens of domestic animals. There was a strong link to European colonization of the Americas (and the associated slave trade to provide workers in plantations) and, from the start, trade in tropical products, such as sugar, to Europe.

Initial introductions were almost random and most introductions probably failed to be economically viable – but many did. For example, Madagascar was the main early source of rice for the USA when in 1686 an English ship sailing from Madagascar docked in Charleston – apparently almost by accident – and from it a local farmer obtained a 'peck' of rice seed (around 7 kg). This gave rise to the variety 'Charleston White'. By 1850 production had risen to 100,000 tons. The next documented introduction of rice to the USA – from Honduras – was almost 200 years later than that from Madagascar (Smith and Dilday, 2002).

Systematic Crop Introduction

Countries with a central responsibility for a great range of territory and agroecological conditions were highly active in the introduction of crops and domestic animals. The British system of colonial botanic gardens specialized in the inter-tropical movement of plantation crops (Brockway, 1979). There was a series of notable successes: cocoa from the Amazon to West Africa; rubber from the Amazon to the Malay Peninsula; oil palm from West Africa to the Malay Peninsula. All these introductions produced globally important trade commodities. Captain Bligh, of HMS *Bounty* fame, introduced breadfruit from the tropical Pacific to the West Indies. He also supplied samples of a new fruit from West Africa to Kew Gardens. This was the akee, which was introduced to the Caribbean, and called *Blighia sapida* in honour of Bligh. There is more to this story: the wife of one of the most famous plant explorers of all (Wilson Popenoe) died after accidentally eating unripe – and therefore poisonous – akee from the economic tree collection at the United Fruit Research station in Honduras.

Even when the origin and destination were tropical, samples were often grown for multiplication in hothouses of botanic gardens in Europe. A particular problem of early tropical introductions was the length of time needed for sea voyages coupled with the inability of many tropical crops to produce dry, storable seed. The Wardian case was a solution to this – a portable greenhouse that could be stored on deck of sailing ships and in which living plants could be watered and grown for months. Of necessity, sea passages with tropical plants via Europe would only by attempted in summer.

The great cost of these early collecting expeditions imposed a correspondingly high value to conserving the collected material as 'back-up' samples in the security of state botanic gardens, if only to avoid the cost of recollecting. A specimen of the first introduction of oil palm (*Elaeis guineensis*) to Java is still growing in the Bogor Botanic Garden after 150 years.

The USA followed the European example when it inherited the Spanish colonial empire, including Puerto Rico, Guam and the Philippines. This dramatically extended the range of conditions for which US agricultural scientists needed to give advice on agricultural production. As usual in colonial agricultural production, work was backed-up by a range of scientists, including botanists and entomologists. Publications valuable to this day ensued, such as Merrill (1945) – who was once the Director of the Philippines Bureau of Science – and Safford (1905). Also, in what was economic colonization in Central America, US companies such as United Fruit developed an excellent network of crop introduction and trials in an attempt to produce tropical crops for US markets. The major success was banana. This had been introduced previously from South-east Asia but large collections of banana varieties and many other tropical fruits and economic plants were maintained in a model system of botanic gardens in Panama (Summit), Costa Rica (Turrialba) and Honduras (Lancetilla).

A major programme of introduction of existing crop varieties was essential for the success of US agriculture as prior to the Columbian exchange North America was a genetic desert with almost nothing of its own. The programme of introduction built up during the 19th century. It was recognized from the start that many introductions would prove unsuccessful but amongst the vast quantities of economic plants introduced were the foundations of US agricultural predominance. If there was a scatter-gun approach to collecting and introduction – with quantity more important than quality – there was also a scatter-gun approach to testing the imported samples. The introduction programme was formally under the US Patent Office and by 1849 the Commissioner was distributing by post over 60,000 seed packages a year to farmers. As described by Kloppenburg (1988), American agriculture was raised on 'the product of thousands of experiments by thousands of farmers committing millions of hours of labor in thousands of diverse ecological niches over a period of many decades.' The apparent over-kill of introduction and distribution for on-farm testing worked well: crops and varieties pre-adapted to a range of conditions were adopted; less useful crops and varieties were quickly discarded. On-farm selection of superior varieties was possible.

The continued reach of the plant introduction programme was substantial: sources were legion. For example, in 1913 the Imperial Research Station in Sokode, Togoland (then a German colony) sent a sample of Kersting's groundnut (now known as *Macrotyloma geocarpum*) to the United States Department of Agriculture (USDA, 1913). This had only be described as a new species 3 years previously, had been grown by the Botanical Centre for the Colonies in the Botanic Garden at Dahlem in Germany, and had just been flagged as of interest by British colonial botanists (Stapf, 1912). In Africa it is now heading for extinction as a crop 'found among the old women who cultivate it on small farms as a "legacy crop"' (Amujoyegbe *et al.*, 2007). Sokode was a typical colonial plant introduction station, with trials of introduced teak and oil palm.

Official US seed distribution direct to farmers reached a peak in 1897 when 22,195,381 packages (each containing five varieties) were distributed (Kloppenburg, 1988). Increasingly, these were not exotic (introduced) varieties, but seed of common varieties. But the pattern of evaluation of samples was changing. In 1887 state agricultural experiment stations (SAESs) were formed. From then on, more of the burden of evaluation of exotic varieties fell to them, although a two-pronged approach was maintained – with farmers and SAESs both applying their different skills in evaluation. But there was a limit to what farmers could achieve in combining valuable characteristics of different varieties: this was the task of the next development – the rise of public and private plant breeding. This was associated with a search for specific valuable characters – in particular, resistances to pests and disease.

The programme of plant introduction as a service to US agriculture was formalized as 'Seed and Plant Introduction' in 1898 under the USDA. This became the most notable global effort in crop introduction, which has now accumulated and maintained for use the largest and most important collection of introduced economic plants ever. The latest inventory (2008) reaches sample number PI 655520 (USDA, 2008), including hundreds of plant genera and thousands of species of economic plants.

Although Kersting's groundnut seems to have failed as an introduction, the success of other crops introduced to the USA could be rapid and highly successful. For example, the spread of soybean from its origin in South-east Asia is the best example of an introduced crop becoming of major importance, first to the USA and now Brazil and Argentina. Lockeretz (1988) called it the 'spectacular rise of the soybean' and wrote: 'The success attending the introduction of the soybean is without parallel in modern US agricultural history. ...The speed and overwhelming success with which this complex and difficult programme achieved was remarkable, so that anyone concerned with other potential new crops should be familiar with this crop introduction par excellence.'

As a direct result of over 400 years of increasingly formalized plant introduction, the USA is now the premier exporter of introduced crops, with soybean and wheat originating in Asia, and maize in Central America.

Soviet Russia covered a region with very wide farming conditions and needs for introduced crops. As with other countries, post-revolution Russia began a wide-ranging programme of crop introduction, associated with an outstanding collector and crop botanist Vavilov (Vavilov, 1951). Vavilov's main claim to fame was the use of his extensive knowledge as a collector on five continents to identify eight Centres of Crop Origin, each with a characteristic suite of crops. For example, the 'Near Eastern' Centre includes wheat, barley, lentils and chickpea, while the Africa Centre includes pearl millet, sorghum and cowpea. While these 'Centres of Origin' have since been extensively questioned and modified – for example, they do not coincide with areas of maximum crop diversity – they are a valuable foundation both to our knowledge of crop origins and evolution, and also to crop introduction.

Importance of Crop Introduction

On the widely accepted view that crops and varieties are 'locally adapted', they would be expected to grow best and produce most in the regions where they were domesticated. Yet this, somewhat surprisingly, seems not generally to be the case. This apparent paradox was commented on by the British colonial botanist, Purseglove, who wrote: 'a striking feature of the present-day distribution of tropical crops is ... that the main areas of production of the major economic crops are usually far removed from the regions in which they originated' (Purseglove, 1968).

Using data only from developing countries produce striking results (Wood, 1988; see Table 4.1): for Africa 70% by value of total crop production is from crops introduced from Asia (26%) and the Americas (44%). At individual country level, it can be as high as 99%. Sugarcane and bananas were first introduced to Africa more than a thousand years ago. Later introductions to Africa included cassava, partly replacing indigenous yams, groundnut replacing (as an oil crop) sesame, maize partly replacing sorghum and *Phaseolus* beans partly replacing cowpea. Therefore, at the unquantifiable cost of losing some of the variation of its indigenous crops, Africa gained entirely new crops, and there has since been a substantial diversification of these (i.e. maize and *Phaseolus* beans). There are no African crops capable of replacing the productivity of introduced sweet potato, *Phaseolus* beans and bananas in, for example, the fertile central highlands of Africa. However, in some regions there was a useful complementarity between introduced crops and local crops, as in the highlands of Ethiopia, where the Near-Eastern crop complement of barley, wheat and chickpea was joined in production by the local domesticates, teff (*Eragrostis tef*), sorghum, noog (*Guizotia abyssinica*), the banana relative *Ensete* and others.

There is a similar figure for developing countries of the Americas, where 32% of production is from indigenous crops, with 18% from African crops (including coffee, oil palm and sorghum) and 50% from Asia (including bananas, sugarcane, citrus and rice). Asia is somewhat different. Seventy per cent of Asian crop production is of Asian origin (mainly Asian rice, the largest production of any crop in the world) with 18% of crops from the Americas and 12% from Africa.

Co-evolved Pests and Diseases and Local Adaptation

Anderson (1954, p. 150) wrote about sunflower: 'the one native American crop. [although] no world crop originated in the area of its modern commercial importance and sunflowers are no exception.' Anderson suggested why: 'In the region where a crop was domesticated there are the maximum number of pests and diseases which have evolved to prey on that particular kind of plant. ... the farther you get from its center of

Table 4.1. Percentage reliance on introduced crops: developing countries in Africa and the Americas[a] (adapted from Wood, 1988).

Country	Introduced crop (%)
Africa	
Malawi	99.1
Zaire	96.5
Mozambique	93.8
Morocco	91.5
Zambia	91.1
Algeria	89.9
Madagascar	88.7
Angola	88.5
Rwanda	88.5
Egypt	87.7
Americas	
Surinam	98.8
Guyana	98.3
Barbados	97.7
Costa Rica	92.8
Cuba	89.4
Panama	84.1
El Salvador	83.6
Uruguay	80.6
Nicaragua	80.4
Chile	80.0

[a]Value of production 1984

origin the more of its pests you can hope to leave behind.'

Hotspots of pests and diseases have been linked to crop centres of origin by Jennings and Cock (1977). These authors demonstrate the higher productivity of introduced crops and recommend that national strategy should emphasize the production of introduced food crops. In contrast, as a result of a high level of biological constraint in centres of origin, they suggest that mixed cropping – as a measure to reduce the spread of pests and diseases – has value for native crops. It is notable that there are no Vavilov Centres in North America and Australia – both countries export large volumes of crops previously introduced from elsewhere and largely grown in monoculture. There are obvious agricultural policy implications from this fact.

These explanations of the value of crop introduction – and cogent arguments that crops faced high pest and disease pressure in their regions of origin – have mainly been ignored by mainstream agriculture. In contrast, assumptions are made that crops somehow do better in their regions of origin – indeed, are 'locally adapted', with the implication that this means 'optimally adapted' – are rife (Altieri and Merrick, 1987; Jarvis and Hodgkin, 1998; Brush, 1999). But evolutionary biologists going back to Darwin have long questioned the idea that 'native is best'. For example, Gould (1997) argued that:

> ... many native plants, evolved by natural selection as adaptive to their regions, fare poorly against introduced species that never experienced the native habitat. If natural selection produced optimality, this most common situation could never arise, for native forms would be 'best' and would prevail in any competition with intruders.

Gould concluded: 'Thus the first order rationale for preferring native plants – that, as locally evolved, they are best adapted – cannot be sustained.'

This questionable view of local adaptation leads to a false development paradigm – that local, indigenous, crops should be the centre of any development programme because they are somehow better. While this may (sometimes) be true for abiotic stress – including climatic adaptation – it may be generally very untrue for biological stress, where 'local' crops are everlastingly constrained by the impact of co-evolved 'local' pests and diseases. The optimum strategy is to move crops away from their pests and diseases to a region with a broadly similar climate: a 'homocline'. Many of these are known and already used for crop introduction, for example, a 'Mediterranean' climate for wine grapes, found around the Mediterranean itself, but also in Australia, South Africa, California and Chile – all now major wine exporters.

Does the 'Enemy Escape Hypothesis' Apply to Crops?

Fortunately, conservation biologists have ridden to the rescue of proponents of crop introduction (and our ideas that 'introduced crops do better'). The setting was the great importance of invasive wild species in dominating and altering global biomes and the associated need to understand just why invasive species were more successful than native species in order to control biological invasions.

Introduced species are the bane of productive land use and of attempts to protect and conserve native species. There are many impacts of introduced plant and animal species, but a general problem is that they can become invasive – spreading fast and widely and competing with native species (Elton, 1958; Simberloff *et al.*, 2005). Very large areas can be dominated by introduced plant species, for example star thistle (*Centaurea solstitialis*) in California, cheatgrass (*Bromus tectorum*) in the intermountain regions of the western USA, and water hyacinth (*Eichhornia crassipes*) in tropical wetlands (Mooney and Cleland, 2001; also see Chapter 8, this volume).

However, an increasing body of research has begun to test hypotheses as to why invasive species actually are invasive. A major topic of research has been around the 'enemy escape (or release) hypothesis', which depends on invasive species escaping the constraints of enemies (including predators, pests and pathogens) in the region of origin.

For example, Mitchell and Power (2003) reviewed the incidence of viruses, rust, smut and powdery mildew fungi that infect 473 plant species naturalized to the USA from Europe. They found that on average, 84% fewer fungi and 24% fewer virus species infect each plant species in its naturalized range than in its native range. Wolfe (2002) surveyed populations of *Silene latifolia* in its native Europe and also as an introduced invasive species in the USA for a range of generalist enemies and specialist diseases. Plants were 17 times more likely to be damaged in Europe than in North America.

Crop Introduction and Plant Pathogens

Re-encounter and new-encounter diseases

During the long history of crop introduction, there has always been a risk of also introducing deleterious biodiversity associated with the crop, that is, its pathogens, pests and weeds. This may happen either at the time of the first introduction or with subsequent introductions. Because pathogens are commonly seed-borne or may be symptomless associates of the crop, they have been frequently moved with their hosts. For example, many of the co-evolved pathogens of major food crops, e.g. wheat rusts, rice blast, potato late blight and soybean rust, to name a few, are now distributed worldwide with their hosts through multiple crop introductions (Wood, 1988). These diseases are often referred to as *re-encounter* diseases, where the host and pathogen are rejoined spatially after separation (Buddenhagen, 1977). One of the most noted historical examples is the introduction of potato late blight (*Phytophthora infestans*) to Europe in the 1840s, about 200 years after the introduction of the potato, which resulted in the Irish potato famine (Allen *et al.*, 1999). Second, *new-encounter* diseases can occur following the reunion of two long-separated components of isolated evolutionary systems, either being a result of intercontinental or regional movement of a crop plant into a new environment, or of the movement of a pathogen into a new region where it infects a related host species (Buddenhagen, 1977; Allen *et al.*, 1999). Wild and weedy relatives can be important sources of inoculum for newly introduced crops (Lenné and Wood, 1991). There are numerous examples of pathogens from wild ecosystems moving to introduced crops (see Thresh, 1981). The potential movement of pathogens into new regions is dealt with below under quarantine.

New-encounter diseases are especially common among virus diseases (Jones, 2009). These include maize streak, rice yellow mottle, groundnut rosette, cocoa swollen shoot, cassava mosaic, peanut clump and black root of common bean, all in Africa on crops of American and/or Asian origin (Thresh, 1980, 1981, 1982; Allen *et al.*, 1998; Jones, 2009). Rice hoja blanca is an example of a new encounter virus disease on an Asian crop in America. Among fungal diseases, red leaf blotch of soybean, common bean scab, cowpea stem rot, Eucalyptus rust, and both maize downy mildew and vascular streak of cocoa are examples of a new-encounter disease in Africa on an Asian crop, in Africa on an American crop, in Australia on an African crop, in South America on an Australian tree, and in Asia on American crops, respectively (Allen *et al.*, 1998; Ploetz, 2007). Vascular wilt of banana is an example of a new-encounter bacterial disease in Africa on an Asian crop.

As crops often originated in one continent as members of large genera with much wider distribution, introduction to other regions has often placed crops in contact with geographically distant but taxonomically related wild relatives and their pathogens with the emergence of such new encounter diseases. We feel that these diseases are more common and important than is generally recognized. With ongoing crop introduction, particularly of fruit and vegetable crops, there are further chances for new-encounter diseases to occur. More attention therefore needs to be paid in the future to the potential of such diseases to cause damaging epidemics, especially in the context of future changing climates (Jones, 2009). At the least, disease risk assessment should be implemented before crop introduction based on a thorough knowledge of the related wild relatives and their associated pathogens.

Importance of quarantine

There is ongoing potential for non-indigenous pathogens and pests to reduce crop yields through crop introduction, climatic events, accidental introduction and deliberate introduction through bio-warfare. Plant quarantine is therefore vital to prevent the introduction of such pathogens into a country, and/or to intercept and eradicate them before they can become widespread and established. Less-developed countries with inadequate quarantine systems are especially vulnerable.

It is only in the past 50 years that such risks have been fully recognized, with the International Plant Protection Convention (IPPC) being adopted in 1952 (FAO, 1999; Roberts, 2009). The IPPC provides a comprehensive framework for preventing and controlling pest spread under which formalized plant quarantine systems have been established in many countries and regions, for example the USDA Animal and Plant Health Inspection Service (APHIS) in the USA, Australian Quarantine and Inspection Service (AQIS) and the European Plant Protection Organization (EPPO). Since then, they have played an important role in restricting the movement of crop-associated biodiversity during the most active period of germplasm movement of major staple food crops through International Standards for Phytosanitary Measures (ISPMs) (Khan, 1977; Neergaard, 1977; Ebbels, 2003). These standards cover risk analysis, treatments to kill or remove pests, systems approaches to pest risk management, and regulatory and certification systems (Roberts, 2009). However, there are many developing countries that lack experienced plant health specialists and/or have inadequate quarantine systems to operationalize the ISPMs.

In spite of effective quarantine systems in countries such as the USA, Australia and in Europe, serious pathogens or new variants of existing pathogens are still occasionally introduced with often considerable effects on crop and food production (Allen *et al.*, 1998). For example, after many years of successful control, potato late blight re-emerged in the USA, Canada and Europe with the introduction of A2 mating type of *Phytophthora infestans* from the Americas (Fry *et al.*, 1993; Allen *et al.*, 1999). Serious blight outbreaks have occurred in the USA and Taiwan but not in Europe where blight continues to be successfully controlled (Pearce, 1997; Allen *et al.*, 1999; Jyan *et al.*, 2004). Similarly, *Ug*99, a new variant of wheat rust (*Puccinia graminis* f.sp. *tritici*) has spread from Uganda north to Kenya, Ethiopia, Sudan, Yemen and Iran and south to Zimbabwe and South Africa in the past decade and now threatens South Asia, one of the world's breadbaskets (The Economist, 2010). The USA remains on high alert with regard to soybean rust, which is widespread and damaging in South America (Madden, 2001; Schneider *et al.*, 2005).

Much more is now known about the distribution of important pathogens of staple food crops. Today's quarantine systems in many countries are highly effective, but not perfect, in preventing the introduction of new, damaging pests. Furthermore, the IPPC provides a comprehensive framework for preventing and controlling pest introduction and spread (Roberts, 2009). However, the framework still needs to be developed into an active, functioning *international* system that makes a real difference to world food security and the economic progress of developing countries.

Lessons for Agrobiodiversity Management

At first sight, some of the discussion above is counterintuitive – for example, how can local crops *not* be locally adapted? The simple answer to this is that crops are subjected to different types of selection pressures. Over long periods of time crops do adapt to local climate, but in so doing, they also become adapted to homologous climates worldwide. In contrast, crops may never adapt closely to local pests and diseases – this is the biological arms race that may never end and may never reach an equilibrium. And very certainly, a crop introduced to another continent will never find the same spectrum of pests and diseases that it may have taken millennia of evolution to co-adapt to in its region of origin.

There are major lessons here to be learned for agrobiodiversity management. First, introduced crops in escaping their co-evolved pests and diseases may escape the need to be grown in complex polycultures – a standard farmers' response to pest and disease pressure, but a response that is labour- and knowledge-intensive. Second, the maintenance of a broad genetic base as a palette for selection under pressure from pests and diseases in regions of origin may be less-needed following introduction to regions with no co-evolved pests and diseases. Farmers can concentrate on quality of variety in its response to biotic restraints, rather than quantity (which necessarily lowers optimum performance). Third, an obvious issue not considered above, the more places on earth we can grow specific crops, the less any local setback of pest, disease or local weather can disrupt global production.

References

Ahmedi, N., Glaszmann, J.C. and Rabary, E. (1991) Traditional highland rices originating from intersubspecific recombination in Madagascar. In: *Rice Genetics II Proceedings of the Second International Rice Genetics Symposium*, 14–18 May 1990, IRRI International Rice Research Institute, the Philippines, pp. 67–79.

Allen, D.J., Lenné, J.M. and Wood, D. (1998) New encounter diseases and allopatric resistance. *Seventh International Congress of Plant Pathology*, Edinburgh, abstract 4.1.5.

Allen, D.J., Lenné, J.M. and Waller, J.M. (1999) Pathogen biodiversity: its nature, characterization and consequences. In: Wood, D. and Lenné, J.M. (eds) *Agrobiodiversity: Characterization, Utilization and Management.* CAB International, Wallingford, UK, pp. 123–153.

Altieri, M.A. and Merrick, L.C. (1987) *In situ* conservation of crop genetic resources through maintenance of traditional farming systems. *Economic Botany* 41, 86–96.

Amujoyegbe, B.J., Obisesan, I.O., Ajayi, A.O. and Aderanti, F.A. (2007) Disappearance of Kersting's groundnut (*Macrotyloma geocarpum* (Harms) Marechal and Baudet) in south-western Nigeria: an indicator of genetic erosion. *Plant Genetic Resources Newsletter* 152, 45–50.

Anderson, E. (1954) *Plants, Man, and Life.* Melrose, London.

Brockway, L. (1979) *Science and Colonial Expansion: The Role of the British Royal Botanic Gardens.* Academic Press, UK.

Brush, S.B. (1999) (ed.) *Genes in the Field.* International Plant Genetic Resources Institute, International Development Research Center, and Lewis Publishers, Rome, Ottawa, Canada and Boca Raton, Florida.

Buddenhagen, I.W. (1977) Resistance and vulnerability of tropical crops in relation to their evolution and breeding. *Annals of the New York Academy of Sciences* 287, 309–326.

Crosby, A. (1972) *The Columbian Exchange: Biological Consequences of 1492.* Greenwood Press, Oxford.

Ebbels, D.L. (2003) *Principles of Plant Health and Quarantine.* CAB International, Wallingford, UK.

Elton, C.S. (1958) *The Ecology of Invasions by Animals and Plants.* Methuen, London.

FAO (1999) *International Plant Protection Convention.* United Nations Food and Agriculture Organization, Rome.

Flad, R.K., Shuicheng, Li, Xiaohong, Wu and Zhijun, Zhao (2010) Early wheat in China: results from new studies at Donghuishan in the Hexi Corridor. *The Holocene* 22 April, doi:10.1177/0959683609358914.

Fry, W.E., Goodwin, S.B., Dyer, A.T., Matussak, J.M., Drenth, A., Tooley, P.W., Sujkowski, L.S., Koh, Y.J., Cohen, B.A., Spielman, L.J., Deahl, K.L., Inglis, D.A. and Sandlan, K.P. (1993) Historical and recent migrations of *Phytophthora infestans*: chronology, pathways and implications. *Plant Disease* 77, 653–661.

Gould, S.J. (1997) An evolutionary perspective on strengths, fallacies, and confusions in the concept of native plants. In: Wolschke-Bulmahn, J. (ed.) *Nature and Ideology: Natural Garden Design in the Twentieth Century.* Dumbarton Oaks, Washington, DC, pp. 11–19.

Jarvis, D.I. and Hodgkin, T. (1998) *Strengthening the scientific basis of* in situ *conservation of agricultural biodiversity on-farm: options for data collecting and analysis.* IPGRI, Rome.

Jennings, P.R. and Cock, J.H. (1977) Centres of origin of crops and their productivity. *Economic Botany* 13, 51–54.

Jones, R.A.C. (2009) Plant virus emergence and evolution: origins, new encounter scenarios, factors driving emergence effects, effect of changing world conditions, and prospects for control. *Virus Research* 141, 113–130.

Jyan, M.H., Liou, R.F., Ann, P.J., Tsai, J.N., Hsih, S.D. and Chang, T.T. (2004) Recent occurrence of *Phytophthora infestans* US-11 as the cause of severe late blight on potato and tomato in Taiwan. *Canadian Journal of Plant Pathology* 26, 188–192.

Khan, R.P. (1977) Plant quarantine: principles, methodology, and suggested approaches. In: Hewitt, W.B. and Chiarappa, L. (eds) *Transfer of Genetic Resources*. CRC Press, Cleveland, Ohio, pp. 289–308.

Kloppenburg, J.R. (1988) *First the Seed: The Political Economy of Plant Biotechnology 1492-2000*. Cambridge University Press, Cambridge.

Lenné, J.M. and Wood, D. (1991) Plant diseases and the use of wild germplasm. *Annual Review of Phytopathology* 29, 35–63.

Lockeretz, W. (1988) Agricultural diversification by crop introduction: the US experience with the soybean. *Food Policy* 13, 154–166.

MacHugh, D.E., Shriver, M.D., Loftus, R.T., Cunningham, P. and Bradley, D.G. (1997) Microsatellite DNA variation and the evolution, domestication and phylogeography of taurine and zebu cattle (*Bos taurus* and *Bos indicus*). *Genetics* 146, 1071–1086.

Madden, L.V. (2001) What are the non-indigenous plant pathogens that threaten US crops and forests? *APSnet Feature Story* October, 2001.

Masefield, G.B. (1972) *A History of the Colonial Agricultural Service*. Clarendon Press, Oxford, 184 pp.

Merrill, E.D. (1945) *Plant Life of the Pacific World*. Macmillan, New York.

Mitchell, C.E. and Power, A.G. (2003) Release of invasive plants from fungal and viral pathogens. *Nature* 421, 625–627.

Mooney, H.A. and Cleland, E.E. (2001) The evolutionary impact of invasive species. *Proceedings of the National Academy of Sciences, USA* 98, 5446–5451.

Neergaard, P. (1977) Quarantine policy for seed in transfer of genetic resources. In: Hewitt, W.B. and Chiarappa, L. (eds) *Plant Health and Quarantine in International Transfer of Genetic Resources*. CRC Press, Cleveland, Ohio, pp. 309–314.

Pearce, F. (1997) The famine fungus. *New Scientist* 2079, 32–36.

Ploetz, R.C. (2007) Diseases of tropical perennial crops: challenging problems in diverse environments. *Plant Disease* 91, 644–663.

Purseglove, J.W. (1968) *Tropical Crops: Dicotyledons*. Longman, London, pp. 12–16.

Roberts, W. (2009) The revised International Plant Protection Convention – a new context for plant quarantine. In: Strange, R.N. and Gullino, M.L. (eds) *The Role of Plant Pathology in Food Safety and Food Security: Plant Pathology in 21st Century*. Springer Science and Business Media, Dordrecht, the Netherlands, pp. 133–136.

Safford, W.E. (1905) Useful plants of the island of Guam. *Contributions from the United States National Herbarium* 9, 1–416.

Schneider, R.W., Hollier, C.A., Whitam, H.K., Palm, M.E., McKemy, J.M., Hernandez, J.R., Levy, L. and De Vries-Paterson, R. (2005) First report of soybean rust caused by *Phakopsora pachyrhizi* in the Continental United States. *Plant Disease* 89, 774.

Simberloff, D., Parker, I.M. and Windle, P.N. (2005) Introduced species policy, management, and future research needs. *Frontiers in Ecology and the Environment* 3, 12–20.

Smith, C.W. and Dilday, R.H. (2002) *Rice: Origin, History, Technology, and Production*. Wiley, Hoboken, New Jersey.

Stapf, O. (1912) A new ground bean. *Bulletin of Miscellaneous Information, Royal Botanic Gardens, Kew* 5, 209–213.

The Economist (2010) Rust in the bread basket. *The Economist*, 1 July 2010.

Thresh, J.M. (1980) The origin and epidemiology of some important plant virus diseases. *Applied Biology* 5, 1–65.

Thresh, J.M. (1981) *Pests, Pathogens and Vegetation*. The Pitman Press, Bath, UK.

Thresh, J.M. (1982) Cropping practices and virus spread. *Annual Review of Phytopathology* 20, 193–218.

USDA (1913) Inventory of Seeds and Plants Imported April 1 to June 30, 1913. Available at: http://ddr.nal.usda.gov/dspace/bitstream/10113/37031/1/pi035.pdf (accessed 2 August 2010).

USDA (2008) Plant Inventory No. 217: Plant Materials Introduced in 2008 (Nos. 652416 – 655520). Available at: www.ars-grin.gov/npgs/pi_books/plant_inv_217_2008.pdf (accessed 2 August 2010).

Vavilov, N.I. (1951) The Origin, Variation, Immunity and Breeding of Cultivated Plants (translated by K. Starr Chester). *Chronica Botanica* 13, 1–366.

Vellend, M., Harmon, L.J., Lockwood, J.L., Mayfield, M.M., Hughes, A.R., Wares, J.P. and Sax, D.F. (2007) Effects of exotic species on evolutionary diversification. *Trends in Ecology & Evolution* 22, 481–488.

Wolfe, L.M. (2002) Why alien invaders succeed: support for the escape-from-enemy hypothesis. *American Naturalist* 160, 705–711.

Wood, D. (1988) Introduced crops in developing countries: a sustainable agriculture? *Food Policy* 3, 167–172.

5 Utilization of Crop Diversity for Food Security

J.M. Lenné and D. Wood

Modern plant breeding therefore greatly increases the potential for broadening the diversity for useful traits in crops locally, regionally and globally and has allowed on-going use of a wealth of crop diversity by millions of farmers.

Wood and Lenné (1999)

Introduction

The most valuable component of agrobiodiversity for food security is, without doubt, crop diversity. Staple crop varieties are the functional units of our food (Wood and Lenné, 1999). This chapter briefly looks at the origin, generation and utility of crop diversity building on several chapters in Wood and Lenné (1999) and complementing Chapters 3 and 4, this volume. We then consider why farmers need crop diversity and how they cultivate it in farming systems. Many examples of the positive impacts on food security from science-based utilization of crop diversity by farmers, especially in developing countries, are next highlighted in the context of meeting the ongoing challenges of achieving food security with less land, water and energy. Emphasis is given to wheat, rice and maize, the world's most important food crops.

The Origin, Generation and Utility of Crop Diversity

Domestication was a key event for crop diversity (Frankel *et al.*, 1995). Early farmers selected from a limited range of plant families, especially grasses and legumes, in nuclear areas of domestication (Wood and Lenné, 1999; see Chapter 3, this volume). Much of the 'wild' genetic diversity excluded from the crop through selection was not needed by the 'crop version' of the species (e.g. shattering, toxins, dormancy etc.) as the primary objective was efficient food production (Harlan, 1975; Simmonds, 1979). However, there remains an evolutionary continuum linking pre-domesticates with present-day varieties (Frankel *et al.*, 1995).

The process of domestication of our major food crops began about 11,000 years ago (Evans, 1998). For example, Asian rice (*Oryza sativa*) is believed to have been first domesticated in China about 10,000 years ago while maize (*Zea mays*) was domesticated in Central America at least 9000 years ago. Two to three millennia after the domestication of early wheats in the Middle East, bread wheat (*Triticum sativum*) appeared abruptly in South-west Asia about 7000 years ago when the already domesticated tetraploid emmer wheat (*Triticum turgidum*) crossed with the diploid weedy goat grass (*Aegilops tauschii*) (Cox and Wood, 1999). This simple event with monumental impact was graphically described by Harlan (1981):

> Some time during the neolithic of the Near East, the genomes of tetraploid wheat combined with that of *Aegilops squarrosa* [now = *Ae. tauschii*]. This little weedy goatgrass is the only member of the genus with a continental distribution and the only one extending into the Central Asian steppes. It transformed a rather ordinary cereal into the most widely grown food crop on earth.

Domestication was followed by up to 10,000 years of natural selection through exposure to a diversity of climates, pests, pathogens and weeds (Frankel *et al.*, 1995); human selection for specific plant traits and dietary and market needs; and wide dispersal. Agriculture spread slowly from primary centres of domestication through the migration of farming people (Evans, 1998). Therefore it allowed crops to spread far beyond the range of their wild ancestors, especially in the last 500 years, exposing them to a great diversity of environments. The redistribution of crops immediately following the voyages of Columbus dwarfs all others in its impact on world food production (Evans, 1998; see Chapter 4, this volume). The combination of natural and human selection and widespread introduction accounts for the remarkable diversity found among and within crop landraces and their extraordinary ranges of adaptation (Wood and Lenné, 1999).

Until the development of modern plant breeding in the late 19th century, all farmers grew landraces. The number of different landraces that could be developed from the crop diversity available was limited only by the ability of farmers to visually distinguish different characters and their efforts in selecting and maintaining varieties (Wood and Lenné, 1999). In spite of this tremendous generation of diversity, Darwin in his 1868 study 'The Variation of Animals and Plants' expressed surprise at how little man has increased the productivity of crop plants by incessant efforts over thousands of years (Evans, 1998). Moreover, there has been a tendency to equate morphological diversity with genetic diversity (Cooper *et al.*, 1992; de Boef *et al.*, 1993; Thrupp, 1998). A mythology has arisen that over-emphasizes the value of morphological diverse, but not necessarily genetically diverse, landraces compared to morphologically uniform but genetically diverse modern varieties (Wood and Lenné, 1997). This is still perpetuated today (Brush, 2004; also see www.croptrust.org).

The development of modern plant breeding demonstrates the striking impact of investment in scientific research on crop productivity and food security. It initiated a process of plant introduction, evaluation and assemblage of collections of crop diversity for current and future use never seen before (Lenné and Wood, 1999). Through targeted hybridization, modern plant breeding allowed the recombination of diversity from widely different backgrounds, countries, climates and cultures in an infinite number of combinations and applied intense selection pressure to remove unwanted characters. The development of modern plant-breeding techniques has therefore greatly facilitated wider use of a wealth of diversity from many sources for increasing crop productivity and, especially, has allowed food production to keep pace with population growth. Investment in crop breeding during the 1940s to 1960s was a key factor in the impact on food production of hybrid maize in the USA and high-yielding varieties of wheat and rice of the Green Revolution in the developing world (Evans, 1998; Reynolds and Borlaug, 2006a,b).

The high-yielding varieties which heralded the Green Revolution were productive and profitable and billions of farmers adopted them (Tripp, 1996; Witcombe *et al.*, 1998). This led to claims of severe loss of landraces (Vellvé and Hobbelink, 1992) and even 'genetic wipe-out' (Fowler and Mooney, 1990). But as many landraces and old varieties were collected and conserved in genebanks for future use, especially during the last half of the 20th century, the actual loss of varietal and, more so, genetic diversity was probably small (Witcombe, 1999; also see Chapter 6, this volume). The diversity located in genebanks – 'diversity in reserve' – has been extensively tapped for breeding programmes during the past 50 years and will continue to be used as needs arise.

Modern plant breeding is supported by gene pools of currently unused cultivars, experimental lines, old varieties, ancestral taxa and wild relatives – any genotype which can be crossed to produce new cultivars, either from within the primary gene pool or from distant relatives through biotechnology (Wood and Lenné, 1999). Therefore, modern plant breeding greatly increases the potential for broadening the diversity for useful traits in crops locally, regionally and globally and has allowed ongoing use of a wealth of crop diversity by millions of farmers.

Many modern varieties of rice, wheat and maize have complex genetic make-up with multiple resistances to diseases, pests and abiotic factors; they are highly genetically diverse (Wood and Lenné, 1999; McNally *et al.*, 2006; Peng *et al.*, 2010). For example, the widely grown rice mega-variety IR64 has more than 50 germplasm sources in its pedigree. The number of landraces in the backgrounds of IRRI rice varieties released from 1966–1994 increased from 4 to 46 (Witcombe, 1999; Table 5.1). A recent analysis of a large, geographically and historically broad dataset has shown that the genetic diversity in rice maintained *in situ* on-farm has in fact continued to survive throughout South and South-east Asia for the 33-year time period covered by the study, notwithstanding the cultivation of IR36 and IR64 over millions of hectares (Ford-Lloyd *et al.*, 2009). A focused study in Nepal showed similar findings (Steele *et al.*, 2009).

Similarly, the number of landraces used in popular CIMMYT wheat varieties increased 10-fold from the 1970s to the 1990s (Frankel *et al.*, 1995; Smale, 1998; Table 5.1). With as many as 70 landraces, from many regions, in the ancestry of CIMMYT wheat varieties bred in the 1990s, their genetic background has never been so wide (Evans, 1998). Analysis of cultivar number, areas, ages, pedigree co-ancestry and genetic distances showed that genetic diversity of modern wheat has *not* decreased over time (Smale *et al.*, 2002). More recent genetic broadening of wheat breeding at CIMMYT has included new sources of spring and winter wheat, wild species, as well as exotic germplasm and landraces from many regions worldwide (Ortiz *et al.*, 2007).

Table 5.1. Increased diversity of landraces used in modern varieties of rice and wheat (Sources: Smale *et al.* (1996); Evans (1998); Witcombe (1999)).

Crop	1950–1960	1990s
IRRI rice	4	45–50
CIMMYT wheat	<10	60

From the modern breeding techniques of the first half of the 20th century, which produced the early, high-yielding, disease- and pest-resistant varieties, crop breeding has evolved through increasingly more sophisticated techniques and biotechnological approaches (e.g. marker assisted selection, genomics and genetic modification technologies) to produce higher yielding, more disease- and pest-resistant, and more abiotic stress-tolerant varieties, as well as hybrids and transgenic or genetically modified (GM) crop varieties (see Chapter 7, this volume). Ongoing methodological improvements have allowed greater access to useful traits from exotic sources (including unrelated plants and microorganisms) continually improving the potential and efficiency for using diversity for increasing food production (Tanksley and McCouch, 1997; Varshney *et al.*, 2006; Moose and Mumm, 2008).

Advances in genome sequencing (for example, sequencing of the rice genome was completed in 2004 (see IRGSP, 2005); the soybean genome in 2008 (see www.jgi.doe.gov); the maize genome in 2009 (see www.maizesequence.org); while the wheat genome is also well advanced (see www.wheatgenome.org)) and wider application of synteny mapping, especially between cereal genomes, have greatly facilitated gene isolation and identification of useful traits (McCouch, 2001; Paterson *et al.*, 2009). Smaller and smaller pieces of genetic material for useful traits are being moved into high-yielding varieties more accurately and efficiently. High-throughput genotyping and phenotyping systems are enabling progeny to be rapidly screened through to advanced breeding lines. The lag time from identifying a useful trait gene to growing the improved variety in farmers' fields has been substantially reduced. Examples of the impacts of successful utilization of crop diversity in feeding millions are given in detail below.

Why do Farmers Cultivate Crop Diversity?

The cultivation of a diversity of crops and landraces by small-scale farmers in many developing countries has been extensively documented (Wood and Lenné, 1993; Thrupp, 1999; Thurston *et al.*, 1999). The reasons why farmers cultivate crop diversity have been summarized for a number of crops in several countries and regions including: cassava in the upper Amazon; maize in Mexico; common beans in East and Central Africa; rice in South-east Asia (Thurston *et al.*, 1999); potatoes in the Andes (Brush *et al.*, 1981); sweet potato in Papua New Guinea (Bourke, 1982) and the Philippines (Conklin, 1957); and yams in Africa, South-east Asia and the Pacific (Clawson, 1986; Thurston, 1992). The reasons include: agronomic (to utilize the diversity of soils and topography on-farm); seasonal (time of seeding, maturity type, temperature and precipitation); cultivation system (mono-culture, inter-crop or mixed crop); economic (access to market with road development, access and affordability of inputs, marketable traits); culinary traits and end use (food or feed); storage quality; indigenous and religious beliefs; social functions; and sentimentality (love of ancestral varieties). Small-scale farmers therefore grow crop diversity for similar and multiple reasons and it is likely that studies of other crops will yield similar results. Although social scientists and anthropologists have extensively studied the socio-economic, culinary, ethnic and religious reasons for growing diversity, very limited study has been done of the biological reasons (Wood and Lenné, 1997). In most cases, the genetic diversity underlying the visual morphological diversity has not been scientifically elucidated, and it is unlikely that most farmers are aware of the extent of genetic diversity contained in their suite of cultivated crop varieties.

Crop landraces and varieties will only be maintained by farmers if they offer an advantage to the farmer and household (Smale and Bellon, 1999). Among the above-listed reasons for maintaining diverse varieties, there will always be trade-offs as some reasons are considered more important than others. This also changes over time as rural development creates and expands marketing opportunities, which impact on the level of crop and varietal diversity cultivated by farmers. For example, in the upper Amazon, the demand of indigenous people for cash and market goods has resulted in increased cultivation of varieties with more marketable traits (Smith, 1996). Similarly, in Mexico, it is common for small farmers to cultivate improved, high-yielding maize varieties under intensive management and traditional landraces under low input management (Thurston *et al.*, 1999). Furthermore, in the Bolivian and Peruvian Andes, farmers often grow intensive plots of potato varieties for market and landrace mixtures for their own use (Zimmerer, 1991; Brush, 2004). It is likely that this occurs in other crops in many developing countries where household food and market demands differ. The socio-economic reasons for growing or not-growing diversity appear to be far more compelling for small farmers than the biological reasons, which are often poorly understood. Very little new research has been done in this area during the past 10 years. Most importantly, even if market pressures lead to less crop diversity being cultivated by farmers, *ex situ* conservation of crop genetic resources in genebanks ensures that this diversity is conserved for future need (see Chapter 10, this volume).

How do Farmers Cultivate Diversity?

Farmers cultivate crop diversity on farm either within the same field/plot and/or between fields/plots (Table 5.2). Within-field diversity includes: (i) monocultures or single species stands with inherent diversity – these include varietal mixtures; (ii) intercrops of structured associations of two crops, e.g. a cereal and a pulse; and (iii) mixed crops or polycultures, which may include many different crops occupying different niches (Wood and Lenné, 1999). Between-field diversity includes: (i) different crops grown in rotation or a crop–livestock system; (ii) different

Table 5.2. Cultivation of crop diversity by farmers.

Within-field diversity	Between-field diversity
Monocultures (e.g. landraces, modern varieties, varietal mixtures, multi-lines, dual-purpose crops)	Mixed farming (e.g. cereals and pastures for livestock; staple food crops and home gardens)
Intercrops	Rotations
Polycultures (home gardens)	Planned varietal deployment

varieties of the same crop grown in different fields; and (iii) staple food crops in the fields and together with homestead gardens. The reasons farmers grow within- or between-field diversity are similar (Wood and Lenné, 1999).

Within-field diversity

Monocultures

Monocultures are single species crop stands. They may be single varieties containing many diverse traits or diverse mixtures of varieties, each with differing genetic makeup. Stands of improved high-yielding crop varieties, hybrids and GM crops are monocultures. Multilines or composite varieties and varietal mixtures of improved varieties are also monocultures. Similarly, pure or mixed stands of traditional varieties or landraces as well as dual-purpose crops cultivated by small-scale farmers in developing countries are also monocultures. In spite of attempts to narrowly redefine the term 'monoculture' to only include single *genotype* stands (see Wolfe, 2000), the scientifically accepted definition is a *mono-specific* crop stand.

Monocultures are the most widely grown type of cropping system on earth. Contrary to some literature (Altieri and Nicholls, 2004) most small-scale farmers in developing countries cultivate monocultures. Most of our staple food from rice, wheat, maize, potato, barley, oilseed crops, pulses and sugarcane is grown in monocultures both in developing and developed countries (Lenné, 1999). Humanity relies on monocultures for food security, and this is unlikely to change for the foreseeable future. Most farmers grow monocultures for ease and economy of management – they are easier to plant, weed, fertilize, harvest, market and process. Although monocultures have become progressively more productive and more resistant to diseases and pests through ongoing advances in agricultural science, they are too often perceived, especially by non-farmers, to be unstable, unsustainable, ecologically dysfunctional and highly vulnerable to pests (Lenné and Wood, 1999). This is a direct result of the 'anti-monoculture' propaganda fomented by NGOs (see Chapter 11, this volume).

Much of the perceived 'vulnerability' of monocultures is based on one major event: the southern corn leaf blight of 1970/71 in the USA due to the use of the T cytoplasm in about 80% of the maize grown (Adams *et al.*, 1971; Ullstrup, 1972). Its susceptibility to a new race of *Bipolaris maydis* resulted in an overall loss of 15% of the total annual maize production in the USA in 1 year only. Although few farmers were affected for more than one season, this event fomented a very extreme view of monoculture 'vulnerability' (Marshall, 1977; Brown, 1983). This has persisted in spite of the success of monoculture agriculture – in both developed and developing countries – to continue to meet the staple food needs of growing populations. In hindsight, this event demonstrated the remarkable response of agricultural research: the susceptible varieties were rapidly replaced, and US maize production exceeded the trend line the following year.

Early reliance of plant breeders on single gene resistances for variable pathogens such as wheat rust and rice blast often resulted in boom–bust cycles and a competition with the pathogens to find new resistance genes (Marshall, 1977; Frankel *et al.*, 1995). Since then, considerable progress has been made in understanding the nature of the most important food crop diseases and how best to successfully manage them. As a result, the past 30 years of crop breeding have been characterized by the increased use of more durable multiple disease and pest resistances including multiple traits stacked into productive varieties (Zhang, 2007; Krattinger *et al.*, 2009). A major strategy of the International Rice Research Institute (IRRI) is to incorporate new genes and traits for resistance to both abiotic and biotic stresses into popular, widely grown mega-varieties such as IR64, continually improving their performance and diversity.

For the foreseeable future, monocultures, especially for staple food crops, will continue to feed the majority of the world's inhabitants (Evans, 1998; Royal Society, 2009). Future investment in agricultural research for food security should therefore give highest priority to the sustainable intensification of monocultures – making them even more productive, resource efficient and environmentally stable.

Varietal mixtures

Farmers in developing countries, especially in subsistence systems, commonly grow crop varietal mixtures, often of landraces but also including improved varieties (Harlan, 1975; Smithson and Lenné, 1996; Thurston *et al.*, 1999). As noted above, these are – by definition – monocultures as only one crop is involved. Varietal mixtures are grown because they prolong harvest and income flow, provide diversity of diet and minimize risk. In spite of their importance, there has been limited scientific research on mixtures in subsistence systems. A few studies on rice in the Philippines (Bonman *et al.*, 1986) and China (Zhu *et al.*, 2000) and common beans in Central and East Africa (Madata, 1989; Pyndji and Trutmann, 1992) have shown some increased yield and decreased disease severity in mixtures.

In contrast, crop varietal mixtures and multilines (genetically similar varieties with varying disease resistances), especially of cereals, have been the subject of considerable attention in temperate, developed countries (Marshall, 1977; Wolfe, 1985, 2000; Smithson and Lenné, 1996; Finckh *et al.*, 2000; Mundt, 2002). From a review of over 120 published studies, mostly in temperate regions under modern agriculture, Smithson and Lenné (1996) showed that improved stability and decreased disease severity were common features of mixtures relative to their components in pure stands. However, in the majority of cases, the yield advantage of mixtures was small, being highest for wheat at 5.4%. A recent meta-analysis of 50 published studies on cereal mixtures confirmed the previous study finding an overall yield advantage of 2.7% (Kiaer *et al.*, 2009). At the same time, a number of studies on soybean, groundnut, barley, maize and wheat found yields of the mixtures to be significantly lower than the poorest component (Smithson and Lenné, 1996). Therefore it is clear that mixtures *per se* do not guarantee yield improvements and, indeed, may produce considerably smaller yields, especially if inappropriate combinations of varieties are used.

In studies where both disease and yield have been measured, spectacular reductions in disease severity (in some cases 80–90% reduction) have not been accompanied by similar yield improvements, which although positive, do not exceed 10% more than the means of their components in pure stands (Smithson and Lenné, 1996). Mixing of varieties with different resistances initiates a complex series of interconnected changes, which affects the pathogens and, in consequence, disease development (Burdon, 1987). A number of reasons have been proposed for the reduction in disease severity in crop mixtures, including dilution and barrier effects as well as induced resistance (Wolfe, 1985; Castilla *et al.*, 2003). However, few of these studies have attempted to understand which mechanisms may have been operating in particular crop–pathogen associations. Most importantly, care must be taken in ascribing yield benefits in mixtures to disease reductions alone (Jeger *et al.*, 1981), as they may be derived from other factors. For example, in a 3-year study of the development of blast in mixtures of upland rice in the Philippines, Bonman *et al.* (1986) found that in the years of greatest blast reduction (>60%) yield increases averaged

3%, while in the year of least reduction (27%), yield in the mixtures increased by 20% over the mean of the components.

One study on blast management in rice mixtures in Yunnan, China (Zhu *et al.*, 2000) has been extensively acclaimed as the model study for disease reduction and yield increases in crop mixtures (Altieri, 2002, 2004; McNeely and Scherr, 2002; Tilman *et al.*, 2002; Pretty *et al.*, 2003; Finckh and Wolfe, 2006; Jackson *et al.*, 2007). Although the blast-susceptible, tall glutinous rice varieties planted in mixtures with the blast-resistant, short hybrid varieties had 94% less severe blast and 89% greater yield than when they were grown in pure stands, the data presented in this study fail to show a consistent association between disease severity and yield. In fact, the site/year – Jianshui/99 – had the highest panicle blast severity and the highest overall yields while site/year – Shiping/99 – showed the greatest mixture effect on yield under the lowest panicle blast severity on both the susceptible varieties Hangkenuo and Zinuo (Zhu *et al.*, 2000).

In their desire to demonstrate a relationship between crop diversity in varietal mixtures, disease reduction and yield increase, Zhu *et al.* (2000) appear to have ignored other factors that may be operating in the Yunnan rice mixtures. Various studies have suggested that complex compensation, competition, complementary and facilitation mechanisms operate in mixtures, accounting for yield increase and stability effects (Fukai and Trenbath, 1993; Castilla *et al.*, 2003). In particular, facilitation is commonly observed in rice mixtures where some components are taller than the others, through prevention of lodging of the tall cultivars. A recent study in Yunnan has clearly shown that prevention of lodging of a tall, blast-susceptible glutinous rice variety was a measurable and important advantage of growing it in a mixture with a resistant hybrid (Revilla-Molina *et al.*, 2009). Prevention of lodging has also been recorded as a positive character in mixtures of barley (Stutzel and Aufhammer, 1989) and wheat (Jackson and Wennig, 1997).

The meta-analysis of Kiaer *et al.* (2009) identified large unexplained variation between mixing effects, indicating that variables such as yield, disease reduction and weed suppression explained only a minority of the differences highlighted in mixture studies. Furthermore, the gains in production from diversity within fields may be countered by the extended and overlapping seasons and the close proximity of neighbours' fields which could exacerbate disease and pest problems. For example, continuous rice cropping (as many as three crops per year) is practised in many fertile areas of Asia. In addition, there will also be problems with ease of harvesting and grain quality with any cereal mixtures used for human food. Much more research is needed to explain why crop mixtures perform better than their components under certain conditions and achieve modest yield increases. Clearly, this should be a requisite before a mixture strategy is widely recommended for improving global food security, especially for poor farmers.

Dual-purpose crops

Dual-purpose crops are often grown by farmers in developing countries as they provide multiple end-products not only from the same crop but also from the same inputs of fertilizer, water and labour. Crop–livestock systems in Asia and sub-Saharan Africa are often based on dual-purpose cereals and legumes (Lenné *et al.*, 2003; Lenné and Thomas, 2005; Herrero *et al.*, 2010). These include maize, wheat, sorghum, soybean, cowpea and groundnut grain used for household consumption and income generation and residues for livestock. Other crops may be grown for food, fuel, thatch and craft products, e.g. baskets. Incorporation of dual-purpose crops into farming systems adds crop diversity. Dual-purpose crops may be traditional landrace or bred varieties. More recently, efforts have been directed at the development of high-yielding food–feed varieties (Lenné *et al.*, 2003).

Mixed crop–livestock systems produce half of the world's food and 50% of the world's cereals (Herrero *et al.*, 2010). For example, maize is widely used as a food–feed crop in intensive smallholder mixed farming systems in East and Southern Africa (Romney *et al.*, 2003). Similarly, in India, sorghum and pearl millet form the backbone of crop–livestock systems in semi-arid areas of India where milk is a major income generator for poor households (Parthasarathy Rao and Hall, 2003). In northern Nigeria, improved, dual-purpose cowpea varieties with higher grain yields and enhanced fodder quality are playing an important role in improving the productivity of traditional crop–livestock systems (Singh *et al.*, 2003).

As the demand for crop residues as feed is very high, improved dual-purpose varieties have had significant impacts on the productivity and efficiency of crop–dairy systems in India (Blümmel and Parthasarathy Rao, 2006). Farmers value the crop residues sometimes as much as the grain owing to their importance as a feed for livestock, particularly in the dry season. Smallholders have been able to increase the milk production of buffalos and cows by up to 50% while at the same time obtaining the same grain output from their crops. This has increased the demand for dual-purpose crops with relatively high-quality crop residues, and burgeoning fodder markets have developed around cities like Hyderabad, India. *Ex ante* impact assessments have predicted high economic returns to the development of dual-purpose sorghum and pearl millet in India and dual-purpose cowpea in Nigeria (Kristjanson and Zerbini, 1999; Kristjanson *et al.*, 2002).

There appears to be considerable potential to further improve both grain yield and residue nutritive value of a number of food–feed crops (Lenné *et al.*, 2003; Blümmel *et al.*, 2007). For example, traits such as brown mid-rib in maize, pearl millet and sorghum and stay-green in maize and sorghum can result in enhancement of many nutritive qualities (Blümmel *et al.*, 2003; Hash *et al.*, 2003; Zerbini and Thomas, 2003). For some legumes, e.g. groundnut, improving the leaf to stem ratio and controlling foliar diseases can greatly enhance nutritive value (CGIAR, 2008). High yielding crop varieties that support both the needs for food security and livestock feed, as well as biofuels, have considerable potential to further intensify agricultural production with the same inputs and to contribute to reducing poverty.

Intercrops

Intercropping is the cultivation of two or more crops in the same field at the same time (Francis, 1986). Commonly, the crops are cultivated in rows or strips or relay cropped. Mixed intercropping is usually practised only in developing country, small-scale agriculture and may increase the productivity of these farming systems. The most commonly cultivated intercrop is cereal–legume for food, feed or both, e.g. maize–beans in East Africa and Central America, maize–pigeon pea in Indonesia, sorghum–pigeon pea in India and millet–cowpea in West Africa.

The major perceived advantage of intercrops is for improving soil fertility, especially through nitrogen fixation by the legume component. However, the fertility benefits will depend on how the crops and their residues are managed. Nitrogen depletion can occur in cereal–legume intercrops when the nutrients taken by the crops are not replaced by manure or fertilizers (Giller, 2001). The other benefits of intercrops include increased yields, improved pest management as well as disease and weed control, and risk spreading. However, the realization and extent of the benefits will depend on the intercrop, the system and the environment (Allen, 1990; Cardona, 1990; Thurston, 1992). One of the main disadvantages of intercrops is competition for water, light and nutrients, which can lead to decreased yield of one or both of the crops. In addition, intercrops can increase labour requirements for weeding, planting and harvesting and usually prevent mechanical harvesting (unless crops are cultivated in strips) (Ransom, 1990). A recent study with canola and wheat intercrops in Canada found that the additional benefits of the intercrops were not sufficient to recommend the system for widespread adoption (Hummel *et al.*, 2009).

Although intercropping has been used in developing countries for thousands of years, it is still poorly understood biologically and agronomically (Lenné and Wood, 1999; Royal Society, 2009). At the local level, intercropping does contribute to food security and improved nutrition through dietary diversity. However, too little is known about the mechanisms that underlie observed effects on yield, pests, diseases and weeds. More research is needed to understand better how intercrops function to enhance their contribution to food security and to develop intercropping systems that are compatible with today's farming systems, management practices and market demands.

Polycultures

Polyculture or multiple-cropping is the cultivation of many crops, both annual and perennial, on the same area of land at the same time (Francis, 1986). As for intercrops, polycultures have been cultivated in developing countries for many thousands of years. At the local level, they provide a diversity of food, feed and other products (fuel, construction materials, medicines etc.) needed by the household as well as a level of household food security and income. Home garden polycultures, for example, consist of an assemblage of trees, shrubs, vines and herbaceous plants, growing in or adjacent to a homestead or home compound (Fernandes and Nair, 1986). Indeed, many of the recommendations for the wider use of polycultures are based on home gardens (as noted above). However, home gardens are not so much determined by ecology as by home economics: the targeted input of nutrients from household waste and small livestock as well as family labour provides a diversity of food for household consumption. Proximity to the home as well as fencing prevents theft of high-value crops such as fruits and tubers. Home gardens can make an important contribution to family nutrition, food security and cash income from surplus production (Landauer and Brazil, 1990) but cannot replace monocultures for most staple food.

Because polycultures are perceived to 'mimic' natural vegetation, facilitate recycling of nutrients, reduce losses due to pests and diseases and achieve high yields, they are considered to be more sustainable and stable than monocultures. However, they share many of the same disadvantages as intercrops (see above). In addition, polycultures do require inputs, as with other agroecosystems, in order to maintain their productivity. The degree of relationship between diversity and increased food (versus biomass) production merits much more study in agroecosystems (Wood and Lenné, 1999). Very little new research has been done on the ecology or biology of polycultures in the past 20 years and even less is known about how they function biologically than is known about intercrops.

Local and under-used crop diversity

Most of our staple food is derived from the widely grown crops – rice, wheat, maize, potato and soybean. This legacy is based on 10,000 years of selection by millions of farmers followed by about 120 years of science-based crop improvement which built on the sound choices of early farmers. It is likely that these crops will continue to play the major role in future global food security. In many countries, especially developing countries, farmers also cultivate a diversity of minor crops at community level for household consumption and income. Although not as productive as staple food crops, many of these crops are highly nutritious, for example: grains such as quinoa from the Andes and finger millet from East Africa and South Asia; a range of roots and tubers from the Andes; and indigenous vegetables and fruit from Asia and Latin America (Crops for the Future, 2009).

These crops are often referred to as 'under-used', 'neglected' and/or 'local'. As under-used, they are perceived to have potential to make a wider contribution to global nutritional and food security through wider promotion. As neglected, many have not been studied by crop scientists and hence their potential for improved productivity through plant breeding is largely unknown. And, as local, especially indigenous fruits and vegetables, are largely unknown outside the area where they are cultivated. One probable reason why local crops have not been used more widely in the locale, country and/or region where they evolved is the presence of their co-evolved pests and diseases (see Chapter 4, this volume). As there could be a brighter future for many of these crops following their introduction to other continents, away from their indigenous pests and diseases, future research should concentrate on this.

Some crops have already become popular in developed countries as 'boutique' foods, such as wild rice, Hopi blue corn and exotic fruits (e.g. mangosteen, dragon fruit, rambutan etc.). However, until developing countries are self-sufficient in basic staple food crops, it is not wise to divert much attention to such crops. It seems unlikely that these crops will reduce our global dependence on increasing yields of staple food crops to match food supply with population growth for the foreseeable future. However, with wider promotion, some of these crops can play a valuable role in dietary diversification and improved nutrition.

Between-field crop diversity

Between-field diversity includes: (i) different crops grown on adjacent fields, which may be part of a rotation or a crop–livestock system; (ii) different varieties of the same crop grown in different fields; and (iii) staple food crops in the fields and horticultural crops in gardens near the house (Lenné and Wood, 1999). Diversity between fields on an individual farm can be planned and controlled by the farmer. Diversity between fields on neighbouring farms may also be planned depending on farm size and type and community networks. Farmer's choices about which crops and varieties to grow will be influenced by climatic and edaphic factors as well as economic considerations.

At community and national level, there is potential for planned varietal deployment between fields, communities and regions, complementing within-field diversification practices, to further reduce risks from pests and disease. Pathogens and pests would be stopped as soon as they encountered resistant varieties. Regional gene deployment strategies have been proposed for: potato late blight (Van der Plank, 1963); for breaking the 'Puccinia pathway' for crown rust of oats in the USA (Browning and Frey, 1969); for wheat stem rust (Knott, 1972); for barley powdery mildew in Europe (Wolfe *et al.*, 1992); to manage rice brown plant hopper migrations in Asia (Roderick, 1994; Horgan, 2009); and, together with synchronized planting, to control rice tungro virus in South-east Asia (Azzam and Chancellor, 2001; Holt and Chancellor, 2002; Tiongco *et al.*, 2008). However, few current examples of the wide-scale use of planned varietal diversity were found in the literature. Of note is the successful deployment of over 30 wheat genotypes with differing res-gene combinations for resistance to leaf rust (*Puccinia triticina*) over 18 million ha in the Gangetic Plain, India (Nagarajan and Saharan, 2007). However, against broad-spectrum pests such as locusts, grasshoppers and army worm, between-field crop diversity strategies will not be effective (see Chapter 8, this volume).

Notable Achievements from Past Investments in Crop Diversity for Food Security

Impacts on food security from science-based utilization of crop diversity

Efforts to increase the global availability of food have led to enormous gains in agricultural productivity, food production and human well-being (Evans, 1998; Evenson and Gollin, 2003; Raudsepp-Hearn *et al.*, 2010). From 1961 to 2007, gross world food production increased from 1.84 to 4.38 billion t (138%) from a land area increase of only 11% (4.51 to 4.93 billion ha (Royal Society, 2009). Great progress has also been made in improving the nutritional quality of food. Importantly, these efforts have done more than just feed millions. The interventions of the past half century have also demonstrated that agriculture can be a key driver of growth and development for many of the world's poorest countries (Byerlee *et al.*, 2009; Hazell, 2009; Spielman and Pandya-Lorch, 2009). Paradoxically, some of these advances in food production that have fed millions have been made during a period of the ongoing erosion of funding for public sector plant breeding. One must question *what* level of advances might have been achieved in increasing crop productivity if funding had continued at the level of the 1970s? Table 5.3 summarizes some outstanding achievements, which are described below.

Table 5.3. Successful examples of feeding millions through science-based utilization of crop diversity.

Crop	Geographical location	Key references
Wheat and rice (Green Revolution)	Asia	Hazell (2009)
Maize	East and southern Africa	Smale and Jayne (2003)
Cassava	West Africa	Nweke (2009)
Wheat	Mexico	Spielman and Pandya-Lorch (2009)
Hybrid rice	Asia	Li *et al.* (2009)
Hybrid sorghum and millet	India	Pray and Nagarajan (2009)
Export horticulture	Kenya	Lenné *et al.* (2005)
Home gardens	Bangladesh	Spielman and Pandya-Lorch (2009)

Rice, wheat and the Green Revolution

In Asia, the Green Revolution resulted in the widespread use of improved rice and wheat varieties in monocultures on high-potential, irrigated land that could be cultivated for two or more seasons annually (Evans, 1998; Kingsbury, 2009; Spielman and Pandya-Lorch, 2009). Yields of staple crops such as rice, wheat and maize have increased several-fold. History records no increase in food production that was remotely comparable in scale, speed, spread and duration (Lipton and Longhust, 1989). The investments in science and technology, along with complementary investments in irrigation systems, road networks, fertilizer production and food price stabilization policies, paid off handsomely (Deane *et al.*, 2010). Millions of small farmers rapidly adopted the new practices and technologies to such a massive extent that between 1965 and 1990, cereal output and yields doubled, pulling many Asian countries back from the brink of famine (Spielman and Pandya-Lorch, 2009). India achieved self-sufficiency in cereals around 1974, a situation widely regarded as inconceivable 15 years previously (Kingsbury, 2009). From 1970 to 1990, an estimated 1.8 billion people benefited from the Green Revolution in terms of improved access to food, increased earnings from agriculture, or both. Large areas of fragile lands were saved from conversion to cropping (Harrington, 1997). Furthermore, the returns to investment were substantial: Raitzer and Kelley (2008) calculated Internal Rates of Return of 34% while Hossain *et al.* (2003) estimated that increased rice production from high-yielding varieties over a 20-year period in Asia was worth $4.3 billion.

In spite of all the evidence to the contrary, anti-development groups have continued to criticize the successes of the Green Revolution on the grounds of inequitable benefits, reduced rural employment opportunities, fostered dependence on agrochemicals and reduced crop diversity (Shiva, 1992; Evans, 1998; see Chapter 11, this volume). The reality is that the Green Revolution allowed billions of people to be fed, increased rural employment opportunities generating income and reducing poverty, benefited both small and large farmers, and resulted in the spread of high-yielding varieties far beyond favourable lands (Thirtle *et al.*, 2002; Hazell, 2009; Deane *et al.*, 2010). Above all, it is strongly correlated with improvements in the Human Development Index (Raudsepp-Hearn *et al.*, 2010). Unfortunately, in spite of this success, the impacts were uneven and significant numbers of poor and hungry remain, but we should not expect agricultural progress to stand proxy for social reform (Evans, 1998).

Maize in East and Southern Africa

Successes in sub-Saharan Africa were less dramatic but still important in addressing the persistent threat of hunger (Spielman and Pandya-Lorch, 2009). In East and Southern Africa, sustained investments in innovative breeding programmes and supportive public policies led to growth in both maize output and yields, mainly from monocultures, which improved the livelihoods of millions of small, resource-poor farmers and their families (Smale and Jayne, 2003). From 1965 to 1990, maize yields in Kenya, Malawi, Zambia and Zimbabwe increased annually between 1 and 5% while annual maize production increases ranged from 1.8% to 3.3% in these same countries, contributing significantly to food security in the region. By 2005, improved maize varieties covered more than 75% of the land under cereal cultivation in the four countries, significantly contributing to food security for millions.

Cassava in West Africa

In West Africa between 1971 and 1989, the application of modern science helped contain the spread of cassava mosaic virus disease (Legg and Thresh, 2000; Nweke, 2009). The virus can cause major losses for cassava, a crop that is central to the food security and incomes of the region's poorest farmers, particularly in times of drought or crisis (Spielman and Pandya-Lorch, 2009). By breeding and disseminating cassava varieties that were resistant to the mosaic disease in Nigeria, Ghana and Uganda, the potential damage posed by this threat was effectively contained. The adoption of disease-resistant cassava varieties, mainly cultivated as monocultures, is estimated to have contributed to making an additional 1.4 million t of cassava flour '*gari*' available per year, enough to feed 29 million people in the region (Nweke, 2009). In addition, as the price of *gari* fell by 40%, millions of poor households benefited. The annual economic rate of return from the investment in the development of resistant varieties was 55%, throughout a 31-year period (Maredia *et al.*, 2000).

Wheat in Mexico

Pioneering efforts in Mexico in the 1950s and 1960s by the late Nobel Prize Laureate Norman Borlaug to breed rust-resistant wheat varieties initiated a global programme to fight a disease that has plagued humanity for thousands of years threatening food security in industrialized and developing countries alike (Spielman and Pandya-Lorch, 2009). This global effort helped protect about 117 million ha of land under wheat monoculture from wheat rusts, directly ensuring the food security of from 60 to 120 million rural households and many more millions of consumers. Through necessity, these efforts are ongoing as new, more virulent rust strains evolve. A new variant of wheat rust (*Puccinia graminis* f.sp. *tritici*) *Ug*99, first identified in Uganda, has spread north to Kenya, Ethiopia, Sudan, Yemen and Iran and south to Zimbabwe and South Africa in the past decade and now threatens South Asia, one of the world's breadbaskets (The Economist, 2010). Scientists have already identified resistance genes that are immediately useful for protecting wheat from *Ug*99 (Fu *et al.*, 2009; Krattinger *et al.*, 2009) and the process of incorporating them into high-yielding wheat varieties has begun (CIMMYT, 2009). This is a very good example of the ability of experienced global research programmes to respond rapidly to new threats to the food security of millions through informed exploitation of crop diversity, and further justifies the critical need for ongoing support.

Comprehensive resistance to biotic factors in high-yielding crop varieties is one of the most valuable contributions that modern crop breeding has made to food security globally (Allen and Lenné, 1998). The widespread adoption of modern, disease-, insect pest- and weed-resistant varieties of staple food crops such as rice, wheat and maize by millions of small-scale, poor farmers in developing countries has also significantly contributed to poverty reduction through increased incomes.

Hybrid rice in Asia

In China, policy reforms promoting private investment in agriculture, along with breakthroughs in rice research, fostered the growth of a vibrant seed industry for hybrid rice (Li *et al.*, 2009). Hybrid rice, cultivated in monocultures, has spread so quickly that it is now on 19 million ha, 70% of all land under rice cultivation in China. Importantly, its yield advantages helped China to feed an additional 60 million people per year during this period (Spielman and Pandya-Lorch, 2009). Other Asian countries are now adopting hybrid rice, including Vietnam (19%), the Philippines (12%), Bangladesh (7%) and India (5%) (see Hybrid Rice Development Consortium, http://hrdc.irri.org). In spite of recent criticisms of hybrid rice technology by NGOs such as GRAIN (GRAIN, 2009), there is no doubt that hybrid rice adoption will be a continuing trend throughout Asia, especially if global rice shortages continue. Countries reliant on rice as a staple food urgently want all the latest technologies to increase rice production for national food security and to avoid the need to import rice from an extremely volatile international market.

Hybrid sorghum and pearl millet in India

In India, similar policy reforms and scientific advances in the mid-1990s encouraged the growth of private investment in the marketing of improved seeds for pearl millet and sorghum, including hybrids. These two crops are commonly cultivated as monocultures in semi-arid regions where nearly 60% of the rural population lives. Hybrids now cover 60–80% of the sorghum and pearl millet area and have increased yields by 60–75% in recent decades (Pray and Nagarajan, 2009).

Export horticulture in Kenya

From relatively humble beginnings, export horticulture has grown steadily in post-independent Kenya, increasing 12-fold in tonnage and 40-fold in value (Lenné *et al.*, 2005). It is the fastest growing agricultural sub-sector and the third largest source of foreign exchange after tourism and tea (Haggblade and Hazell, 2010). Kenya is the largest exporter of vegetables to the EU, and the UK is its major customer. About 70% of exported vegetables are grown by smallholders, with up to 50,000 smallholders alone growing French beans (Lenné *et al.*, 2005). Smallholders producing export vegetables have average annual household incomes almost five times higher than non-export smallholders. The export sub-sector also employs hundreds of thousands of semi-skilled and unskilled Kenyans who would struggle to find alternative employment. Continued growth of the export vegetable sub-sector will therefore beneficially support the food security and livelihoods of export company employees and smallholders as well as the Kenyan economy.

Home gardens in Bangladesh

In Bangladesh, Helen Keller International has worked in partnership with more than 70 local organizations and the Government of Bangladesh to promote home gardening, small livestock production and nutritional education for home consumption and the market (Spielman and Pandya-Lorch, 2009), to supplement staple rice production. These homestead food production programmes have reached 5 million poor people and contributed to combating micronutrient deficiencies that can be major causes of diseases among women and children. This model has great potential to spill over to other developing countries where home gardens are an appropriate strategy for improving household nutrition and food security.

Transgenic crops increase crop diversity and decrease agriculture's environmental footprint

Transgenic approaches to the development of disease, pest and weed management in crops are becoming increasingly important technologies for boosting agricultural production, reducing production costs thus improving input use efficiency and generating profits for small farmers (James, 2009). More than 13 million farmers in 25 countries planted 125 million ha of transgenic or genetically modified (GM) crops in 2008. The main crops/traits were herbicide-resistant soybean, maize, canola and cotton; stacked traits (herbicide-resistant/Bt-crops) followed by Bt-crops. The incorporation of both traits into crops has increased their diversity. All are planted as monocultures. Three new countries, including Egypt and Burkina Faso, and 1.3 million new farmers experienced the benefits associated with such crops in 2008. In Burkina Faso in 2009, the GM cotton area soared from 8500 ha to 115,000 ha – which is 29% of the national area (James, 2010). GM crops have increased production by 141 million t in the past 12 years, thus contributing to increased food availability and affordability.

In spite of many NGO claims, peer-reviewed surveys have clearly shown a positive impact of commercialized GM crops (Carpenter, 2010). Average yield increases range from 16% for Bt maize to 30% for Bt cotton, with an average 85% yield increase in one study for herbicide-resistant maize. And, importantly, GM crops decrease agriculture's environmental footprint by reducing pesticides, saving on fossil fuel use and decreasing carbon dioxide emissions and soil loss through reduced cultivation. From 1996 to 2007, GM crops saved 360,000 t of pesticides (James, 2009). Results from 12 countries indicate, with few exceptions, that GM crops have benefited farmers. The benefits, especially in terms of increased yields, are greatest for small farmers in developing countries, who have taken advantage of spill-over technologies originally targeted at larger farmers in developed countries (Carpenter, 2010). The role of transgenic or GM crops for food security is covered in more detail in Chapter 7, this volume, while the impact and value of spill-overs is considered further in Chapter 13, this volume.

The most spectacular and rapid adoption of a GM crop has been for Bt-cotton in India (James, 2010). From 2002, the year of its release, until 2007, the area under Bt-cotton has increased by more than 210 times and the number of Bt-farmers by 190 times. In 2009, 5.6 million small farmers in India grew Bt-cotton, mainly in monocultures, on 8.4 million ha, which is 87% of the national area. From 2002–2008, Indian cotton production doubled from 15 million to 31 million bales, mainly due to the rapid adoption of Bt-cotton hybrids and some new conventional hybrids (Campbell *et al.*, 2010).

Small farmers in India have benefited from: (i) 63% increase in cotton yield; (ii) 55% reduction in chemical sprays; and (iii) 110% increase in profits, equivalent to about US$ 250/ha over the non-Bt cotton (Gandhi and Namboodiri, 2006; Qaim, 2006). Average cotton yields increased from 308 kg/ha in 2002 with non-Bt cotton to 560 kg/ha in 2007 with Bt-cotton, at least 50% of the increase being attributed to Bt technology. From 2005 to 2007, exports of raw cotton increased from 0.9 to 4.8 million bales, making an important contribution to the Indian economy. Further, in 2008, Bt-cotton contributed US$1.8 million to the national farm economy and reduced insecticide use by 50%. Thus, there have been huge social and economic benefits as well as intangible environmental benefits. The ever-increasing demand for Bt-cotton seed is a clear reflection of farmers' confidence in this technology and its benefits.

Paradoxically, with over 87% of the entire Indian cotton area cultivated to Bt-cotton, NGOs in India continue to vehemently oppose it (Herring, 2006). The most vocal spokesperson for the movement 'Operation Cremate Monsanto' was Shiva (cited in Herring, 2006), who stated:

> Pushed into deepening debt and penury by Monsanto-Mahyco and other genetic-engineering multinationals, the introduction of Bt-cotton heralds the death of thousands of farmers ... High costs of cultivation and low returns have trapped Indian peasants in a debt trap from which they have no other escape but to take their own lives.

Clearly, this anti-Bt cotton movement, which began in 1998, has been a failure. Rather than asking why there has been spectacular and rapid adoption of Bt-cotton by small farmers and seed companies, a nearly doubling of yields, and a five-times increase in cotton exports, activists continue to declare 'the failure of Bt-cotton'. The Cremate Monsanto's assumption that small farmers are hapless before the powers of corporations illustrates a key weakness of elite interpretation of rural dynamics: the urban, educated class is culturally, politically and economically superior to the peasantry (Herring, 2006). Farmers cultivating Bt-cotton believed that they were reducing toxification of soil, water and people and reducing expenditure on pesticides, thus risk of indebtedness. To the activists, this outcome was inconceivable (Herring, 2006). This misreading has been largely hegemonic among the NGO activists. This issue is explored in more detail in Chapter 11, this volume.

Other GM crops due for commercialization in the next few years are Bt-rice in China and β-carotene enriched Golden Rice in South-east Asia (James, 2009, 2010). In November 2009, China approved its first GM rice for commercial production, which will enable China to further increase yields by reducing pest damage. Similarly, β-carotene enriched Golden Rice will be tested in farmers' fields in several countries in South-east Asia in 2010 as an important strategy to address chronic vitamin A deficiency in women and children (see www.goldenrice.org). Future targets include drought tolerance, tolerance to other abiotic stresses, a range of disease- and pest-resistant traits as well as additional nutritional traits. All of these traits will increase crop diversity for food security. The world is poised for a second wave of growth in GM crops as developing countries recog nize their contribution to food security and prosperity (James, 2010).

Due to their demonstrated potential for producing more affordable food and for mitigating challenges associated with climate change, GM crops are gradually gaining increased political support (James, 2009):

- G8 members meeting in Hokkaido, Japan, in July 2008 recognized for the first time the significance of the important role that GM crops can play in food security. The G8 leaders' statement on biotech crops reads: 'We will accelerate research and development and increase access to new agricultural technologies to boost agriculture production; we will promote science-based risk analysis, including on the contribution of seed varieties developed through biotechnology.'
- The European Commission stated that 'GM crops can play an important role in mitigating the effects of the food crisis'.
- The World Health Organization (WHO) has emphasized the importance of GM crops because of their potential to benefit the public health sector by providing more nutritious food, decreasing its allergenic potential and also improving the efficiency of production systems.

Future Utilization of Crop Diversity for Food Security

Just as past technologies have already successfully fed millions, science-based tools and technologies have great potential to continue to utilize crop diversity to improve future food production. However, scientists face far greater challenges today to feed the next 3 billion people: using crop diversity to substantially increase crop productivity from less land and more efficient use of water and energy (Evans, 1998; World Bank, 2008; Royal Society, 2009). These challenges must also be placed in the context of the recent slowing in the rate of yield increases in major food crops and the unpredictable effects of climate change on global crop productivity. However, these challenges are not new to science. Not only have they been known by scientists for more than 10 years (Evans, 1998), ongoing research is already producing more stress-tolerant crops. And, there is still considerable scope for using crop diversity to reduce yield losses from pests to close the yield gap (Evans, 2003).

As in the past, many future approaches to utilizing crop diversity will build on and extend existing knowledge and technologies, continuing to make a major contribution to food security. Others will be novel and require further research (Royal Society, 2009). Improvements in crop management including more targeted and efficient fertilizer use, improved irrigation systems and water use efficiency, improved pest control strategies and reduced tillage systems are also likely to contribute to closing the yield gap, especially if applied synergistically with genetic improvements (Evans, 1998, 2003; Royal Society, 2009). Ongoing investment in technologies and approaches that deliver modest cumulative benefits and new investments in novel tools and approaches with potential to significantly improve crop productivity will *both* be needed. Some of the emerging and future targets for investment in utilization of crop diversity are listed in Table 5.4 and briefly discussed below in the context of meeting the ongoing challenges of achieving food security with less land, water and energy.

Table 5.4. Examples of emerging and novel biological tools and technologies and their applications.

Tools and technologies	Applications
Genetic and phenotype analysis	
• Genome sequencing and genomics for sequencing entire crop genomes and identifying genes affecting crop production • Marker technology and marker-assisted selection (MAS) for identifying and monitoring desired genes in breeding progeny • Genetic modification (GM) for introducing desired novel genes into crop plants • Virus-induced gene-silencing • Phenotyping platforms for effectively revealing sets of genes that influence agronomically significant phenotypes, e.g. drought tolerance • High throughput analysis for chemical profiling, e.g. abiotic stress responses and novel crop protection chemicals • Isotopic analysis for drought resistance and water use efficiency • Modelling for predicting how genes will respond in different environments	• Improving genetic yield potential by harnessing hybrid vigour • Development of apomixis in hybrid crops • MAS for selection of desirable traits governed by multiple genetic loci, e.g. drought tolerance • MAS for development of submergence-tolerant rice varieties • Vitamin A biofortification, e.g. Golden rice and orange-fleshed sweet potato • Modification of photosynthetic efficiency, e.g. converting C3 crops such as rice to C4 photosynthesis for up to 50% yield increase • GM rice for drought and salinity tolerance • GM rice for aluminium toxicity • MAS and GM applications for variable pathogens such as wheat stem rust and potato late blight • Parasite-derived resistance for plant viruses, e.g. virus coat protein resistance for papaya ring spot
Crop management practices	
• Crop protection chemicals	• Novel chemicals that mimic plant resistance compounds • Herbicide seed coating on herbicide-resistant crops for weed control
• Genetic control of post-harvest losses	• Ripening resistant tomatoes etc. • Improved processing, storage and packaging of foods to ensure food safety
• Increased water-use efficiency and reduced water loss	• Regulated deficit irrigation and mulching to increase water-use efficiency and decrease water loss
• Reduced erosion	• Conservation tillage systems
• Improved nutrient and water uptake	• Manipulation of the rhizosphere
• Precision farming	• Remote sensing to inform management decisions
• Soil pathogen control	• Fostering disease suppressive soils

Emerging and novel technologies

The science underpinning food crop production is being revolutionized by new technological developments (NRC, 2008; Royal Society, 2009). These include highly sensitive imaging and powerful and informative biochemical analysis (genome sequencing), which can now be applied to high throughput systems. As many thousands of plants can be analysed in a single experiment, plants with desired traits are rapidly and accurately identified. Widespread use of improved computing technologies that can handle large datasets is also creating unprecedented opportunities for genetic improvements in crops and/or in crop management.

Continued genetic improvement of staple cereals is crucial to meeting future global food security. However, genetic manipulation of wheat has been greatly impeded by the size and complexity of its genomes (almost fully sequenced now) and also aspects of its biology that prevent the easy application of advanced technologies developed in model plants. New techniques such as Virus-Induced Gene Silencing are opening new avenues for functional genomics in wheat (Cakir *et al.*, 2010). The ability to generate knockdown phenotypes without having to perform the difficult and time-consuming process of transformation and regeneration is a highly significant advantage, as is the ability to silence all copies of a gene present in complex genomes.

Table 5.4 gives some examples of emerging and novel biological tools and technologies and their applications that are already contributing and/or likely to contribute to future increased food production. Some have already reached farmers' fields and are already being promoted, e.g. submergence-tolerant rice in Asia. Others are near-field technologies that are likely to achieve substantial impact in the next 5 years, e.g. vitamin A-fortified orange-fleshed sweet potato and Golden Rice. Even more are under development, e.g. GM rice for drought and salinity tolerance and converting C3 crops to C4 photosynthesis for significant yield increase, which may be the best way of substantially increasing yields of rice and wheat in future (Evans, 1998; NRC, 2008; Royal Society, 2009; IRRI, 2010).

The major advantages of many of the genetic tools and technologies is their rapidity (breeding programmes are accelerated), their accuracy (desired genes are inserted without linkage drag of deleterious genes), their reliability and their cost-effectiveness. Many of these tools and technologies will enable improved, higher-yielding varieties to be cultivated in farmers' fields in 5–10 years – twice as fast as 50 years ago. In addition, their growing potential to be widely used in the research that underpins crop management practices should result in significant advances in mitigating stresses caused by abiotic and biotic factors (Royal Society, 2009).

The ongoing development of novel tools and technologies has also opened up possibilities to solve complex, difficult problems. Although resolving these problems will require longer-term genetic strategies, the potential has been enabled by these revolutionary developments (Royal Society, 2009), for example: the potential to understand and engineer non-host resistance to crop pathogens (Jones and Dangl, 2006); crop management for enhanced mycorrhizal function (for phosphorus uptake) (Belimov *et al.*, 2009); genetic improvement of root architecture for improved phosphorus acquisition (Lynch, 2007) and of nitrogen fixation capability for nitrogen use (Markmann and Parniske, 2009); and crop management to improve grain and nutritional quality (Bruulsema *et al.*, 2008). These pioneering and, in some cases, radical approaches may result in dramatic increases in productivity associated with reduced need to utilize more land and greater efficiencies in water and energy use for future food security.

References

Adams, M.W., Ellingboe, A.H. and Rossman, E.C. (1971) Biological uniformity and disease epidemics. *BioScience* 21, 1067–1070.

Allen, D.J. (1990) The influence of intercropping with cereals on disease development in legumes. In: Waddington, S.R., Palmer, A.F.E. and Edje, O.T. (eds) Workshop on Research Methods for Cereal/Legume Intercropping in Eastern and Southern Africa, *CIMMYT Eastern and Southern Africa On-Farm Research Report* No. 17, pp. 62–67.

Allen, D.J. and Lenné, J.M. (eds) (1998) *The Pathology of Food and Pasture Legumes.* CAB International, Wallingford, UK.

Altieri, M.A. (2002) Agroecology: the science of natural resource management for poor farmers in marginal environments. *Agriculture, Ecosystems and Environment* 93, 1–24.

Altieri, M.A. (2004) Linking ecologists and traditional farmers in the search for sustainable agriculture. *Frontiers in Ecology and Environment* 2, 35–42.

Altieri, M.A. and Nicholls, C.I. (2004) *Biodiversity and Pest Management in Agroecosystems*, 2nd edn. Food Products Press, New York.
Azzam, O. and Chancellor, T.C.B. (2001) The biology, epidemiology and management of tungro virus in Asia. *Plant Disease* 86, 88–100.
Belimov, A.A., Dodd, I.C., Hontzeas, N., Theobald, J.C., Safronova, V.I. and Davies, W.J. (2009) Rhizosphere bacteria containing 1-aminocyclopropane-1-carboxylate deaminase increase yield of plants grown in drying soil via both local and systemic hormone signalling. *New Phytologist* 181, 413–423.
Blümmel, M. and Parthasarathy Rao, P. (2006) Economic value of sorghum stover traded as fodder for urban and peri-urban dairy production in Hyderabad, India. *International Sorghum and Millets Newsletter* 47, 97–100.
Blümmel, M., Zerbini, E., Reddy, B.V.S., Hash, C.T., Bidinger, F. and Khan, A.A. (2003) Improving the production and utilization of sorghum and pearl millet as livestock feed: progress towards dual-purpose genotypes. *Field Crops Research* 84, 143–158.
Blümmel, M., Bidinger, F.R. and Hash, C.T. (2007) Management and cultivar effects on ruminant nutritional quality of pearl millet (*Pennisetum glaucum* (L.) R. Br.) stover: II. Effects of cultivar choice on stover quality and productivity. *Field Crops Research* 103, 129–138.
Bonman, J.M., Estrada, B.A. and Denton, R.I. (1986) Blast management in upland rice cultivar mixtures. In: *Proceedings of the 1985 Jakarta conference: Progress in upland rice research*, International Rice Research Institute, Los Baños, the Philippines, pp. 375–382.
Bourke, R.M. (1982) Sweet potato in Papua New Guinea. In: *Sweet Potato, Proceedings of the 1st International Symposium*, AVRDC, Shanhua, Taiwan.
Brown, W.L. (1983) Genetic diversity and genetic vulnerability: an appraisal. *Economic Botany* 37, 4–12.
Browning, J.A. and Frey, K.J. (1969) Multiline cultivars as a means of disease control. *Annual Review of Phytopathology* 14, 355–382.
Brush, S.B. (2004) *Farmers' Bounty: Locating Crop Diversity in the Contemporary World.* Yale University Press, New Haven, Connecticut.
Brush, S.B., Carney, H.J. and Huaman, Z. (1981) Dynamics of Andean potato agriculture. *Economic Botany* 35, 70–88.
Bruulsema, T.W., Witt, C., Garcia, F., Shutian, L., Nagenda Rao, T., Chen, F. and Ivanova, S. (2008) A global framework for fertilizer best management practices (FBMPs). *Better Crops* 92, 13–15.
Burdon, J.J. (1987) *Diseases and Plant Population Biology*. Cambridge University Press, Cambridge.
Byerlee, D., de Janvry, A. and Sadoulet, E. (2009) Agriculture for development: toward a new paradigm. *Annual Review of Resource Economics* 1, 15–31.
Cakir, C., Gillespie, M.E. and Scofield, S.R. (2010) Rapid determination of gene function by Virus-induced Gene Silencing in wheat and barley. *Crop Science* 50, S-77–S-84.
Campbell, B.T., Saha, S., Percy, R., Frelichowski, J., Jenkins, J.N., Park, W., Mayee, C.D. and Gotmare, V. (2010) Status of the global cotton germplasm resources. *Crop Science* 50, 1161–1179.
Cardona, C. (1990) Effect of intercropping on insect populations: the case of beans. In: Waddington, S.R., Palmer, A.F.E. and Edje, O.T. (eds) Workshop on Research Methods for Cereal/Legume Intercropping in Eastern and Southern Africa, *CIMMYT Eastern and Southern Africa On-Farm Research Report* No. 17, pp. 56–61.
Carpenter, J.E. (2010) Peer-reviewed surveys indicate positive impact of commercialised GM crops. *Nature Biotechnology* 28, 319–321.
Castilla, N.P., Vera Cruz, C.M., Mew, T.W. and Zhu, Y. (2003) Using rice cultivar mixtures: a sustainable approach for managing diseases and increasing yield. *International Rice Research Notes* 28, 5–11.
CGIAR (2008) Superior nut. In: Global Recommitment to Agriculture, *CGIAR Annual Report 2008.* Washington, DC, p. 27.
CIMMYT (2009) Ug 99 wheat stem rust. Available at: www.ug99.info/?m=200908 (accessed 24 November 2009).
Clawson, D.L. (1986) Harvest security and interspecific diversity in traditional tropical agriculture. *Economic Botany* 39, 56–67.
Conklin, H.C. (1957) Hanunoo Agriculture: A Report on an Integral System of Shifting Cultivation on the Philippines. *FAO Forestry Development Paper* 12. FAO, Rome.
Cooper, D., Vellve, R. and Hobbelink, H. (eds) (1992) *Growing Diversity: Genetic Resources and Local Food Security*. Intermediate Technology Publications, London.
Cox, S. and Wood, D. (1999) The nature and role of crop biodiversity. In: Wood, D. and Lenné, J.M. (eds)

(1999) *Agrobiodiversity: Characterization, Utilization and Management.* CAB International, Wallingford, UK, pp. 35–58.
Crops for the Future (2009) *Crops for the Future: Paths out of Poverty.* Strategic Plan 2009–2013, Selangor, Malaysia. Available at: www.cropsforthefuture.org (accessed 24 November 2009).
Deane, C., Ejeta, G., Rabbinge, R. and Sayer, J. (2010) Science for development: mobilizing global partnerships. *Crop Science* 50, v–viii.
De Boef, W., Amanor, K., Wellard, K. and Bebbington, A. (eds) (1993) *Cultivating Knowledge: genetic diversity, farmer experimentation and crop research.* Intermediate Technology Publications, London.
Evans, L.T. (1998) *Feeding the Ten Billion: Plants and population growth.* Cambridge University Press, Cambridge.
Evans, L.T. (2003) Agricultural intensification and sustainability. *Outlook on Agriculture* 32, 83–89.
Evenson, R.E. and Gollin, D. (eds) (2003) *Crop Variety Improvement and its Effect on Productivity: The Impact of International Agricultural Research.* CAB International, Wallingford, UK.
Fernandes, E.C.M. and Nair, P.K.R. (1986) An evaluation of the structure and function of tropical homegardens. *Agricultural Systems* 21, 279–310.
Finckh, M.R. and Wolfe, M.S. (2006) Diversification strategies. In: Savary, S. and Cooke, B.M. (eds) *Epidemiology of Plant Diseases: Facing the Changes of the 21st Century.* Springer, Dordrecht, the Netherlands, pp. 269–308.
Finckh, M.R., Gacek, E.S., Goyeau, H., Lannou, C., Merz, U., Mundt, C.C., Munk, L., Nadziak, J., Newton, A.C., Vallavielle-Pope, C. and Wolfe, M.S. (2000) Cereal variety and species mixtures in practice, with emphasis on disease resistance. *Agronomie* 20, 813–837.
Ford-Lloyd, B.V., Brar, D., Khush, G.S., Jackson, M.T. and Virk, P.S. (2009) Genetic erosion over time of rice landrace agrobiodiversity. *Plant Genetic Resources: Characterization and Utilization* 7, 163–168.
Fowler, C. and Mooney, P. (1990) *Shattering: Food, Politics and Loss of Genetic Diversity.* University of Arizona Press, Tucson, Arizona.
Francis, C.A. (ed.) (1986) *Multiple Cropping Systems.* Macmillan, New York.
Frankel, O.H., Brown, A.H.D. and Burdon, J.J. (1995) *The Conservation of Plant Biodiversity.* Cambridge University Press, Cambridge.
Fu, D., Uauy, C., Distelfeld, A., Blechl, A., Epstein, L., Chen, X., Sela, H., Fahima, T. and Dubcovsky, J. (2009) A kinase-START gene confers temperature-dependent resistance to wheat stripe rust. *Science* 323, 1357–1360.
Fukai, S. and Trenbath, B.R. (1993) Processes determining intercrop productivity and yields of component crops. *Field Crops Research* 34, 247–271.
Gandhi, V. and Namboodiri, N.V. (2006) The adoption and economics of Bt-cotton in India: Preliminary results from a study. Indian Institute of Management (IIM), Ahmedabad, India. *Working paper No. 2006-09-04,* pp. 1–27, September 2006.
Giller, K. (2001) *Nitrogen Fixation in Tropical Cropping Systems.* CAB International, Wallingford, UK.
GRAIN (2009) Oryza hybrida blog. Available at: www.grain.org/hybridrice/?docs (accessed 4 November 2009).
Haggblade, S. and Hazell, P.B.R. (2010) Successes in African agriculture: lessons for the future. *International Food Policy Institute Issue Brief* 63.
Harlan, J.R. (1975) *Crops and Man.* American Society of Agronomy and Crop Science Society of America, Madison, Wisconsin.
Harlan, J.R. (1981) Ecological settings for the emergence of agriculture. In: Thresh, J.M. (ed.) *Pests, Pathogens and Vegetation.* Pitman, London, pp. 3-22.
Harrington, L. (1997) Diversity by design. *CGIAR News* 4, 5–8.
Hash, C.T., Bhasker, A.G., Raj, S.L., Sharma, A., Beniwal, C.R., Folkertsma, R.T., Mahalakshmi, V., Zerbini, E. and Blimmel, M. (2003) Opportunities for market-assisted selection (MAS) to improve feed quality of crop residues in pearl millet and sorghum. *Field Crops Research* 84, 79–88.
Hazell, P.B.R. (2009) Transforming agriculture: the Green Revolution in Asia. *IFPRI Discussion Paper* 00911, 2020 Vision Initiative. IFPRI, Washington, DC.
Herrero, M., Thornton, P.K., Notenbaert, A., Msangi, S., Wood, S., Kruska, R., Dixon, J., Bossio, D., van de Steeg, Freeman, H.A., Li, X. and Parthasarathy Rao, P. (2010) *Drivers of change in crop-livestock systems and their potential impacts on agro-ecosystem services and human well-being to 2030.* Study commissioned by the Systemwide Livestock Programme, ILRI, Addis Ababa, Ethiopia.
Herring, R.J. (2006) Why did 'Operation Cremate Monsanto' fail? *Critical Asian Studies* 38, 467–493.
Holt, J. and Chancellor, T.C.B. (2002) Modelling the spatio-temporal deployment of resistant varieties to

reduce the incidence of rice tungro disease in a dynamic cropping system. *Plant Pathology* 48, 453–461.

Horgan, F. (2009) Will farmers rely on varietal resistance to combat insect attack? Available at: http://ricehoppers.net/2009/06/will-farmers-rely-on-varietal-resistance-to-combat-insect-attack (accessed 8 August 2009).

Hossain, M., Gollin, D., Cabanilla, V., Cabrera, E., Johnson, N., Khush, G.S. and McLaren, G. (2003) International research and genetic improvement in rice: evidence from Asia and Latin America. In: Evenson, R.E. and Gollin, D. (eds) *Crop Variety Improvement and its Effect on Productivity: The Impact of International Agricultural Research*. CAB International, Wallingford, UK.

Hummel, J.D., Dosdall, L.M., Clayton, G.W., Turkington, T.K., Lupwayi, N.Z., Harker, K.N. and O'Donovan, J.T. (2009) Canola-wheat intercrops for improved agronomic performance and pest management. *Agronomy Journal* 101, 1190–1197.

IRGSP (2005) The map-based sequence of the rice genome. *Nature* 436, 793–800.

IRRI (2010) GRiSP International Rice Research Institute (IRRI), Los Baños, the Philippines. Available at: www.irri.org (accessed 20 July 2010).

Jackson, L.E., Pascual, U. and Hodgkin, T. (2007) Utilizing and conserving agrobiodiversity in agricultural landscapes. *Agriculture, Ecosystems and Environment* 121, 196–210.

Jackson, L.J. and Wennig, R.W. (1997) Use of wheat cultivar blends to improve grain yield and quality and reduce disease and lodging. *Field Crops Research* 52, 261–269.

James, C. (2009) Global status of commercialized biotech/GM crops in 2008: ISAAA Brief 39. Available at: www.isaaa.org/resources/publications/briefs/39/executivesummary (accessed 30 October 2009).

James, C. (2010) Global status of commercialized biotech/GM crops in 2009: ISAAA Brief 41. Available at: www.isaaa.org/resources/publications/briefs/41/executivesummary (accessed 26 January 2010).

Jeger, M.J., Jones, D.G. and Griffiths, E. (1981) Disease progress of non-specialised fungal pathogens in intra-specific mixed stands of cereal cultivars. II. Field experiments. *Annals of Applied Biology* 98, 199–210.

Jones, J.D.G. and Dangl, J.L. (2006) The plant immune system. *Nature* 444, 323–329.

Kiaer, L.P., Skovgaard, I.M. and Ostergard, H. (2009) Grain yield increase in cereal variety mixtures: a meta-analysis of field trials. *Field Crops Research* 114, 361–373.

Kingsbury, N. (2009) *Hybrid: The History and Science of Plant Breeding*. The University of Chicago Press, Chicago, Illinois.

Knott, D.R. (1972) Using race-specific resistance to manage the evolution of plant pathogens. *Journal of Environmental Quality* 1, 227–231.

Krattinger, S.G., Lagudah, E.S., Spielmeyer, W., Singh, R.P., Huerta-Espino, J., McFadden, H., Bossolini, E., Selter, L.L. and Keller, B. (2009) A putative ABC transporter confers durable resistance to multiple fungal pathogens in wheat. *Science* 323, 1360–1363.

Kristjanson, P.M. and Zerbini, E. (1999) Genetic Enhancement of Sorghum and Millet Residues Fed to Ruminants. *ILRI Impact Assessment Series 3*. International Livestock Research Centre, Nairobi, Kenya.

Kristjanson, P.M., Tarawali, S., Okike, I., Singh, B.B., Thornton, P.K., Manyong, V.M., Kruska, R.L. and Hoogenboom, G. (2002) Genetically improved dual-purpose cowpea: assessment of adoption and impact in the dry savannah of West Africa. *ILRI Impact Assessment Series 9*. International Livestock Research Centre, Nairobi, Kenya.

Landauer, K. and Brazil, M. (eds) (1990) *Tropical Home Gardens*. United Nations, University Press, Tokyo.

Legg, J.P. and Thresh, J.M. (2000) Cassava mosaic virus in East Africa: a dynamic disease in a changing environment. *Virus Research* 71, 135–149.

Lenné, J.M. (1999) *Conservation of Agrobiodiversity for Global Food Security*. An Inaugural Professorial Lecture delivered at the University of Greenwich, Chatham Maritime, UK, 27 January 1999.

Lenné, J.M. and Thomas, D. (2005) Integrating crop-livestock research and development in sub-Saharan Africa: option, imperative or impossible? *Outlook on Agriculture* 35, 167–175.

Lenné, J.M. and Wood, D. (1999) Vegetational diversity in agroecosystems: a mixed blessing for successful pest management? In: Terry, P.J. (eds) *International Crop Protection: Achievements and Ambitions*. 1999 British Crop Protection Council Proceedings No. 73, pp. 75–98.

Lenné, J.M., Fernandez-Rivera, S. and Blümmel, M. (eds) (2003) Approaches to improve the utilization of food-feed crops. *Field Crops Research* 84, 1–227.

Lenné, J.M., Pink, D.A.C., Spence, N.J., Ward, A.F., Njuki, J. and Ota, M. (2005) The vegetable export system: a role model for local vegetable production in Kenya. *Outlook on Agriculture* 34, 225-232.

Li, Jiming, Xin, Yeyun, Yuan, Longping (2009) Hybrid rice technology development: ensuring China's food security. *IFPRI Discussion Paper 00918*, 2020 Vision Initiative. IFPRI, Washington, DC.

Lipton, M. and Longhurst, R. (1989) *New Seeds and Poor People.* Unwin Hyman, London.
Lynch, J.P. (2007) Roots of the second green revolution. *Australian Journal of Botany* 55, 493–512.
Madata, C.S. (1989) Effect of cultivar mixtures on yield of common bean (*Phaseolus vulgaris* L.) and on the development of anthracnose, angular leaf spot and halo-blight. PhD thesis. Michigan State University, East Lansing, Michigan.
Maredia, M.K., Byerlee, D. and Pee, P. (2000) Impacts of food crop improvement research: evidence from Sub-Saharan Africa. *Food Policy* 25, 531–559.
Markmann, K. and Parniske, M. (2009) Evolution of root endosymbiosis with bacteria: how novel are nodules? *Trends in Plant Science* 14, 77–86.
Marshall, D.R. (1977) The advantages and hazards of genetic homogeneity. *Annals of the New York Academy of Sciences* 287, 1–20.
McCouch, S.R. (2001) Genomics and synteny. *Plant Physiology* 125, 152–155.
McNally, K.L., Bruskiewich, R., Mackill, D., Buell, C.R., Leach, J.E. and Leung, H. (2006) Sequencing multiple and diverse rice varieties. Connecting whole-genome variation with phenotypes. *Plant Physiology* 141, 26–31.
McNeely, J.A. and Scherr, S.J. (2002) *Ecoagriculture: strategies to feed the world and save wild biodiversity.* Island Press, Washington, DC.
Moose, S.P. and Mumm, R.H. (2008) Molecular plant breeding as the foundation of 21st century crop improvement. *Plant Physiology* 147, 969–977.
Mundt, C.C. (2002) Use of multiline cultivars and cultivar mixtures for disease management. *Phytopathology* 40, 381–410.
Nagarajan, S. and Saharan, M. (2007) Epidemiology of *Puccinia triticina* in the Gangetic Plain and planned containment of crop losses. *Developments in Plant Breeding* 12, 71–76.
NRC (2008) *Emerging Technologies to Benefit Farmers in Sub-Saharan Africa and South Asia.* National Academy of Sciences, National Academies Press, Washington, DC.
Nweke, F. (2009) Controlling cassava mosaic virus and cassava mealybug in Sub-Saharan Africa. *IFPRI Discussion Paper* 912.
Ortiz, R., Trethowan, R., Ortiz Ferrara, G., Iwanaga, M., Dodds, J.H., Crouch, J.H., Crossa, J. and Braun, H.-J. (2007) High yield potential, shuttle breeding, genetic diversity, and a new international wheat improvement strategy. *Euphytica* 157, 365–384.
Parthasarathy Rao, P. and Hall, A.J. (2003) Importance of crop residues in crop-livestock systems in India and farmers' perceptions of fodder quality. *Field Crops Research* 84, 189–198.
Paterson, A.H., Bowers, J.E., Bruggmann, R., Dubchak, I., Grimwood, J., Gundlach, H., Haberer, G., Hellsten, U., Mitros, T., Poliakov, A., Schmutz, J., Spannagl, M., Tang, H., Wang, X., Wicker, T., Bharti, A.K., Chapman, J., Feltus, F.A., Gowik, U., Grigoriev, I.V., Lyons, E., Maher, C.A., Martis, M., Narechania, A., Otillar, R.P., Penning, B.W., Salamov, A.A., Wang, Y., Zhang, L., Carpita, N.C., Freeling, M., Gingle, A.R., Hash, C.T., Keller, B., Klein, P., Kresovich, S., McCann, M.C., Ming, R., Peterson, D.G., Mehboob-ur-Rahman, Ware, D., Westhoff, P., Mayer, K.F., Messing, J. and Rokhsar, D.S. (2009) The Sorghum bicolour genome and the diversification of grasses. *Nature* 457, 551–556.
Peng, S., Huang, J., Cassman, K.G., Laza, R.C., Visperas, R.M. and Khush, G.S. (2010) The importance of maintenance breeding: a case study of the first miracle rice variety-IR8. *Field Crops Research* 119, 342–347.
Pray, C.E. and Nagarajan, L. (2009) Pearl millet and sorghum improvement in India. *IFPRI Discussion Paper* 00919. 2020 Vision Initiative, IFPRI, Washington, DC.
Pretty, J.N., Morison, J.I.L. and Hine, R.E. (2003) Reducing food poverty by increasing agricultural sustainability in developing countries. *Agriculture, Ecosystems and Environment* 95, 217–234.
Pyndji, M.M. and Trutmann, P. (1992) Managing angular leaf spot of common bean in Africa by supplementing farmer mixtures with resistant varieties. *Plant Disease* 76, 1144–1147.
Qaim, M. (2006) Adoption of Bt cotton and impact variability: Insights from India. *Review of Agricultural Economics* 28, 59–71.
Raitzer, D.A. and Kelley, T.G. (2008) Benefit–cost meta-analysis of investment in the International Agricultural Research Centers of the CGIAR. *Agricultural Systems* 96, 108–123.
Ransom, J.K. (1990) Weed control in maize/legume intercrops. In: Waddington, S.R., Palmer, A.F.E. and Edje, O.T. (eds) Workshop on Research Methods for Cereal/Legume Intercropping in Eastern and Southern Africa. *CIMMYT Eastern and Southern Africa On-Farm Research Report No. 17*, pp. 41–44.
Raudsepp-Hearne, C., Peterson, G.D., Tengo, M., Bennett, E.M., Holland, T., Benessaiah, K., MacDonald,

G.K. and Pfeifer, L. (2010) Untangling the environmentalist's paradox: why is human well-being increasing as ecosystems degrade? *BioScience* 60, 576–589.
Revilla-Molina, I.M., Bastiaans, L., Van Keulen, H., Kropff, M.J., Hui, F., Castilla, N.P., Mew, T.W., Zhu, Y.Y. and Leung, H. (2009) Does resource complementarity or prevention of lodging contribute to the increased productivity of rice varietal mixtures in Yunnan, China? *Field Crops Research* 111, 303–307.
Reynolds, M.P. and Borlaug, N.E. (2006a) Centenary Review. Impacts of breeding on international collaborative wheat improvement. *Journal of Agricultural Science, Cambridge* 144, 3–17.
Reynolds, M.P. and Borlaug, N.E. (2006b) Centenary Review. Applying innovations and new technologies for international collaborative wheat improvement. *Journal of Agricultural Science, Cambridge* 144, 95–110.
Roderick, G.K. (1994) Genetics of host adaptation in plant hoppers and their ecology and management. In: Denno, R.F. and Perfect, T.J. (eds) *The Planthoppers*. Chapman Hall, London, pp. 551–570.
Romney, D.L., Thorne, P., Lukuyu, B. and Thornton, P.K. (2003) Maize as food and feed in intensive smallholder systems: management options for improved integration in mixed farming systems of east and southern Africa. *Field Crops Research* 84, 159–168.
Royal Society (2009) *Science and the sustainable intensification of global agriculture*. Royal Society Policy Document 11/09.
Shiva, V. (1992) *The Violence of the Green Revolution: Ecological degradation and political conflict in Punjab*. Zed Press, New Delhi, India.
Simmonds, N.W. (1979) *Principles of Crop Improvement*. Longman, London.
Singh, B.B., Ajeigbe, H.A., Tarawali, S.A., Fernandez-Rivera, S. and Abubakar, M. (2003) Improving the production and utilization of cowpea as food and fodder. *Field Crops Research* 84, 169–178.
Smale, M. (1998) *Farmers, Gene Banks and Crop Breeding: economic analyses of diversity in wheat, maize, and rice*. Kluwer Academic Publishers, Boston, Massachusetts.
Smale, M. and Bellon, M.R. (1999) A conceptual framework for valuing on-farm genetic resources. In: Wood, D. and Lenné, J.M. (eds) *Agrobiodiversity: Characterization, Utilization and Management*, CAB International, Wallingford, UK, pp. 387–408.
Smale, M. and Jayne, T.S. (2003) Maize in Eastern and Southern Africa: 'seeds' of success in retrospective. *EPTD Discussion Papers 97*. International Food Policy Research Institute (IFPRI), Washington, DC.
Smale, M., Aquino, P., Cross, J., del Toro, E., Dubin, J., Fischer, T., Fox, P., Khairallah, M., Mujeeb-Kazi, A., Nightingale, K.J., Ortiz-Monasteria, I., Rajaram, S., Sing, R., Skovmand, B., van Ginkel, M., Varughese, G. and Ward, R. (1996) *Understanding Global Trends in the use of Wheat Diversity and International Flows of Wheat Genetic Resources*. International Maize and Wheat Improvement Centre (CIMMYT), Mexico DF, Mexico.
Smale, M., Reynolds, M.P., Warburton, M., Skovmand, B., Trethowan, R., Singh, R.P., Ortiz-Monasterio, I. and Crossa, J. (2002) Dimensions of diversity in modern spring bread wheat in developing countries from 1965. *Crop Science* 42, 1766–1779.
Smith, R.C. (1996) Biodiversity won't feed our children. In: Redford, K.H. and Mansour, J.A. (eds) *Traditional Peoples and Biodiversity Conservation in Large Tropical Landscapes*. The Nature Conservancy, Arlington, Virginia, pp. 197–218.
Smithson, J.B. and Lenné, J.M. (1996) Varietal mixtures: a viable strategy for sustainable productivity in subsistence agriculture. *Annals of Applied Biology* 128, 127–158.
Spielman, D.J. and Pandya-Lorch, R. (2009) *Millions Fed: proven success in agricultural development*. International Food Policy Research Institute, Washington, DC.
Steele, K.A., Gyawali, S., Joshi, K.D., Shrestha, P., Sthapit, B.R. and Witcombe, J.R. (2009) Has the introduction of modern rice varieties changed rice genetic diversity in a high-altitude region of Nepal? *Field Crops Research* 113, 24–30.
Stutzel, H. and Aufhammer, W. (1989) Effects of winter barley cultivar mixtures on lodging. *Journal of Agricultural Science* 112, 47–55.
Tanksley, S.D. and McCouch, S.R. (1997) Seed banks and molecular maps: unlocking genetic potential from the wild. *Science* 277, 1063–1066.
The Economist (2010) The Grim Reaper: Rust in the Breadbasket. Available at: www.economist.com (accessed 30 July 2010).
Thirtle, C., Beyers, L., Lin, L., McKenzie-Hill, V., Irz, X., Wiggins, S. and Piesse, J. (2002) The impact of changes in agricultural productivity on the incidence of poverty in developing countries. *DFID Report No. 7946*. London.

Thrupp L. (1998) *Cultivating Diversity: Agrobiodiversity and Food Security*. World Resources Institute, Washington, DC.
Thurston, H.D. (1992) *Sustainable Practices for Plant Disease Management in Traditional Systems*. Westview Press, Boulder, Colorado.
Thurston, H.D., Salick, J., Smith, E.M., Trutmann, P., Pham, J.-L. and McDowell, R. (1999) Traditional management of agrobiodiversity. In: Wood, D. and Lenné, J.M. (eds) *Agrobiodiversity: Characterization, Utilization and Management*. CAB International, Wallingford, UK, pp. 211–244.
Tilman, D., Cassman, K.G., Matson, P.A. and Naylor, R. (2002) Agricultural sustainability and intensive production practices. *Nature* 418, 671–677.
Tiongco, E.R., Angeles, E.R. and Sebastian, L.S. (2008) *The Rice Tungro Virus Disease: A Paradigm in Disease Management*. Philippines Rice Research Institute and Honda Research Japan Co. Ltd, Nueva Ecija, the Philippines.
Tripp, R. (1996) Biodiversity and modern crop varieties: sharpening the debate. *Agriculture and Human Values* 13, 48–63.
Ullstrup, A.J. (1972) The impacts of the southern corn leaf blight epidemic of 1970–1971. *Annual Review of Phytopathology* 10, 48–63.
Van der Plank, J.E. (1963) *Plant Disease, Epidemics and Control*. Academic Press, New York.
Varshney, R.K., Hoisington, D.A. and Tyagi, A.K. (2006) Advances in cereal genomics and applications in crop breeding. *Trends in Biotechnology* 24, 490–499.
Vellvé, R. and Hobbelink, H. (1992) Intellectual property rights on life forms: opportunities and concerns. In: *Biotechnology and Development: Expanding the Capacity to Produce Food*. Advanced Technology Assessment System, Issue 9, United National Department of Economic and Social Development, New York, pp. 245–249.
Witcombe, J.R. (1999) Does plant breeding lead to a loss of genetic diversity? In: Wood, D. and Lenné, J.M. (eds) *Agrobiodiversity: Characterization, Utilization and Management*. CAB International, Wallingford, UK, pp. 245–272.
Witcombe, J.R., Virk, D.S. and Farrington, J. (1998) *Seeds of Choice. Making the Most of New Varieties for Small Farmers*. Oxford and IBH Publishing Company, New Delhi, India.
Wolfe, M.S. (1985) The current status and prospects of multiline cultivars and variety mixtures for disease resistance. *Annual Review of Phytopathology* 23, 251–273.
Wolfe, M.S. (2000) Crop strength through diversity. *Nature* 406, 681–682.
Wolfe, M.S., Brandel, U., Koller, B., Limpert, E., McDermott, J.M., Muller, K. and Schaffner, D. (1992) Barley mildew in Europe: population biology and resistance. In: Johnson, R. and Jellis, G.J. (eds) *Breeding for Disease Resistance*. Kluwer Academic Publications, the Netherlands, pp. 125–129.
Wood, D. and Lenné, J.M. (1993) Dynamic management of domesticated biodiversity by farming communities. In: *Proceedings of the Norway/UNEP Expert Conference on Biodiversity*, Trondheim, Norway, May 1993, pp. 84–98.
Wood, D. and Lenné, J.M. (1997) The conservation of agrobiodiversity on-farm: questioning the emerging paradigm. *Biodiversity and Conservation* 6, 109–129.
Wood, D. and Lenné, J.M. (eds) (1999) *Agrobiodiversity: Characterization, Utilization and Management*. CAB International, Wallingford, UK.
World Bank (2008) *Meeting Growing Demand for Agriculture through Innovations in Science and Technology*. World Development Report 2008, World Bank, Washington, DC.
Zerbini, E. and Thomas, D. (2003) Opportunities for improvement of nutritive value of sorghum and pearl millet residues in South Asia through genetic enhancement. *Field Crops Research* 84, 3–16.
Zhang, Z. (2007) Strategies for developing Green Super Rice. *Proceedings of the National Academy of Sciences* 104, 16402–16409.
Zhu, Y., Chen, H., Fan, J., Wang, Y., Li, Y., Chen, J., Fan, J., Yang, S., Hu, L., Leung, H., Mew, T.W., Teng, P.S., Wang, Z. and Mundt, C.C. (2000) Genetic diversity and disease control in rice. *Nature* 406, 718–722.
Zimmerer, K.S. (1991) The regional biogeography of native potato cultivars in highland Peru. *Journal of Biogeography* 18, 165–178.

6 Impact of Introduction of Modern Varieties on Crop Diversity

J.R. Witcombe, K.D. Joshi, D.S. Virk and B.R. Sthapit

Introduction

When modern varieties (MVs) are grown by farmers for the first time they can only replace landraces and hence will reduce the extent of their cultivation. Such a reduction in the cultivation of traditional varieties is considered to be a disadvantage and some NGOs intervene to establish village seed banks to preserve traditional cultivars (e.g. Satheesh, 1996) and recommend active measures to conserve traditional varieties to prevent or slow the introduction of MVs (e.g. Ravishankar and Selvam, 1996). However, preventing access to new varieties is both technically difficult and also ethically dubious as it prevents farmers – often those in greatest need – from reaping the economic benefits from cultivating new varieties. For example, from surveys in eastern India, described below, we found that farmers who replaced rice landraces with modern varieties increased their rice self-sufficiency by 20% on average.

A contrary, more realistic, viewpoint to finding ways of preventing the adoption of MVs is that they have an essential role to play in the agricultural system and in improving livelihoods and food security. The negative impacts on agrobiodiversity can be overestimated. The reduction in cultivation of traditional varieties does not lead to a reduction in several measures of agrobiodiversity (reviewed by Witcombe, 1999) and available diversity or reserve diversity may increase. Loss of agrobiodiversity can also be moderated by using more participatory, client-oriented breeding and seed delivery methods that can maintain greater genetic diversity by more quickly delivering a greater choice of varieties to farmers.

Speed of delivery of new varieties is important because diversity can be measured over time as well as space. Perhaps changes over time are the most important in helping farmers maintain yields. The vulnerability of cultivars increases over time as pests and diseases have longer to adapt to the host. Rapid turnover of cultivars reduces this risk and this process can be driven by farmers having to constantly replace cultivars as they become disease susceptible (Souza *et al.*, 1994) or by plant breeding and varietal extension methods that consistently provide access to newer and better varieties at frequent intervals (Witcombe *et al.*, 1996).

Many studies have shown that in developing countries temporal diversity is low because cultivars are replaced only after long periods and, even in favourable agricultural areas, farmers cultivate varieties that are decades old (Witcombe *et al.*, 1996). This slow replacement is due to inefficient methods of plant breeding and varietal extension. In developing countries, farmers are mainly dependent on public-sector breeding efforts for important crops such as rice, wheat and grain legumes. Unfortunately the incentives for plant breeders in this sector lie in officially releasing varieties, almost entirely on the basis of data from on-station trials, and not in the actual delivery of new varieties to farmers. There is a disconnection in the linear transfer of technology model between the role of the plant breeder, whose work finishes when a variety is released, other than the need to produce breeder seed, and the role of agricultural extension, which only starts when the variety is released and has to be promoted. The breeder who is not responsible for the extension of a variety can find ready excuses for any lack of adoption on extension system inefficiencies and the lack of innovation by farmers. More participatory approaches – client-oriented breeding (COB) or participatory plant breeding (PPB) – bring the client to the forefront and recognize that plant breeding is pointless if its products are not grown by farmers (Witcombe *et al.*, 2005). The breeder becomes more concerned with actual delivery; only varieties that are adopted can be considered a success, and hence the rate of adoption of new varieties and temporal diversity can be increased.

We review here how client-oriented methods do indeed deliver varieties to farmers more quickly than conventional methods. Indeed, in some agricultural systems they are the only methods that have had demonstrable impact. The studies we report here were selected only on the basis of the existence of data on the varietal change process over years and not as typical examples. They have taken place in specific and contrasting agricultural environments in South Asia. This is the region of the world with the highest number of people living in poverty and suffering from malnutrition. Here, many farming households cultivate small areas of land and increases in crop yield and the stability of yield are vital for improved food security.

Measuring Changes in Varietal Diversity Over Time

Case study 1. Cultivar replacement in high-altitude rice in Nepal

The breeding of the high altitude rice Machhapuchhre 3 (M3) was the first peer-reviewed report of a successful Participatory Plant Breeding (PPB) programme. Since its release in 1996 its adoption, and those of two other unreleased lines i.e. Machhapuchhre 9 (M9) and Lumle 2 from the same programme, has been monitored over years (Joshi *et al.*, 2001; Joshi and Witcombe, 2003; Steele *et al.*, 2009). The most recent survey was for 2004 and the adoption of the varieties by household was high and differed greatly between villages (Fig. 6.1). The adoption pattern was related to altitude. In the villages Maramche

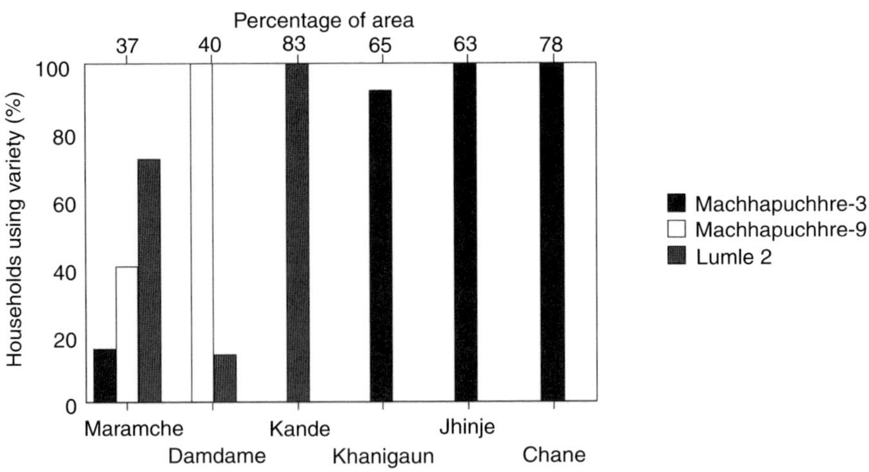

Fig. 6.1. Adoption by households of three rice varieties from COB in 2004 in six villages in Kaski district, Nepal, from a survey of 131 farmers. Adoption by area is also shown above the bars for each village.

and Damdame, at elevations between 1400 and 1600 m, mainly Lumle 2 and M9 were grown as they are better adapted than M3 to these altitudes. In Khanigaun, Jhinje and Chane, which are at higher altitudes above 1600 m, farmers only adopted M3 from among the new varieties as it was the only one with sufficient chilling tolerance to be grown at such altitudes. Kande was more similar to Maramche and Damdame in altitude. Only Lumle 2 was adopted there and this was probably due to farmers not having access to the seed of M9. Clearly, environmental heterogeneity in the form of altitude maintains diversity through the differing adaptation to altitude of the three modern varieties.

However, within-village diversity appears to have been reduced when adoption by household is considered as nearly all, or all, of the sampled households grew the same MV. However, these farmers did not grow the varieties on all of their land (Figs 6.1 and 6.2). Overall, although nearly 100% of the sampled farmers across the six villages had adopted at least one of the MVs the adoption was incomplete. Although about half of the farmers had adopted the variety on 100% of their land (Fig. 6.2) the remainder used them on only some of their land and, in most cases, this was on less than half of it. Farmers in the same village had made very different decisions on how widely they grew the new introductions. This could have been because adoption ceilings had not yet been reached because limited seed availability currently restricts the area or have many other socio-economic explanations. This uneven pattern of adoption created spatial diversity and reduced loss of agrobiodiversity when measured at a between-farm level.

This variation in adoption decisions had a varying impact at a between-farm level according to altitude. The overall proportion of the rice area on which the MVs were grown was higher in the four high-altitude villages and varied from 63% to 83% (Fig. 6.2). In all of these villages initial diversity, i.e. before the introduction of the MVs, was already low with only two or three landraces being recorded. In the two lower altitude villages, Maramche and Damdame, the modern varieties were adopted on less than 50% of the land. Although in these cases, weighted diversity could not be determined in the absence of data on the individual varietal portfolios of the farmers, the impact on agrobiodiversity was likely to have been favourable. Steele *et al.* (2009) showed that (taking Kaski region as a whole and assuming

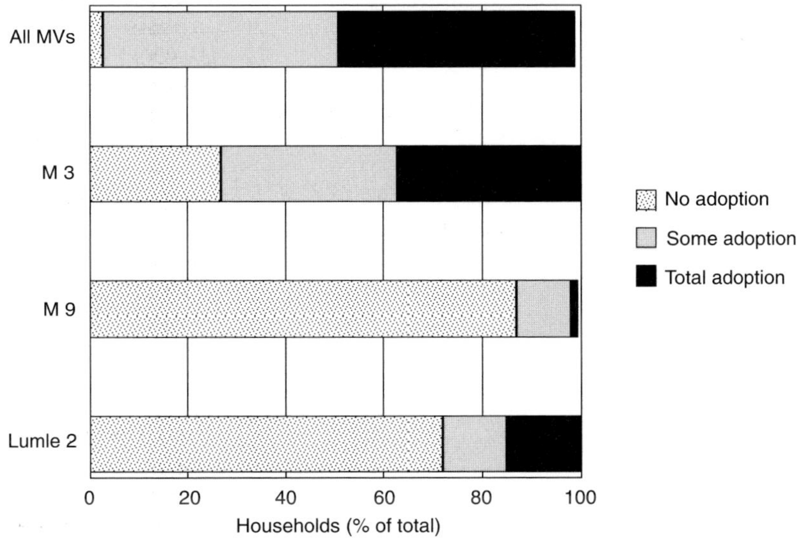

Fig. 6.2. Extent of adoption of modern varieties by the 131 farmers, according to the amount of land they devoted to the varieties.

the three new varieties were evenly adopted) only if they were grown on more than 60% of the area was there any loss in the weighted diversity. This limit varies and will be higher still with any increase in the number of new varieties or the diversity among them.

Farmers in Maramche village made the most diverse decisions on the adoption of the new varieties and is also the one for which we have data to examine changes across years (Fig. 6.3). The process of varietal replacement takes place as Lumle 2 becomes more popular and ousts M3 and M9. However, the rate of decline in M9 is much lower than the increase in Lumle 2 as 39% of the Lumle 2 adopters continue to grow M9. This has led to an increase in diversity and illustrates how the continuing introduction of new varieties adds not just to temporal diversity but also to spatial diversity. If the cultivation of M9 continues to decline there may eventually be a decrease in spatial diversity but only if newer varieties fail to start replacing M9.

Case study 2. Cultivar replacement in upland rice in eastern India

A client-oriented breeding (COB) programme in upland rice (Virk *et al.*, 2003) using the few cross, large population size approach of Witcombe and Virk (2001), was undertaken in India. The improvement of Kalinga III was targeted as it was the best of the upland varieties in participatory varietal selection (PVS) trials in western India (Witcombe *et al.*, 1996) and was also liked by farmers in Jharkhand, India. It was crossed with IR64, the most popular lowland variety in Jharkhand. The breeding programme started in 1996 and by 2001 two varieties had been identified for release, Ashoka 200F (Birsa Vikas Dhan 109) and Ashoka 228 (Birsa Vikas Dhan 110). These varieties yielded from 18% to 25% more than Kalinga III in farmers' fields and about 20% more on research stations (Virk *et al.*, 2003). From 2001, seed of these varieties was distributed to farmers, with the

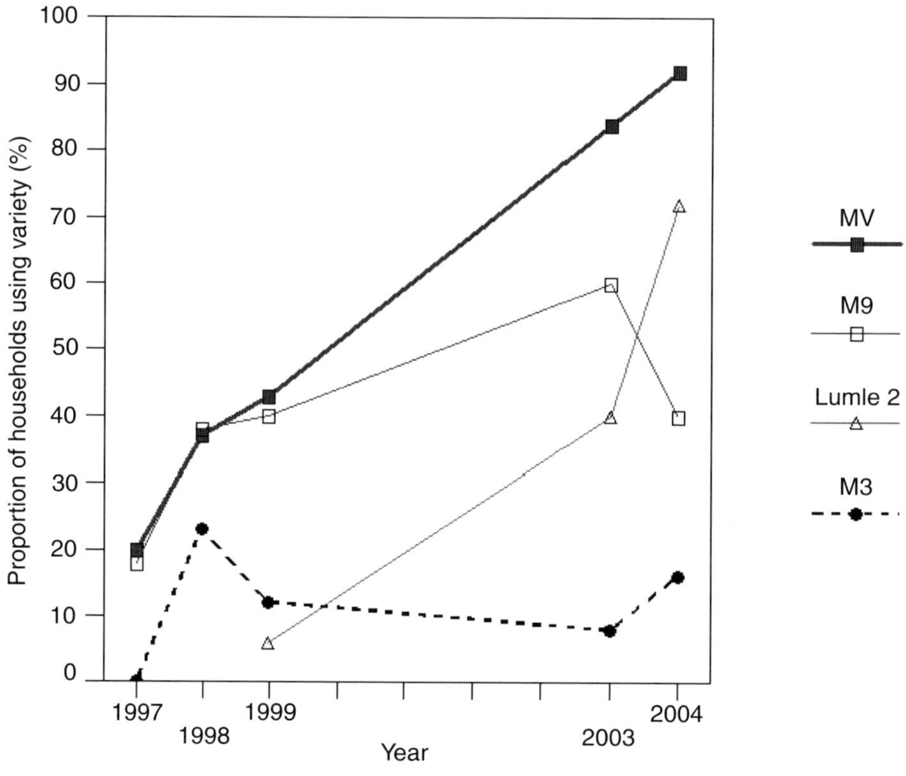

Fig. 6.3. Adoption of three new modern varieties in Maramche village from 1997 to 2004.

active involvement of the plant breeders, in eastern and western Indian states. It was done through NGOs in externally funded development projects and research projects. In 2004, the DFID Plant Sciences Research Programme funded an impact assessment that surveyed over 150 farmers who had been given small quantities of seed in 2001 or 2002. This survey comprised 150 households in: Ranchi (4), Hazaribag (23) and Saraikela (30) districts of Jharkhand state; West Midnapur (10) and Purulia (50) districts of West Bengal state; and Dhenkanal (10), Keonjhar (6) and Mayurbhanjh (8) districts of Orissa state.

A major finding was that the COB varieties were highly accepted and about 98% of farmers given seed adopted the varieties. The areas that these farmers devoted to the two Ashoka varieties in all three states, starting from a very low base, had increased to between 80% and 90% of their suitable rice land (Figs 6.4, 6.5 and 6.6).

In the study villages in all three states the Ashoka varieties were the most successful of all the upland MVs. The only modern upland varieties farmers grew before the introduction of the Ashoka varieties were Kalinga III (in Jharkhand and West Bengal) and Vandana (in Jharkhand). These were only adopted because of decades of seed supply by development projects and:

- They were quickly replaced by the Ashoka varieties; and
- They were far less used (<30% of the area after many years compared with >80% for the Ashoka varieties after only a few years).

Orissa was the only state of those studied where the Ashoka varieties were adopted in medium land and – since agrobiodiversity is higher in more favourable environments in the absence of genetic erosion – this was the district where the greatest varietal diversity was found. In Orissa, the reduction in varietal diversity – as measured by richness, i.e. the

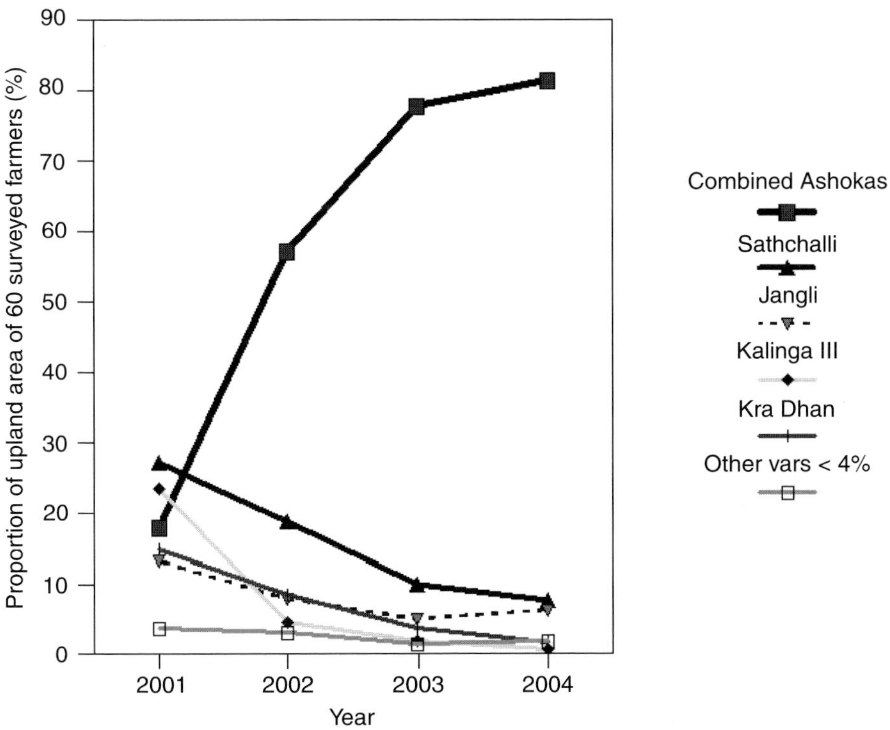

Fig. 6.4. Adoption of Ashoka varieties and other landraces and varieties from 2001 to 2004 by 60 farmers in West Bengal who were first given seed in 2001.

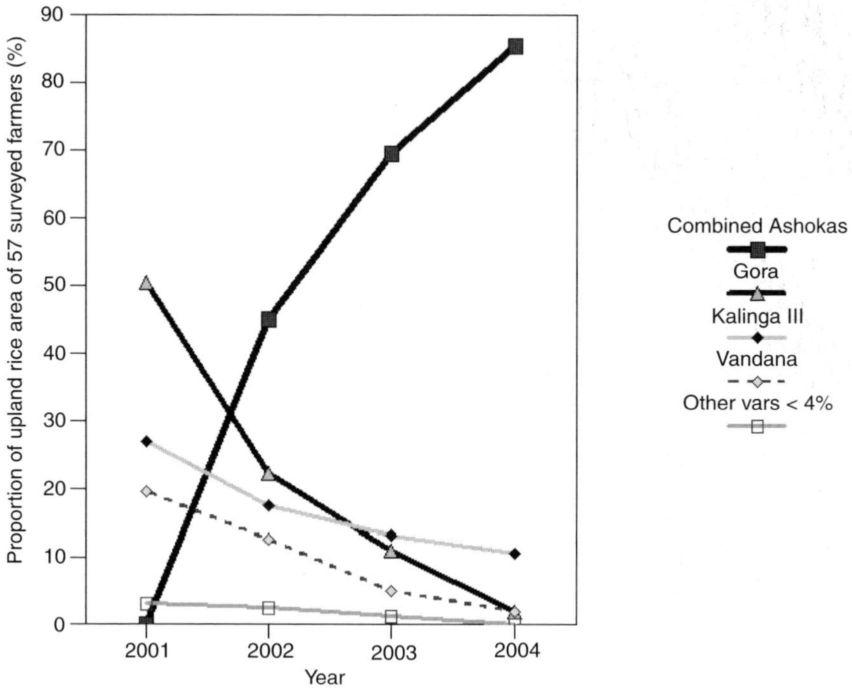

Fig. 6.5. Adoption of Ashoka varieties and other landraces and varieties from 2001 to 2004 by 57 farmers in Jharkhand who were first given seed in 2001.

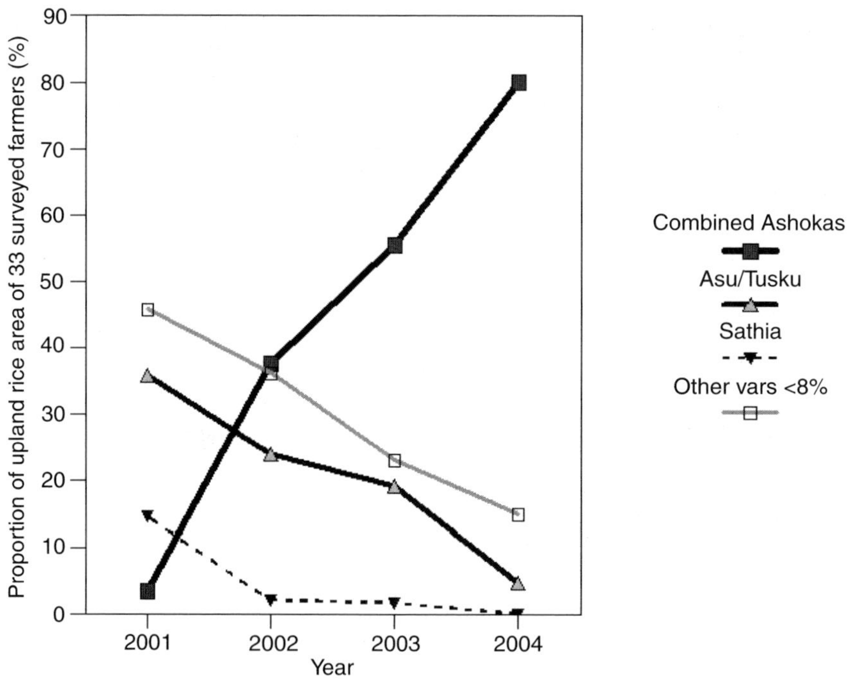

Fig. 6.6. Adoption of Ashoka varieties and other landraces and varieties from 2001 to 2004 by 33 farmers in Orissa who were first given seed in 2001.

total number of genotypes present (Frankel *et al.*, 1995) – was profound (Fig. 6.7) in the medium rice land of the adopting farmers and the number of landraces they grew decreased from 20 in 2001 to six in 2004 (the two Ashoka varieties, Annapurna, Kalinga III and Heera are MVs while all others are assumed to be landraces). The mean count of varieties per household declined from 2.4 to 1.5 from 2001 to 2004 resulting in a decline in the Shannon-Wiener index (which takes into account the frequency at which each variety is grown) from 1.2 to 0.7 (Virk and Witcombe, 2007).

There are several important caveats to this decline in diversity.

- All the landraces would still be grown if the farmers also include those who did not adopt the Ashoka varieties. Hence, reserve diversity and richness has not declined.
- Overall two additional varieties (Ashoka 200F and Ashoka 228) are grown. This adds to the varietal richness. However, this amounts to essentially only one addition as the two varieties are very similar at a molecular marker level (Steele *et al.*, 2004) but,

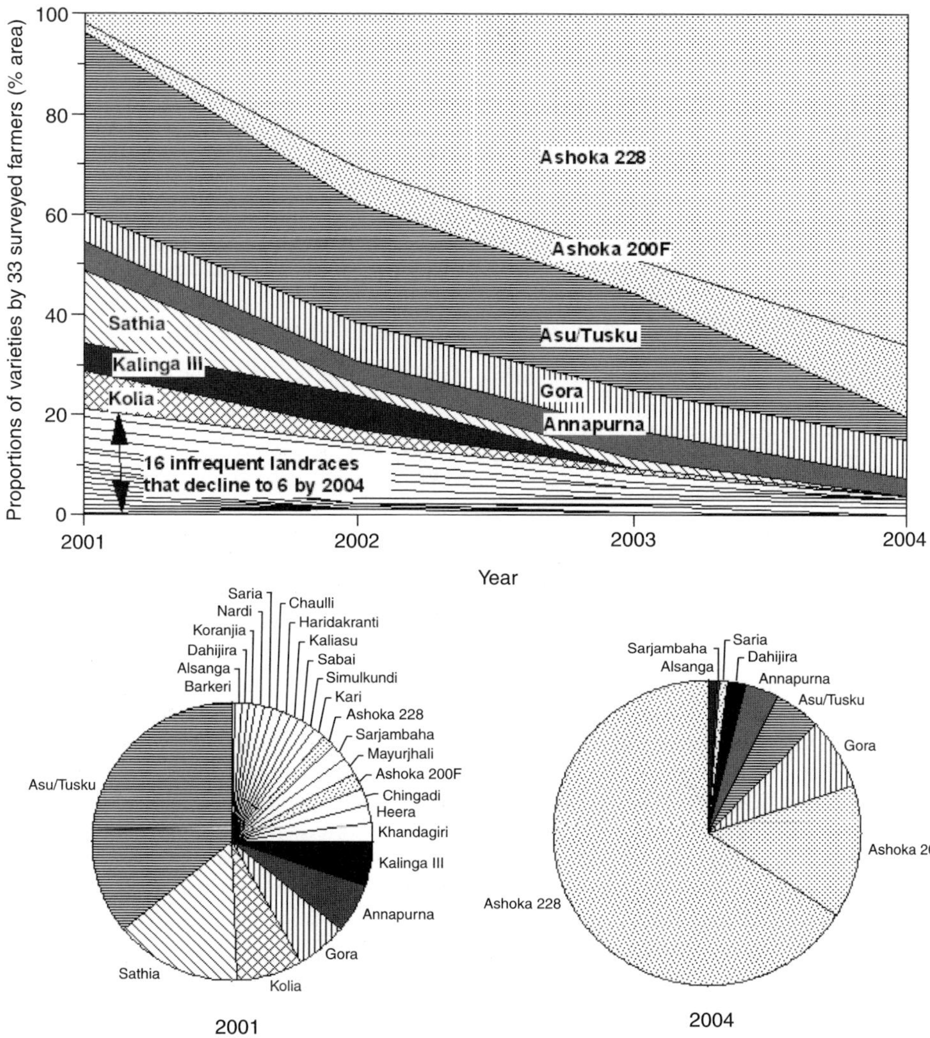

Fig. 6.7. Change in the varietal composition of the rice grown by 33 farmers in Orissa over time (top), in 2001 (bottom left) and in 2004 (bottom right).

when phenotype is considered, they do differ significantly in their flowering time.

- The impact on weighted diversity is sensitive to the scale that is used. On a larger scale (all villages in the area) weighted diversity probably increases because of the addition of the new varieties. Weighted diversity declines only when the measure is restricted to the farmers that have adopted the Ashoka varieties.
- Among the farmers who adopted the Ashoka varieties temporal diversity greatly increased between 2001 and 2004.
- There has been an increase in diversity at a phenotypic level (called apparent diversity, see Witcombe, 1999). The Ashoka varieties have a rare combination of traits: drought tolerance and adaptation to the upland environment was combined with fine grains whereas all previous upland varieties, with the exception of the lower yielding Kalinga III, were coarse grained.

Case study 3. Client-oriented breeding in rice in low altitude areas of Nepal

The client-oriented breeding (COB) programme in rice for the low-altitude regions in the Nepal Terai commenced in 1997 and followed the same few cross, large population size approach of Witcombe and Virk (2001). We report here on progress by 2008. By that time, only three crosses had sufficient time to produce a variety that could have been grown by farmers.

- The first cross, Kalinga III/IR64, was made in IRRI in 1996 at the request of CAZS-NR and seed was brought into Nepal from India in 1998 at the F_3 generation.
- A second cross, Radha 32/Kalinga III, was made in Nepal in 1998. It was chosen as both of the parents were liked by farmers in PVS trials in Chitwan. Radha 32 had poor grain quality and Kalinga III poor lodging resistance and these weaknesses could be eliminated through the complementary phenotypes of the parents.
- A mutation breeding programme in Pusa Basmati 1 (PB 1), an aromatic, dwarf, rice variety from India, was equivalent in effort to that of a third cross. Seed of PB 1 was irradiated in 1998 to produce mutations, but it was later found that the seed that was irradiated had been harvested from PB 1 plants that had naturally crossed to other varieties in the field. This out-crossing, almost certainly, produced more variation than the mutations and there was much diversity in the material.

The PPB programme used only a few crosses because it fits well with the particular constraints and advantages of working with farmers, and because the approach is soundly based in theory. Because few parents were employed their choice is crucial. Participatory varietal selection (PVS) helped greatly in this process because it identified both parents and important target traits. Large population sizes were maintained in subsequent generations, all of which were grown in farmers' fields in different production systems, from upland to lowland, in both the early (*Chaite*) and main seasons. In contrast to commonly used pedigree breeding methods that start single plant selection in the F_2 we used bulk population breeding methods (Witcombe and Virk, 2001). These delay strong selection until a bulk population in the F_4 generation when between-plant heritability is considerably higher than in the F_2.

In accordance with the client-oriented programme the breeders also promoted the varieties that emerged from the breeding programme. This was why it was feasible to examine adoption of varieties as early as 2008 from a breeding programme, most of which began in 1998. In conventional public-sector breeding no variety would have reached farmers by this stage. A survey was made in a total of 36 villages distributed evenly across six widely dispersed districts of those in the Nepal Terai. In these villages over 2000 rice-growing households were identified and over 300 farmers from the stratum of farmers identified as growing some variety related to the programme (including varieties outside of the COB programme but introduced through PVS; Table 6.1). The proportion of households growing a variety from the COB programme was already high at this early stage (17%) and eight varieties were grown at a frequency above 0.8%. (There were several other varieties that were reported as well but at lower frequencies.) Six varieties were popular at a more local level, i.e. in individual villages where from 32% to 71% of farmers used them.

Moreover, the few crosses from which these varieties were derived did not reduce genetic diversity. Steele *et al.* (2004) showed that diversity among the varieties derived from the first cross was high. There were large genetic differences between the rice varieties for different ecosystems with the upland adapted varieties having most alleles from Kalinga III and the lowland varieties most from IR64. For example, upland variety Barkhe 1027 had about 80% of its alleles from Kalinga III and the lowland variety Barkhe 3004 had 67% from IR64.

A diversity analysis using DArT markers also showed that diversity was high between the varieties irrespective of whether the varieties were from the same or different crosses (Table 6.2). Indeed, the greatest similarity was between Barkhe 2014 (cross 1) and Sugandha 1 (cross 3). The high frequency of non-parental alleles in Barkhe 2014 and the fact that only one of the two parents of Sugandha 1 is known (the irradiated parent had out-crossed) would explain this apparent anomaly.

The impact on agrobiodiversity in the Nepal case study was highly positive with many new varieties added to farmers' options over the short period of the COB programme. The relatively high proportion of users (17%) did so on a small proportion of their land (averaging about 12% of their total rice land). Since the Nepal Terai is not a marginal agricultural environment there has been a high adoption of MVs, and diversity in some of the districts is very low with the predominant modern variety occupying most of the rice area. Hence, the introduction of new varieties for diverse rice domains that are both aromatic and non-aromatic adds to the available diversity and, at least at these early stages of adoption, increases the weighted diversity if, as in the most probable scenario, they mainly replace more common varieties.

Table 6.1. Proportion of farmers who grew a COB variety in the main season among all of the 2222 rice-growing households (hh) in the 36 villages in the six terai districts.

Variety	Derived from cross	For rice domain[a]	Mean use (% all hh)	Maximum use in any village (% hh)	District
Barkhe 1027	1	Up & med up	2.8	61	Banke
Barkhe 2014	1	Med	2.0	44	Kanchanpur
Barkhe 3004	1	Med & low	2.4	33	Nawalparasi
Judi 572	2	Up	2.4	61	Banke
Barkhe 2001	3	Med	1.4	32	Banke
Barkhe 2024	3	Med	0.8	4.4	Chitwan
Sugandha 1	3	Up (aromatic)	1.0	4.3	Chitwan
Sunaulo sugandha	3	Med & low (aromatic)	7.5	71	Kanchanpur
COB user			17.4		

[a]Up = upland, med = medium land, low = lowland.

Table 6.2. Jaccard's coefficient of similarity based on 373 DArT (Diversity Arrays Technology, www.diversityarrays.com) loci among six varieties bred by COB in Nepal.

	Barkhe 2014	Barkhe 1027	Barkhe 3004	Judi 572	Sunaulo sugandha
Barkhe 1027	0.61				
Barkhe 3004	0.39	0.37			
Judi 572	0.37	0.34	0.63		
Sunaulo sugandha	0.40	0.42	0.25	0.34	
Sugandha 1	0.66	0.54	0.39	0.46	0.47

Discussion

The case studies show that client-oriented breeding can rapidly produce new, highly accepted varieties and can do so in situations where prior breeding efforts have met with much less success. The impact these successes have had on varietal diversity is mixed and both declines and increases in spatial diversity, as measured by weighted diversity, were found depending on the agricultural system and the existing diversity into which the new varieties are introduced. However, there is always, at least in the initial stage, an increased temporal diversity and the new varieties added to the richness, i.e. total available diversity among modern cultivars.

Farmers continued to grow landraces in several of the case studies. This was most likely because there were no suitable modern varieties because of a lack of plant breeding for the target domain or because plant breeding efforts were insufficiently client oriented. Less likely is that suitable varieties actually existed but farmers had not had access to them. This could be true for recent varieties but older varieties would have diffused through the farmers' innovation system (e.g. Maurya, 1989). When landraces were found there was often a richness of diversity that could exist for several reasons. Farmers may be following a risk aversion strategy and consider a portfolio of varieties will reduce the risk of poor production years or will sustain production under adverse environmental changes (Rosenzweig and Stark, 1989). Brush and Meng (1998) also describe the role of differing grain quality and environmental heterogeneity in maintaining diversity. Other explanations are that different varieties have different bundles of traits and farmers choose the varieties for different purposes or because farmers experiment with partial adoption over extended periods before deciding on more complete adoption (a factor that may be important in all of three cases reviewed here).

Gollin and Smale (1999) point out that if farmers continue to grow landraces when MVs are available they are willing to sacrifice higher yield to gain some other characteristics. The extent of this lost yield is then a measure of the farmer's willingness to pay for genetic diversity. In the case of the Ashoka upland rice it was not the first time that farmers had had access to modern varieties but only a minority of surveyed households had adopted varieties such as Kalinga III and Vandana and they did so only on a minority of their land. Farmers traded-off the higher yield of these older improved varieties against their inferior traits – the lower drought tolerance of Kalinga III and the poorer grain quality of Vandana. However, when COB was successful the need for these trade-offs was much reduced. The Ashoka varieties had higher yields, better drought tolerance and better grain quality (Virk and Witcombe, 2007) and hence the Ashoka varieties were widely used whether measured by the proportion of households or the proportion of the land on which farmers grew them. Given the benefits of the Ashoka varieties, the cost to farmers (of preserving the prior landrace diversity by forgoing them) would be high given the 20% increase in rice self sufficiency reported by users (see Introduction above). Hence, from a functional viewpoint, the reduction in the diversity of the farmers' portfolio did not reduce the perceived major benefit of biodiversity – stability of production – as the Ashoka varieties provided this through their inherent higher yield and drought tolerance. Similarly, in Nepal the high and stable yield of the new varieties also provided the same benefit of yield stability that was provided by a diversity of landraces.

The only clear example of increased agrobiodiversity at all levels (spatial, temporal and available diversity) was the example of low-altitude rice domains in Nepal. Seven diverse varieties have been adopted by farmers since 2002 and there are several other new varieties at the early stage of adoption. The COB programme has added to varietal choice and hence varietal richness (Witcombe, 1999). Moreover, it is in an area, unlike the example of the medium rice land in Orissa, where in some districts varietal diversity was very low because a single variety predominated. However, partial adoption may be because of limited quantities of seed in the early stages of the adoption process and over time some of the new varieties are likely, in their turn, to predominate. The most important factor in making sure that any such predominance is temporary is the breeding of newer, better varieties and this is most likely to happen rapidly through client-oriented methods.

No evidence was found that any landraces have been discarded by all the farmers in any of the agricultural environments we consider, so the genetic resources available for crop improvement – the reserve diversity – have not declined. Moreover, another measure of diversity – temporal diversity – increased. If the COB programmes are institutionalized and continue then temporal diversity will continue to be high. COB can, through its client orientation and emphasis on seed delivery, continue to provide a continuous stream of new varieties. For example, in Jharkhand two new varieties were released from the COB programme in 2009 that can be grown where farmers are currently cultivating the Ashoka varieties – PY 84 (Birsa Vikas Dhan 111) and Sugandha 1 (Birsa Vikas Dhan Sugandha 1).

However, the sustainability of COB programmes is in doubt as, despite successes, they have not been institutionalized. Despite many efforts at seeking funds our COB programmes have not attracted sustainable funding from international donors. Instead these donors continue to fund centralized international breeding programmes of the International Agricultural Research Centres (IARCs) while governments continue to fund the transfer of the technology model of public-sector breeding programmes. For future food security there has to be a shift to also provide funding to research-oriented NGOs for client-oriented plant breeding. Such NGOs have a comparative advantage for more farmer-oriented field research over the more centralized and research-station-centred approach of the IARCs and the public sector. There is no justification to rely on just one conventional, centralized plant breeding paradigm when COB has proven to be very effective both in marginal environments and in more productive ones that provide most of the world's food.

Our argument that there is no evidence that any landraces have been made extinct by the new varieties is supported by the evidence of Ford-Lloyd *et al.* (2009). They assessed the molecular diversity in germplasm collections of rice landraces made from 1962 to 1995. They found that the allelic diversity available on farm in later years was no lower than in earlier ones. They concluded that the genetic diversity on farm had survived in South and South-east Asia over the 33 years of the study. Hence, available diversity had not declined. However, the actual area over which these are growing could not be assessed. A decline in diversity would certainly be found if weighted diversity were to be used; the authors give examples of areas where IR36 and IR64 were adopted and diversity fell to zero.

In other areas where there has been no sustainable effort on client-oriented breeding, landrace diversity persists almost undiminished. In 1974 the senior author collected wheat and barley genetic resources in the Gilgit and Skardu regions in northern Pakistan (Witcombe, 1975) and he revisited these areas 25 years later. The landraces were still grown, and were still diverse. Breeding efforts had been limited to the introduction of dwarf plant varieties. Their lack of adaptation to local conditions and their much lower straw yields, in an area where straw yield was highly valued, had prevented their adoption. The local farmers may have preserved diversity but at a high cost; they have forgone the considerable benefits that a more client-oriented breeding programme could have provided.

Acknowledgements

Thanks are due to Vikas Kumar and Dr S.C. Prasad of Gramin Vikas Trust, Ranchi, India for their help during the survey in eastern India. Bikash Paudel, Rachana Devkota and Dhruba Neupane of LI-BIRD (Local Initiatives for Biodiversity, Research and Development), Nepal, were collaborators in the Nepal COB study and their contributions are gratefully acknowledged. We thank Dr Katherine Steele of CARIAD, Bangor University, for the use of her DArT data.

References

Brush, S.B. and Meng, E. (1998) Farmers' valuation and conservation of crop genetic resources. *Genetic Resources and Crop Evolution* 45, 139–150.

Ford-Lloyd, B.V., Brar, D., Khush, G.S., Jackson, M.T. and Virk, P.S. (2009) Genetic erosion over time of rice landrace agrobiodiversity. *Plant Genetic Resources* 7, 163–168.

Frankel, O.H., Brown, A.D.H. and Burdon, J.J. (1995) *The Conservation of Plant Diversity*. Cambridge University Press, Cambridge.

Gollin, D. and Smale, M. (1999) Valuing genetic diversity: crop plants and agroecosystems. In: Collins, W.W. and Qualset, C.O. (eds) *Biodiversity in Agroecosystems*. CRC Press, Boca Raton, Florida, pp. 237–265.

Joshi, K.D. and Witcombe, J.R. (2003) The impact of participatory plant breeding (PPB) on landrace diversity: a case study for high-altitude rice in Nepal. *Euphytica* 134, 117–125.

Joshi, K.D., Sthapit, B.R. and Witcombe, J.R. (2001) How narrowly adapted are the products of decentralised breeding? The spread of rice varieties from a participatory plant breeding programme in Nepal. *Euphytica* 122, 589–597.

Maurya, D.M. (1989) The innovative approach of Indian farmers. In: Chambers, R., Pacey, A. and Thrupp, L.A. (eds) *Farmer First: Farmer Innovation and Agricultural Research*. Intermediate Technology Publications, London, pp. 9–13.

Ravishankar, T. and Selvam, V. (1996) Contributions of tribal communities in the conservation of traditional cultivars. In: Sperling, L. and Loevinsohn, M. (eds) *Proceedings of Conference on Using Diversity and Maintaining Genetic Resources on Farm*, New Delhi, June 1995. International Development Research Centre (IDRC), pp. 268–274.

Rosenzweig, M.R. and Stark, O. (1989) Consumption smoothing, migration, and marriage: evidence from rural India. *Journal of Political Economy* 97, 905–926.

Satheesh, P.V. (1996) Genes, gender and biodiversity: Deccan Development Society's community seed banks. In: Sperling, L. and Loevinsohn, M. (eds) *Proceedings of Conference on Using Diversity and Maintaining Genetic Resources on Farm*, New Delhi, June 1995. International Development Research Centre (IDRC), pp. 268–274.

Souza, E., Fox, P.N., Byerlee, D. and Skovmand, B. (1994) Spring wheat diversity in irrigated areas of two developing countries. *Crop Science* 34, 774–783.

Steele, K.A., Edwards, G., Zhu, J. and Witcombe, J.R. (2004) Marker evaluated selection in rice: shifts in allele frequency among bulks selected in contrasting agricultural environments identify genomic regions of importance to rice adaptation and breeding. *Theoretical and Applied Genetics* 109, 1247–1260.

Steele, K.A., Gyawali, S., Joshi, K.D., Shrestha, P., Sthapit, B.R. and Witcombe, J.R. (2009) Has the introduction of modern rice varieties changed rice genetic diversity in a high-altitude region of Nepal? *Field Crops Research* 113, 24–30.

Virk, D.S. and Witcombe, J.R. (2007) Trade-offs between on-farm varietal diversity and highly client-oriented breeding – a case study of upland rice in India. *Genetic Resources and Crop Evolution* 54, 823–835.

Virk, D.S., Singh, D.N., Kumar, R., Prasad, S.C., Gangwar, J.S. and Witcombe, J.R. (2003) Collaborative and consultative participatory plant breeding or rice for the rainfed uplands of eastern India. *Euphytica* 132, 95–108.

Witcombe, J.R. (1975) Wheat and barley from two Himalayan regions. *Euphytica* 24, 431–434.

Witcombe, J.R. (1999) Does plant breeding lead to a loss of genetic diversity? In: Wood, D. and Lenné, J.M. (eds) *Agrobiodiversity: Characterization, Utilization and Management*. CAB International, Wallingford, UK, pp. 245–272.

Witcombe, J.R. and Virk, D.S. (2001) Number of crosses and population size for participatory and classical plant breeding. *Euphytica* 122, 451–462.

Witcombe, J.R., Joshi, A., Joshi, K.D. and Sthapit, B.R. (1996) Farmer participatory crop improvement. I: Varietal selection and breeding methods and their impact on biodiversity. *Experimental Agriculture* 32, 445–460.

Witcombe, J.R., Joshi, K.D., Gyawali, S., Musa, A., Johansen, C., Virk, D.S. and Sthapit, B.R. (2005) Participatory plant breeding is better described as highly client-oriented plant breeding. I. Four indicators of client-orientation in plant breeding. *Experimental Agriculture* 41, 299–319.

7 Transgenics Can Enhance Crop Diversity – Under Certain Circumstances

J. Gressel

The Need to Breach the Genetic Glass Ceiling

There has been a major loss of crop biodiversity, both in the number of crops cultivated, and especially in the proportion of our food supply produced by four crops. These four crops, wheat, rice, maize and soybean, now supply 80% of the necessary calories for humans and their domesticated animals. How did we deteriorate to such a situation where a pandemic attack on one crop could have dire consequences on world food security? Are the farmers stupid to have brought us to this situation?

These issues are discussed at great length in a recent book by the author (Gressel, 2008a). This chapter distils and condenses some of the issues described in that book, except for the sections that deal with the implications deriving from new methodologies coming on line.

Farmers' concerns, just as those of most people, are to provide the wherewithal to support themselves and their families, and their short-term considerations are to cultivate the crops that repeatedly provide them with a profit. They have to make tough choices about which crops to cultivate and despite the misconception that farmers are very conservative, they will rapidly adopt new crops or crop varieties that better fit their agricultural ecosystem. When a novel adoption seems absurd to us, we should look more closely at the reasons that the farmers used to make that choice. Many wonder why African farmers rapidly adopted maize over far more transient-drought-tolerant sorghum, clearly better adapted to Africa than maize. Flavour was a small part of the issue, but the major reason was the African weaver bird (*Quelea quelea*) that decimates the crop in huge flocks, considered to be 'feathered locusts' (Doggett, 1988). Breeding made the sorghum taste worse to bird and man, but the birds persevered and ate bitter varieties after they had eaten the tastier ones. The farmers thus chose and adopted maize with its covered ears that the birds have yet to discover.

In general it can be stated that the big four, wheat, rice, maize and soybean, have taken over because they have the greater inherent diversity in their genomes. In this author's lifetime maize has been domesticated to complete its lifecycle many hundreds of kilometres north and south of where it could be cultivated when he was a child. Wheat originated in the arid subtropics and over millennia had the inherent diversity that allowed it to be adapted to far more agro-ecological areas than oats, which originated nearby.

Far too often breeders do not realize that each crop has its own 'genetic glass ceiling' of inherent diversity. If it or its interbreeding relatives do not possess a trait, the trait cannot appear from thin air, no matter how many crosses are made. A genetic glass ceiling is a ceiling, lower in some species and clearly quite high in others. Thus, if we want to increase crop biodiversity we must find and insert the genes the breeder cannot find in a given species, or use other means to deal with the impediment to cultivation. For example, two methods are currently used to keep weaver birds from attacking sorghum: spraying nesting sites with an organophosphate insecticide or exploding dynamite in the nesting areas (Mundy, 2000). Transgenically breaching the glass ceiling by adding maize genes to sorghum so that it produces a covered ear (Gressel, 2008a) may sound like science fiction, but is more appealing than dynamiting or poisoning birds.

Constraints to crop biodiversity

Both high ceiling and low genetic glass ceiling crops may lack the inherent genetic variability to deal with many of the constraints facing crop cultivation, such as diseases, insects and weeds, despite the widespread use of pesticides, as well as drought, flooding, cold, heat, etc. In the highly developed world pesticides are used to deal with many constraints, along with irrigation, desalination and cooled or heated greenhouses, allowing some crops such as tomatoes to be grown at all times, precluding the need for vast genomic diversity. This is a drop in the bucket though when dealing with enhancing crop biodiversity – it is a reduction in the proportion of arable land devoted to four crops, by increasing crop biodiversity that matters, providing that this can be done without negatively affecting food security. Indeed, if other crops are adequately domesticated, this should enhance food security by lessening the dependence on the big four.

Much of the developing world is plagued with intractable problems: weeds related to crops and thus unable to be selectively controlled with herbicides; underground root-parasitic weeds out of reach of herbicides; and fungal, viral and bacterial diseases as well as insects and other arthropods, especially those of stored grain, that both eat the grain and vector mycotoxin-producing fungi (Gressel *et al.*, 2004).

Solving these problems and increasing yields in major crops could increase crop biodiversity, as absurd as it may sound. When market forces of overproduction limit area under cultivation, land is available for other crops that perhaps the consumer can afford to buy because of lower prices of the major commodity crops. For example, solving the yield reduction caused by parasitic weeds in Africa would at least double maize production, lowering the price and forcing farmers to reduce the area planted to maize (Ejeta and Gressel, 2007). They could then plant legumes on the newly available land, increasing crop biodiversity, and enriching the diet in missing protein. If the above constraints can be solved with minor crops before solving them for major crops, then the minor crops might be cultivated more widely, increasing crop biodiversity.

One type of enhancing of biodiversity will decrease crop biodiversity: increasing the biodiversity of weed, pathogen and insect pests as promoted by some misguided urban 'experts' on the environment can only lower crop yields, resulting in higher prices for staples and requiring more land under cultivation. Agroecosystems should be devoted to crops, not to yield-reducing pests, even in the name of increasing biodiversity.

Crop biodiversity can be gained or lost for many reasons

Efforts are currently underway to rapidly domesticate a number of crops to produce biofuels (Gressel, 2008b; Vega-Sanchez and Ronald, 2010). It is often forgotten that 80 years ago about 20% of temperate agriculture was devoted to one biofuel crop, oats, fulfilling much of the energy requirements of draught animals on farm, in cities and also for farm labourers. The need disappeared, despite experts at the time assuredly stating that no farmer would purchase a tractor because you had to buy fuel, when fuel could be cultivated. If a crop is no longer needed, it will disappear, as have the many fibre crops for rope manufacture.

Temperate and subtropical root and tuber crops that could not be stored disappeared and the potato remained. Some might be resuscitated transgenically by enhancing their shelf life. Groundnuts (peanuts) from South America tasted better than African bambara nuts (*Vigna subterranea*) (Heller *et al.*, 1997) hence are widely cultivated throughout Africa. Experts in increasing biodiversity are often surprised that a crop is abandoned: 'Apparently a range of positive traits such as rusticity, pest tolerance and high productivity under low levels of inputs cannot counterbalance the lack of market interest' in the tuber mashua (*Tropaeolum tuberosum*) (Grau *et al.*, 2003). If domestication of mashua were to continue as it has with so many other species, breeding out the pungent taste and odour might also increase the insect and disease problems requiring pesticide use. Consumer tastes change in this world of 'instant' with working couples and less time to prepare food; wheat in prepared bread is gaining over rice and other carbohydrate crops requiring even a short period of cooking, despite what this author and many readers may claim to prefer to eat.

The Use of Transgenics to Breach the Genetic Glass Ceiling

Classical genetics has been an excellent tool to exploit the genetic potential of crops and their interbreeding relatives, and adapting the crops to various agroecosystems. Problems arise when one must bring in traits from interbreeding relatives, as when crossing one brings a whole genome of undesirable traits along with desirable ones. Some undesirable traits can be backcrossed away quickly and easily, especially with molecular marker-assisted breeding. Still, some undesirable traits are closely linked near the desired trait on the same chromosome as the desirable trait, and getting them to recombine away by chromosome crossing-over is laborious. Sometimes one wishes to keep a trait in most of the plant, but not the edible portion (e.g. the insect and human poisons that were in hallucinogenic 'love apples' before they were domesticated into edible tomatoes). In molecular terms, this requires mutating a general promoter on the toxin genes to be tissue specific, which is not an easy breeding task.

Transgenics have the ability to bring single isolated chosen genes from wherever they may exist into the crop that needs them. They can do so without bringing the deleterious baggage of other genes from the source. Genetic engineering also allows one to selectively suppress plant genes in a tissue-specific manner, e.g. the poisons from the tomato progenitor fruits, or the pungent flavour chemicals from mashua tubers, etc. One can stack a whole variety of genes in the same crop either by crossing various transformants with each other or by 'co-transformation' of a group of genes at one time. Whole metabolic pathways have been recently co-transformed (Zhu *et al.*, 2008).

There is far less randomness with transgenesis than with breeding, as >99.9% of the genome remains unaltered. The level of transgene expression and its stability are a function of where the gene was inserted. When a transgene inserts into a vital gene, the transformant dies or is sufficiently 'wimpy' to be discarded in favour of healthier cohorts.

To preclude the baggage brought from wild interbreeding relatives of the crop, it has been proposed to bring needed traits as isolated single genes only from such relatives, using the same molecular tools but under the guise of cisgenics (from interbreeding species) (Jacobsen and Schouten, 2009) versus transgenics from wherever. This limits trait hunting to wild relatives, which may not have the best traits; indeed, if the gene is for disease or insect resistance, the trait from wild relatives might just confer back the obnoxious tasting or poisonous compounds that our ancestors bred out of the crop.

The World is Using Transgenics, Mainly in the Wrong Direction

Some rare uses of transgenics helped preserve crop biodiversity; e.g. the Hawaiian papaya industry was saved from complete devastation from ringspot virus disease by engineering a virus coat protein gene into high quality papaya varieties (Ferreira *et al.*, 2002). This is a major exception. Mainly the crops with the greatest genetic variability have had their genetic glass ceiling raised by transgenics. The dominant genetic engineered crops in commercial cultivation are soybean, maize, cotton and oilseed rape (canola), some of the most widespread crops in cultivation. Over 80% of the world's soybeans are derived from backcrossing a single successful herbicide-resistant transformant ('event' in regulatory parlance) into local varieties. This decreases soybean diversity, albeit to a small extent because there is a 'linkage disequilibrium': the genes closest to the herbicide-resistant gene will remain linked to the resistance gene, decreasing variability in that region. Only after more than a decade of cultivation has a new resistance gene been introduced, with a different chromosomal location.

Very few herbicide-resistant and insect-resistant genes are distributed among soybean, maize, oilseed rape and cotton, i.e. a very low genetic diversity of genes is used. This poor diversity of genes increases the likelihood that the weed or insect pests they control will more rapidly evolve resistance than if a wider diversity of genes were used. Other genes and more multi-stacked traits are appearing, but most are going into the same crops. This has been excellent (so far) for the human and agricultural environment. Where Bt insect resistance has been used there has been an extreme reduction in the amount of insecticides used, far lowering pesticide poisonings of farm applicators as well as lessening the impact on non-target insects (Brookes and Barfoot, 2010). Some point to cases where transgenics seem to have zero impact on yield, but this is a polemic. One must also compare how the transgenic crop affects the farmer's balance sheet, and look at (typically non-costed) effects to current farmers' practice. It is also clear that the environmental impact is currently orders of magnitude less than the insecticides used in conventional agriculture as well as those used in organic farming. The trend in insecticide development has been to find narrow-spectrum compounds with minimum effect on non-target insect groups. All those who care about the environment endorse this trend. Separate Bt genes were chosen for engineering into crops in the same manner to control either lepidopteran (moth), coleopteran (beetle), or dipteran (flies and mosquitoes) pests. This has led to mirid bugs purportedly becoming secondary pests on cotton (Lu *et al.*, 2010). Even though this problem would be expected with any insecticide, as nature abhors vacuums, it is being used to claim that Bt transgenes are 'bad', and the bad old days of broad-spectrum insecticides are better. But flawed studies (Lu *et al.*, 2010) based on 1 year's data on a mirid outbreak in China have been brought to the public eye. The problems from mirid bugs were less severe in subsequent years despite increased plantings of Bt cotton, contradicting the dire predictions in that paper. Still, a greater biodiversity of insect resistance genes is needed as part of the breeders' arsenal, just as chemical diversity is needed by those spraying conventional insecticides.

The main herbicide resistance gene commercialized is to the broad spectrum, low mammalian toxicity and low environmental persistence, very inexpensive glyphosate. The rapid adoption brought an environmental revolution: heavy, high energy-using ploughs and discs were left to rust and minimum tillage practices were instituted. The concomitant savings from far less soil erosion as well as the benefits to soil structure due to less compaction have been incomparable. Because the herbicide glyphosate was so cheap, the technology so flexible and easy to use compared to other herbicides, farmers threw all caution about resistance management to the wind, and relied almost entirely on glyphosate. This was abetted by industry who claimed that it was nigh impossible for resistance to evolve (Bradshaw *et al.*, 1997), as well as government regulators who removed many excellent alternative herbicides from permitted use. Nature abhors scientists who claim that evolution of resistance is unlikely, and glyphosate resistance evolved in some of the most pernicious weeds, the worst being widespread resistance in *Sorghum halepense* in Argentina, *Lolium* species worldwide and *Amaranthus* and *Conyza* species mainly in

North America (Heap, 2010). Glyphosate was one of the few herbicides that could control *S. halepense*, as it is systemic and can penetrate to the underground rhizomes. *Sorghum halepense* evolved a 'phoenix' mechanism of resistance: the herbicide burns off the leaves, and the new shoots arise from the rhizomes through the ashes of the leaves, due to an evolved lack of herbicide translocation to the underground rhizomes. Companies are now 'stacking' the glyphosate resistance gene with resistances to other herbicides, one which kills only broadleaf weeds, or one to which the grass weeds such as *Lolium* and *S. halepense* have already evolved resistance. Not only does agriculture need crop biodiversity, it needs chemidiversity of herbicides. In the last 30 years only one new herbicide has been developed that acts on a novel target site.

Thus we have a situation where modern transgenics have helped agriculture, and have increased the genetic diversity in the crops that needed it the least. There are so many lesser cultivated crops that need new genes so that their cultivation might be expanded, or that they may be cultivated with less external inputs, lowering production costs thereby rendering them more competitive. Clearly we are heading in the wrong direction. Is no one putting the genes in the right crops? Clearly not enough is being done. But a perusal of the scientific literature shows that most of the needed genes are already known and in the databases, often put into some of the needed crops. A check of the various governmental web sites shows that many have been field-tested. Why are they not in the fields and on our tables? What can be wrong about having less mycotoxins in our food, having cheaper food, having food requiring less refrigeration and a longer shelf life? Every major medical association, many national and international academies of science have endorsed these products as safe for humans and the environment and beneficial to agriculture. Some countries have collective amnesia about their own history. *Phytophthora* attacking potatoes starved a goodly proportion of the Irish population to death and an equal number emigrated to avoid death, yet the recently produced transgenic potatoes resistant to this blight are not cultivated in Ireland. Instead, Irish farmers pour huge amounts of fungicides on potatoes, yet the chemicals hardly affect the fungus. Can the Irish be so addicted to chemicals?

Regulatory Impediments to Enhancing Agrobiodiversity

Regulatory regimes are needed to assure safety as a general rule; human nature is to cut corners in safety, whether in driving, deepwater drilling for oil, or whatever. Logical regulatory systems identify hazards and then assess the risks that the hazard can cause damage. When there is no hazard, a product can achieve 'GRAS' (generally regarded as safe) status and the product is exempt from further regulatory scrutiny. Historically, regulatory scrutiny came into play only after actual risk was demonstrated. This has not been the case with transgenics. When the first transgenic organisms were generated the scientists performing the experiments considered the possibilities to be so unknown that they themselves decided on a moratorium to assess risks, after which they understood that a transgenic organism could be no more risky than the transgene product. Obviously an organism transformed with a toxin or allergen-encoding gene could be more risky than the wild type if the gene is expressed. If transgenics are used to suppress endogenous toxins or allergens in an organism, the transgenic organism is inherently less risky than the wild type.

Few nations have regulatory systems that assess hazards and risks in a multi-tiered manner, fast-tracking transgenics that obviously pose no risks. Few systems compare the risks of presently used agronomic procedures versus transgenics. Thus, in much of the world you can cultivate castor bean or *Jatropha*, producing the highly toxic ricin or curcin, respectively, with impunity and dump the toxic residues on soil as 'manure' without environmental impact studies (Gressel, 2008b). If the toxin-encoding genes were suppressed or excised using the tools of recombinant DNA, there would be a requirement for extensive and expensive toxicity and environmental impact studies. For that reason the wild types of these species are being planted despite the biosafety and biosecurity risks; ricin is a toxin of choice of bioterrorists as a Bulgarian diplomat discovered on the streets of London, after being approached from behind by a Russian agent with a hollow-pointed umbrella.

Likewise, no regulators compare the risks from multiple applications of human-toxic insecticides with risks so low as to be unknown for the Bt gene in crops such as aubergine (also known as eggplant or brinjal). Many advocates of hyper-regulation do not hear the 'so low as to be' and just hear 'unknown' and demand absolute safety, knowing full well that nothing can be proven to be absolutely safe.

The present regulatory regimes thus favour blockbuster crops such as maize, cotton and soybean and only large multinational companies can afford the costs required for regulatory approval. As regulation is 'event based', a single transformant ('event') undergoes registration and is then backcrossed into other varieties of the crop, bringing adjacent genes with it, i.e. 'linkage disequilibrium' as discussed in an earlier section. If the same gene, in the same construct is engineered into other crops or even into other varieties of the same crop, regulation starts from the beginning, as they are different 'events'. 'Familiarity', an important concept that allows regulators in other areas to focus on real hazards and risks and not waste time with familiar non-hazards or hazards with negligent risks, does not come into play with transgenics. Due to this, even a large multinational will wait more than a decade to change 'events', as was seen with glyphosate herbicide-resistant soybean, because of the regulatory costs. It is too expensive to replace 'events', even with the same gene expressed on a different chromosomal location. With conventional breeding, better varieties appear almost yearly.

Thus the pile in front of the regulator includes files for transgenics with novel proteins, for transgenics with suppressed toxin genes and for transgenics carrying a gene that is already in four other crops. The files are equally thick and at least, in theory, equal time must be given to all.

Science-based changes have been proposed by many unbiased scientists with no axe to grind. They call for use of familiarity and use of a level of scrutiny that is appropriate for both hazard and risk (Bradford *et al.*, 2005). Unless such changes are made it will be impossible to use transgenic techniques to attain crop biodiversity by inserting genes missing in the underutilized crops needing further domestication, or even for suppressing endogenous genes that are deleterious to the cultivation of such crops.

The high regulatory thresholds play into the hands of an unholy alliance – the large multinationals that can afford the cost, and those who oppose transgenics. This is most peculiar if not illogical; a large proportion of the activists denigrate transgenics as a cover for their dislike of globalization and multinational corporations that they fear will corner the seed market. Targeting transgenics *per se* actually prevents public sector and small biotech company research, and from getting their products to market. Those products are typically the crops and genes that would expand crop biodiversity. Thus the call for greater use of transgenics to increase biodiversity and to logically regulate them is joined by ex-ideologues of the environmental (Lewis, 1992; Brand, 2009) and organic (Ronald and Adamchak, 2008) movements. These thinkers have performed the risk benefit analyses for the general case and see that the benefits of transgenics far outweigh the perceived or unknown risks.

Regulatory regimes are mandated by politicians who determine the terms of reference. Where politicians have cared about food security and farmer productivity (e.g. Canada), all novel traits, whether introduced transgenically or by breeding, are regulated, but familiarity and logic are part of the process. Some regulatory regimes focus mainly on what the gene does (USA, Argentina), others more on where it came from and where it becomes localized in the genome (Europe). Claims have been made that the USA is dangerously moving towards Europe (Davison, 2010). There is little coherence among the regimes; sovereignty is promoted as a way to make it harder to register a product. Eventually this could backfire: countries where it is too complicated to cost-effectively register products will not have the products their neighbours are using to increase crop productivity and agrobiodiversity.

New Molecular Methods That Could Assist Enhancing Crop Biodiversity

The first generation of heavily commercialized transgenic crops is quite analogous to the first generation of widely purchased automobiles, the Ford Model T. The Model T uptake was amazingly rapid and it revolutionized personal transportation in the same way as the first generation of transgenics revolutionized weed and insect control in agriculture. In both cases, a limited number of versions was released, were inexpensive, and widely appreciated by users. There were opponents who tried to pass laws to prevent uptake (a flagman must run in front of the Model T to prevent scaring horses) but the voices of Luddites were eventually stifled by a populace that soon saw through their illogical motivation. Like the Model T Ford, with its breakable suspension, no electric starter, poor steering, inefficient motor, poor tyres, etc. which make that revolution quite different from today's vehicles, so the first generation of transgenics will be replaced by far better and more diverse models, as excellent and revolutionary as modern vehicles. There can be many more genes encoding a wider variety of agronomic and consumer traits, in many more species. This can assist in further domesticating species outside of the four that provide us with most of our calories. Generating the newer models that will bear these traits is being facilitated by the new technologies that have been/are coming into play and are briefly discussed below to give a taste of how transgenics are developing from their Model T days.

Promoting timed and tissue-specific gene expression

Most Model T transgenic crops have their genes under the control of non-specific promoters, the sequences that actually control when and where a gene is expressed in the life of a plant. A case in point is the 35S promoter sequence that originated from a virus that attacks plants. Genes under its control are expressed in most tissues, much of the time, whether needed or not. Besides the waste of energy in expressing genes when not needed, it can preclude full use of crops. For example, castor bean and *Jatropha* are being touted as biofuel crops, and the meal after oil removal would make excellent cheap, high protein animal feed if it were not for the inconvenient presence of small amounts of ricin and curcin, exceedingly toxic proteins that render the meal a biosecurity problem both in normal handling and in the hands of terrorists (Gressel, 2008b). If antisense or RNA interference (RNAi) were used to suppress ricin or curcin synthesis under a non-specific promoter, the leaves, roots, stems and seedcoats would be subject to insect attack. If the RNAi or antisense system were under a strong seed-specific promoter, the level might be reduced sufficiently to allow the meal to be used as feed. Similarly, if the Bt gene for stem borer was under the control of a stem-specific promoter, there would have been no expression in maize pollen. This would have saved considerable amounts of research funds dedicated to ascertaining that even though Monarch butterflies force-fed Bt maize pollen died (Losey *et al.*, 1999), the pollen is innocuous because in nature Monarch butterflies do not eat maize pollen, nor could they, as they arrive after maize has shed its pollen (Stanley-Horn *et al.*, 2001).

Thus the new technologies allow transgenes to be promoted only when/where needed as:

1. Tissue specific promoters – as described above;
2. Temporal specific promoters – e.g. senescence or ripening specific promoters that could degrade non-palatable or unwanted products just before or after harvest; and
3. Inducible promoters – e.g. promoters that turn on energy-expensive protective pathways following incipient stress so that the plant can cope when the stress is acute; having such pathways operative at all times is a waste of resources.

Targeted gene insertions

The presently commercialized Model T transgenic crops resulted from tedious transformation protocols that randomly introduced the gene in one or more copies at any old position in the genome. Some insertion events lethally disrupted genes and others inserted into areas where expression is greatly blocked, some less blocked, and some positions where expression is at the desired level. Some transgenes later became inactive after a few generations. Thus, a considerable amount of screening and testing has to be performed before an 'event' can be released. If it is later decided to add (stack) other transgenes with the first trait, the process begins anew, or the newly desired transgene(s) is/are crossed in from other events. The different transformed traits are probably on different chromosomes and will segregate from each other during backcrossing. Thus, achieving homozygosity takes time and patience, and transferring the multiple traits to other varieties takes even more time as each genetically segregates its own way. This 'herding cats' can be precluded using a few novel techniques.

Zinc finger nucleases (Srivastava and Gidoni 2010; Weinthal *et al.*, 2010; Zhang *et al.*, 2010) and meganucleases (sometimes called homing nucleases) (Puchta, 2005) are endonucleases that recognize specific long stretches of DNA, and cause double strand breaks. If they are introduced together with transgene constructs that have as borders sequences that are homologous to the ends of the broken DNA, the transgene goes to that site by 'homologous recombination' using the plant's own genome repair system. If the site proves to allow good and stable expression, other transgenes can be added to the same site. All transgenes targeted to the site are genetically 'linked' and will be inherited as if a single gene, facilitating backcrossing into multiple varieties. Many regulatory authorities insist on knowing precisely where a gene has been inserted, requiring sequences of the flanking regions. Because of the nature of endonuclease insertion these data need be gathered only once.

If the endonuclease binding site is within an expressed gene, that gene will be silenced. This is good if the gene expression is unwanted (e.g. encodes a toxin or encodes fruit softening). Similarly, intergenic sites can be chosen for targeting, when one is found that allows excellent stable gene expression. The homologous recombination systems also allow deleting a stretch of DNA and replacing it with another that encodes one or more amino acids not in the original gene. This type of site-directed mutagenesis allows changes that would not occur with the one nucleotide at a time mutagenesis that occurs in nature.

The sequence specificity of the zinc finger nucleases and the meganucleases is also their biggest drawback. They either have to be designed, synthesized and optimized for the particular sequence of the gene to be disrupted in each species, as the same gene rarely has sufficiently long stretches with the same precise nucleotide sequence in more than one species. Thus, the intragenic site that can be disrupted in one species may not exist in another. In the case of meganucleases, one company has a library of tens of thousands of meganucleases, each recognizing different sequences. These can be matched with a whole gene sequence for the crop in question to see whether they have a matching meganuclease (Cellectis, www.cellectis.com). If not, one can be custom synthesized, at a cost. Conversely, for *Arabidopsis* at least, a genome browser is available that displays the zinc finger sites that can be targeted by reagents available in the public domain OPEN (Oligomerized Pool ENgineering) platform. It is claimed that the platform is 'sufficiently robust to target most *Arabidopsis* genes' at a high frequency (Zhang *et al.*, 2010).

Minichromosomes

A cluster of transgenes can be arranged as a single 'minichromosome' that can be engineered into a crop and be inherited in further generations as a separate minichromosome, in parallel to their big native brethren (Yu *et al.*, 2007). The cluster can contain a complex of traits: insect, fungus, herbicide, drought, heat, frost, flooding resistance along with enhanced nutritional and flavour traits, etc. They will be inherited as a single dominant gene in backcrosses into other varieties of the same species. The same minichromosome, if found effective in one crop species, can be transformed into others as well. In essence, this is as if a chromosome was backcrossed from one species to another, akin to backcrossing to another variety, where the whole minichromosome is moved. Whether regulators will accept both familiarity and the analogy to regulatory exemption of backcrosses, and then lower the level of scrutiny on multiple species use of the same minichromosome, is an open question.

Dealing with crop to weed transgene flow

Transgene movement from crop to wild species is thought to have the possibility of leading to dire environmental consequences (Ellstrand, 2003). This is often mooted as a reason to prevent the release of transgenic crops. Interestingly, the examples given where this may happen are cases where native genes have introgressed into related weeds or ruderal species, not into wild species in natural habitats. It is telling that so many supposed environmentalists, who claim possibilities of ecological disasters, do not understand and distinguish between weeds and wild species, or natural ecosystems and agroecosystems. Few crops can interbreed with wild species due to genetic incompatibilities and, even if the genetic barriers did not exist, proximity is a problem. Thus, the 'dire consequences' could only be in the farmers' own fields. Not all transgenic traits would confer an advantage to weeds related to crops, and the farmer would be more than happy if the transgene would confer disadvantages, as the weeds related to crops cannot usually be controlled by crop-selective herbicides. Still, there are transgenic traits that one would not want transferred to weeds. Foremost among such traits is herbicide resistance. By engineering herbicide resistance into the crop, the farmer suddenly has the tool to control the related weed. The last thing desired would be for the transgene controlling herbicide resistance to move into the weed, losing the technology. This has already happened when non-transgenic herbicide-resistant rice was achieved by mutagenesis. The herbicide was exceedingly effective in controlling the feral form of rice (often called red rice or weedy rice), but only for a few years in many parts of the world. The mutated gene rapidly crossed into the weedy rice and backcrossed with it, and a good technology was lost. While maize, soybean and cotton do not have weedy relatives in much of the world, rice, sorghum and oilseed rape do. Sunflowers, carrots, rye, beets, radishes, oats as well as wheat also have pernicious weedy relatives living adjacent to them in agro-ecosystems. Many transgenic traits being used would have little effect on their weediness, but surely herbicide resistance does and, in many cases, abiotic and biotic stress tolerances might increase their weediness, if the weeds do not possess those traits already.

While such gene movement of non-transgenic traits cannot be prevented in open agriculture, it can be dealt with transgenically. Various solutions have been proposed to either attempt to 'contain' transgenes within the crop or to mitigate any positive effect should they cross into related weeds. Most of the containment methods proposed, e.g. chloroplast genome transformation, male sterility and genetic use restriction technologies (GURTs) or 'terminator' technologies, are at best unidirectional, preventing the crop from pollinating the weed but not the weed pollinating the crop. Even in the direction that they work, they are typically 'leaky' containers at best (Chapter 4 in Gressel, 2008a). Once a transgene has leaked to a weed, it will rapidly spread if it has a selective advantage. Thus, 'Transgenic Mitigation' (TM) technologies were proposed to ensure that any transgene that leaks out to weeds (or wild populations) cannot establish and compete in its ecosystem (Gressel, 1999). Transgenic mitigation is attained by tandemly linking the desired (problematic) transgene with other transgenes that are either of positive or neutral value to the crop, but would render the hybrid with the weed (or wild species) as well as its backcross progeny with the weed (or wild species) uncompetitive. Because the transgenic traits are tandemly linked, the 'problem' trait and the mitigation trait are genetically linked and will be inherited together. In cases where special care is needed (e.g. where the 'problem' transgene encodes a pharmaceutical trait), the 'problem' transgene can be flanked on either side to preclude ultra-rare mutation in the mitigator or its being separated by even more rare crossing-over events.

The mitigator traits must be tailored to the crop in question. Ideal mitigator traits (anti-weediness traits) for many field crops are transgenes that increase crop yield by reducing plant height (classical Green Revolution traits), genes that prevent seed 'shatter' (seed drop) that replenishes the weed seed bank, or genes that induce uniform germination or super-sensitivity to other herbicides used in a rotational regime (Gressel and Valverde, 2009). Transgenes that prevent pollen and flower formation can be used as mitigator transgenes for vegetatively-propagated crops such as potatoes or some tree species used in forest plantations. Anti-bolting (premature flower stalk formation) genes (RNAi or antisense of genes on the pathway to gibberellic acid) can be used for root crops such as carrots, radishes and beets to the benefit of the crop and to the detriment of the weed. A special modification has been suggested for wheat, a crop with three distinct genomes that are similar but not quite homologous to the genomes of many weedy *Aegilops* species, and 'homoeologous' recombination can transfer genes. Wheat has the *phi*1 gene located on the long arm of chromosome 4B that prevents such homoeologous recombination in hybrids, and such recombination occurs only after chromosome 4BL is lost in subsequent progeny. In this case, it was proposed to insert the transgenes of choice to a nearby site on chromosome 4BL (Weissmann *et al.*, 2008) such that they cannot integrate into the weed chromosomes.

Transgenic mitigation has been demonstrated to fulfil its promise in screenhouse (Al-Ahmad and Gressel, 2006; Al-Ahmad *et al.*, 2006) and field (Rose *et al.*, 2009) experiments with oilseed rape. Thus, after further experimentation with each species that has interbreeding weeds, it should be possible to determine that useful transgenes can be put into crops with weedy or wild relatives without fear that there may be detrimental effects on the agroecosystem or natural ecosystem, respectively, as long as they are transgenically mitigated.

Concluding Remarks

Crop biodiversity can be enhanced by introducing new genes not found in the crop genome nor in the genome of interbreeding relatives by transgenic technology. Diversity can also be enhanced by 'surgically' suppressing deleterious genes from the tissues where they are not wanted. The former cannot be performed by breeding: genes do not appear from thin air by crossing. Transgenic technologies are 'cleaner' than breeding when it comes to bringing genes from related species, as only the 'gene' is moved and not a whole genome that brings many undesirable traits. Breeding can result in tissue-specific suppression of genes; that is how tomatoes went from being a poisonous or hallucinogenic 'love apple' (depending on dose) to what we eat today. This would have happened much more quickly using tissue-specific suppression of the toxin pathway.

Breeding is still and will always remain a very necessary tool. Once transgenic traits are in a crop, they must be moved into a large number of geographically and ecosystem adapted varieties; there is no 'one size fits all' with crops.

Until now, the multinational private sector seed companies have only been interested in blockbuster products in major crops reducing crop biodiversity. The farmer will ask: Why cultivate other legumes when herbicide-resistant soybeans are so easy to grow? Why cultivate drought-tolerant sorghum when maize has resistance to herbicides and insects? It will be a while until these large multinational companies come around to understand the need and market for increasing crop biodiversity. This void must be filled by the public sector along with small, smart, rapidly-acting local biotechnology companies as well as local seed companies. The technologies are becoming easier to use, many are off-patent and in the public domain.

In some cases it is to the common good to

increase the biodiversity of genes in crops and the biodiversity of crops: for example, when it comes to reducing mycotoxins, enhancing nutritional value, etc. Here, public sector involvement is imperative and an effort must be made to convince those that delegitimize transgenics as a whole because they do not like multinational corporations or globalization, to endorse transgenics as a way to increase crop biodiversity and local diversity of agriculture.

Moore's law pertaining to the exponential increase of computer chip capacity and price decrease with time has been extrapolated to both the cost of DNA sequencing and DNA syntheses (Carlson, 2003). The cost is continually being greatly reduced. The transgenic Moore's law can be extrapolated to the costs of finding the genes you need and transforming them into the desired crop for enhancing crop biodiversity. Thus, the scientific aspects of increasing crop biodiversity through transgenics are becoming economically feasible with a profit incentive. Unfortunately this is being fought by a non-scientific risk analysis that does not balance risks with benefits and is instead political and emotional. The upshot of this is denying the farmer and the consumer the choice to cultivate and consume the transgenics that could be produced. It is fascinating to see how those that are pro-choice in so many other areas of our lives work so hard to prevent us from having the choice to increase crop biodiversity through the only means that can breach the genetic glass ceilings of crops that could and should be cultivated to lessen our unhealthy dependence on so few species.

References

Al-Ahmad, H. and Gressel, J. (2006) Mitigation using a tandem construct containing a selectively unfit gene precludes establishment of *Brassica napus* transgenes in hybrids and backcrosses with weedy *Brassica rapa*. *Plant Biotechnology Journal* 4, 23–33.

Al-Ahmad, H., Dwyer, J., Moloney, M.M. and Gressel, J. (2006) Mitigation of establishment of *Brassica napus* transgenes in volunteers using a tandem construct containing a selectively unfit gene. *Plant Biotechnology Journal* 4, 7–21.

Bradford, K.J., Van Deynze, A., Gutterson, N., Parrott, W. and Strauss, S.H. (2005) Regulating transgenic crops sensibly: lessons from plant breeding, biotechnology and genomics. *Nature Biotechnology* 23, 439–444.

Bradshaw, L.D., Padgette, S.R., Kimball, S.L. and Wells, B.H. (1997) Perspectives on glyphosate resistance. *Weed Technology* 11, 189–198.

Brand, S. (2009) *Whole Earth Discipline: An Ecopragmatist Manifesto.* Viking Press, New York.

Brookes, G. and Barfoot, P. (2010) Global impact of biotech crops: environmental effects, 1996-2008. *AgBioForum* 13(1), #6.

Carlson, R. (2003) The pace and proliferation of biological technologies. *Biosecurity and Bioterrorism: Biodefense Strategy, Practice, and Science* 1, 203–214.

Davison, J. (2010) GM plants: science, politics and EC regulations. *Plant Science* 178, 94–98.

Doggett, H. (1988) *Sorghum*, 2nd edn. Longman, Harlow, UK.

Ejeta, G. and Gressel, J. (eds) (2007) *Integrating New Technologies for Striga Control: Ending the Witch-hunt.* World Scientific, Singapore.

Ellstrand, N.C. (2003) *Dangerous Liaisons – when cultivated plants mate with their wild relatives.* Johns Hopkins University Press, Baltimore, Maryland.

Ferreira, S.A., Pitz, K.Y., Manshardt, R., Zee, F., Fitch, M. and Gonsalves, D. (2002) Virus coat protein transgenic papaya provides practical control of papaya ringspot virus in Hawaii. *Plant Disease* 86, 101–105.

Grau, A., Ortega Dueñas, R., Nieto Cabrera, C. and Hermann, M. (eds) (2003) *Mashua (*Tropaeolum tuberosum *Ruíz & Pav.).* International Plant Genetic Resources Institute, Rome.

Gressel, J. (1999) Tandem constructs: preventing the rise of superweeds. *Trends in Biotechnology* 17, 361–366.

Gressel, J. (2008a) *Genetic Glass Ceilings: Transgenics for Crop Biodiversity.* Johns Hopkins University Press, Baltimore, Maryland.

Gressel, J. (2008b) Transgenics are imperative for biofuel crops. *Plant Science* 174, 246–263.

Gressel, J. and Valverde, B.E. (2009) A strategy to provide long-term control of weedy rice while mitigating herbicide resistance transgene flow, and its potential use for other crops with related weeds. *Pest Management Science* 65, 723–731.

Gressel, J., Hanafi, A., Head, G., Marasas, W., Obilana, B., Ochanda, J., Souissi, T. and Tzotzos, G. (2004) Major heretofore intractable biotic constraints to African food security that may be amenable to novel biotechnological solutions. *Crop Protection* 23, 661–689.

Heap, I.M. (2010) International survey of herbicide-resistant weeds. Available at: www.weedscience.org (accessed 8 November 2010).

Heller, J., Begemann, F. and Mushonga, J. (eds) (1997) *Bambara groundnut* Vigna subterranea *(L.) Verdc.* International Plant Genetic Resources Institute, Rome.

Jacobsen, E. and Schouten, H.J. (2009) Cisgenesis: an important sub-invention for traditional plant breeding companies. *Euphytica* 170, 235–247.

Lewis, M.W. (1992) *Green delusions: An Environmentalist Critique of Radical Environmentalism.* Duke University Press, Durham, North Carolina.

Losey, J.E., Rayor, L.S. and Carter, M.E. (1999) Transgenic pollen harms monarch larvae. *Nature* 399, 214.

Lu, Y., Wu, K., Jiang, Y., Xia, B., Li, P., Feng, H., Wyckhuys, K.A.G. and Guo, Y. (2010) Mirid bug outbreaks in multiple crops correlated with wide-scale adoption of Bt cotton in China. *Science* 328, 1151–1154.

Mundy, P.J. (2000) Red-billed queleas in Zimbabwe. In: Cheke, R.A., Rosenberg, L.J. and Kieser, M.E. (eds) *Workshop on Research Priorities for Migrant Pests of Agriculture in Southern Africa.* Natural Resources Institute, Chatham, UK.

Puchta, H. (2005) The repair of double-strand breaks in plants: mechanisms and consequences for genome evolution. *Journal of Experimental Botany* 56, 1–14.

Ronald, P.C. and Adamchak, R.W. (2008) *Tomorrow's Table – Organic Farming, Genetics and the Future of Food.* Oxford University Press, New York.

Rose, C.W., Millwood, R.J., Moon, H.S., Rao, M.R., Halfhill, M.D., Raymer, P.L., Warwick, S.I., Al-Ahmad, H., Gressel, J. and Stewart, C.N.J. (2009) Genetic load and transgenic mitigating genes in transgenic *Brassica rapa* (field mustard) × *Brassica napus* (oilseed rape) hybrid populations. *BMC Biotechnology* 9, 93. Available at: www.biomedcentral.com/1472-6750/9/93 (accessed 8 November 2010).

Srivastava, V. and Gidoni, D. (2010) Site specific gene integration technologies for crop improvement. *In Vitro Cellular and Developmental Biology – Plant* 46, 219–232.

Stanley-Horn, D.E., Dively, G.P., Hellmich, R.L., Mattila, H.R., Sears, M.K., Rose, R., Jesse, L.C.H., Losey, J.E., Obrycki, J.J. and Lewis, L. (2001) Assessing the impact of Cry1Ab-expressing corn pollen on Monarch butterfly larvae in field studies. *Proceedings of the National Academy of Science USA* 98, 11931–11936.

Vega-Sanchez, M.E. and Ronald, P.C. (2010) Genetic and biotechnological approaches for biofuel crop improvement. *Current Opinion in Biotechnology* 21, 218–224.

Weinthal, D., Tovkach, A., Zeevi, V. and Tzfira, T. (2010) Genome editing in plant cells by zinc finger nucleases. *Trends in Plant Science* 15, 308–321.

Weissmann, S., Feldman, M. and Gressel, J. (2008) Hypothesis: transgene establishment in wild relatives of wheat can be prevented by utilizing the *Ph*1 gene as a *senso stricto* chaperon to prevent homoeologous recombination. *Plant Science* 175, 410–414.

Yu, W., Han, F., Gao, Z., Vega, J.M. and Birchler, J.A. (2007) Construction and behavior of engineered minichromosomes in maize. *Proceedings of the National Academy of Science USA* 104, 8924–8929.

Zhang, F., Maeder, M.L., Unger-Wallace, E., Hoshaw, J.P., Reyon, D., Christian, M., Li, X., Pierick, C.J., Dobbs, D., Peterson, T., Joung, J.K. and Voytas, D.F. (2010) High frequency targeted mutagenesis in *Arabidopsis thaliana* using zinc finger nucleases. *Proceedings of the National Academy of Science USA,* in press doi:10.1073/pnas.0914991107.

Zhu, C., Naqvi, S., Breitenbach, J., Sandmann, G., Christou, P. and Capell, T. (2008) Combinatorial genetic transformation generates a library of metabolic phenotypes for the carotenoid pathway in maize. *Proceedings of the National Academy of Science USA* 105, 18232–18237.

8 Management of Crop-associated Biodiversity Above-ground

J.M. Lenné

Biological pest control can benefit the pocket, health and the environment

Neuenschwander (2004)

While some of the major twists in the Gordian knot of vegetational diversity can be perceived, we are a long way from unravelling its complexity

Andow (1991)

Introduction

Crop-associated biodiversity (C-AB) includes all of the interacting species of weeds, natural vegetation, pollinators, pathogens, arthropod and insect pests, parasitoids and predators associated with a crop in an agroecosystem. For convenience, we have separated above-ground C-AB from below-ground C-AB, which is covered in the following chapter, accepting that there is potential for interactions between both groups. In Wood and Lenné (1999), fungal, bacterial and viral biodiversity – with emphasis on harmful disease-causing organisms – and insect biodiversity – with emphasis on beneficial associates including pollinators, parasitoids and predators – were comprehensively described and discussed (Allen *et al.*, 1999; La Salle, 1999). Furthermore, Polaszek *et al.* (1999) analysed some of the effects of pest management strategies on pathogens, insect pests and weeds in agroecosystems. This chapter extends some of the main findings of these three chapters into the context of management for food security.

Our main focus is the management of harmful above-ground C-AB – pathogens, insect pests and weeds – for enhanced food production to meet food security needs in an environmentally benign manner. Due to the breadth of this topic, we must be highly selective, concentrating on major issues and successful initiatives. Most emphasis will be given to initiatives where beneficial C-AB has been successfully manipulated and used to manage harmful C-AB to reduce crop losses. The most successful examples are for biological control of insect pests and weeds, *key agroecosystem services*. We will also consider beneficial C-AB such as pollinators. In addition, we will look at the impact of GM crops on associated beneficial C-AB and discuss the role of associated vegetation in managing harmful C-AB. Where the unqualified term 'pest' is used, it collectively includes pathogens, insect pests and weeds.

In contrast to the management of crop biodiversity for food security where the principal objective is to increase yields and productivity, in most cases, the main aim of management of harmful C-AB is to preserve existing yields by reducing losses or for protecting incremental yield gains. Significant yield gains are usually a bonus. This important difference is frequently misunderstood by the non-scientific community and policy makers.

Their expectations are often for significantly increased yields from successful pest management strategies. When yield increases are not achieved, the technology may be judged 'a failure'. This is clearly illustrated by criticisms of lack of significant yield gains from genetically modified (GM) crops for controlling insect pests and weeds such as herbicide-tolerant and Bt soybean, maize, cotton and rape (canola) (e.g. see Shiva and Jafri, 2004). Although only modest yield increases are common, such crops more than meet farmers' expectations by reducing crop losses and use of costly inputs such as toxic insecticides and labour for weeding (McIntyre *et al.*, 2009). As a result, farmers recognize and appreciate the increased profits, reduced workload and added health benefits due to reduced exposure to insecticides. The benefits are often greatest for small farmers in developing countries (Carpenter, 2010).

Importance of Pests in Agroecosystems

Pathogens, insect pests and weeds (including invasive species) are critically important components of farming systems globally, for biological and economic reasons (Lenné and Wood, 1999). Damaging pests can have a significant impact on the stability and sustainability of food production and food security by substantially reducing crop yields. And, in spite of ongoing scientific advances and successes in managing many important pests of staple food crops, significant crop losses still occur globally.

Two comprehensive surveys covering many crops and countries almost 30 years apart estimated losses of about 42% (Cramer, 1967; Oerke *et al.*, 1994; Evans, 2003; Royal Society, 2009). Table 8.1 presents losses and their value for rice, wheat and maize (adapted from Teng, 1999 and based on Oerke *et al.*, 1994). Worldwide pre-harvest crop losses due to weeds, insect pests and diseases for major food crops such as wheat, rice, maize and potatoes have been estimated at from 37% to 51%. Postharvest losses can add a further 20% to pre-harvest losses. The major insect pests of rice, wheat and maize responsible for these losses are listed in Royal Society (2009; Table 2.1, p. 17). Overall, in the absence of management measures such as resistant varieties, pesticides, biological control and integrated management, losses could be as high as 50–80% (Oerke and Dehne, 2004). Such losses severely compromise the efficiency of production, wasting often scarce and costly inputs of energy, water, nutrients and labour. There is little wonder that farmers have laboured for millennia and agricultural science has devoted more than 100 years to developing improved methods for controlling pests.

Pest Management Strategies

Modern pest management strategies have evolved and developed considerably during the past century. From a limited number of moderately effective crop protection chemicals and a rudimentary understanding of host-plant resistance, multiple and integrated

Table 8.1. Global production and annual estimated yield losses due to pests of four major food crops 1988–1990 (Source: Teng (1999) adapted from Oerke *et al.* (1994)).

		Losses (US$ billion)					
Crop	Actual production (US$ billion)	Diseases	Insects	Weeds	Total	Total attainable production[a]	Loss (%)
Rice	106.4	33.0	45.4	34.2	112.5	218.9	51
Wheat	64.6	14.0	10.5	14.0	38.5	103.1	37
Maize	44.0	7.8	10.4	9.3	27.4	71.4	38
Potatoes	35.1	9.8	9.6	5.3	24.8	59.9	41

[a]Actual production plus total losses equals total attainable production

strategies are now commonly used for most major insect pests, diseases and weed problems of staple food crops. Although all strategies are potentially available for all pest groups, experience has shown that specific management strategies are more appropriate and successful for different pest groups (Lenné and Wood, 1999; Polaszek *et al.*, 1999).

For pathogens, crop diversity through host-plant resistance as well as crop protection chemicals has been a widely successful strategy (Allen *et al.*, 1999). Large-scale breeding programmes generating high yielding, disease-resistant staple food crop varieties initiated in the 1960s and 1970s continue to be the strategy of choice for managing important diseases of staple food crops and enhancing production for food security as was demonstrated in Chapter 5, this volume. It is not often necessary to seek alternative strategies such as biological control for above-ground harmful pathogens.

In sharp contrast, for arthropod and insect pests, although insecticides are still widely and often over-used, there have been many notable successes from the manipulation of C-AB, especially through biological control and integrated pest management (IPM: combinations of resistance with biological, chemical and cultural control). Biological control is a key agroecosystem service provided by beneficial C-AB. At the same time, it should be noted that host-plant resistance through GM Bt crops such as maize, soybean, rape/canola and cotton has been increasingly successful for controlling harmful C-AB in the past 10 years (see Chapters 5 and 7, this volume). With the development of improved methods and tools, host-plant resistance is likely to become more important for managing insect pests in future.

Finally, for weeds and invasive plant species, herbicides, integrated management and biological control have been commonly and successfully used. More recently, GM herbicide tolerance is showing increasing success globally. For example, in 2008, herbicide tolerance deployed in soybean, maize, canola, cotton and lucerne occupied 63% or 79 million ha of the global GM crop area of 125 million ha (McIntyre *et al.*, 2009). The potential for alien pests to reduce crop yields through accidental or deliberate introduction has led to the development of quarantine systems, discussed in Chapter 4, this volume.

Role and Impact of Beneficial Crop-associated Biodiversity in Managing Pests in Agroecosystems

Biological control of arthropod pests in the field

There is a wealth of examples of the successful use of parasitoids, predators and pathogens for managing insect and arthropod pests. Tables 8.2 to 8.4 give some notable examples, mainly in important food crops and cropping systems. Selected examples are discussed in more detail, with emphasis on developing countries and food security. For each C-AB group, we also assess the advantages, limitations and, where information is available, the economic impact.

Parasitoids

Parasitoids are extremely important elements in farming systems (La Salle, 1999) and have been the most common type of natural enemy used against insect pests (Van Driesche *et al.*, 2008). They comprise a diverse range of insects that lay their egg on or in the body of an insect host, especially caterpillars, which is then used as food for developing larvae. Most insect parasitoids are wasps (Hymenoptera) or flies (Diptera) including Ichneumonid, Braconid and Chalcid wasps and Tacinid flies as well as some Coleoptera, Neuroptera and Lepidoptera (Van Driesche *et al.*, 2008). Greathead (1986) recorded 393 species of parasitoids which have been used in biological control programmes and noted that they have been effective more than twice as often as predators. Numerous cases of successful and complete biological control using parasitoids have been reviewed by DeBach and Rosen (1991) and in Africa by Neuenschwander *et al.* (2003). Table 8.2 lists some of the successful examples on food crops.

Table 8.2. Successful examples of biological control of major crop pests by parasitoids.

Crop/Pest	Biological control agent	Geographical location	Impact	Key references
Cassava mealybug	Encyrtid wasp *Anagyrus lopezi*	Africa	Reduced losses: 95% Estimated savings: US$20 billion B:C ratio 200:1	Neuenschwander (2004); Neuenschwander *et al.* (2003); Zeddies *et al.* (2001)
Mango mealybug	Encyrtid wasp *Gyranusoidea tebygi*	Benin	Reduced losses: >36% Estimated savings: US$530 million	Neuenschwander (2004); Neuenschwander *et al.* (2003)
Cereal leaf beetle	Several parasitic wasps	North-western USA	Reduced losses: significant Estimated savings: US$6.75 million annually	DeBach and Rosen (1991); Evans *et al.* (2006)
Cereal stemborers	Several parasitic wasps	East and Southern Africa	Reduced losses: >40% Estimated savings: US$183 million	Neuenschwander *et al.* (2003); Kipkoech *et al.* (2006)
Potato tuber moth	Several parasitic wasps	South Africa, Zambia and Zimbabwe	Reduced losses: 1.8 million fewer bags of potatoes rejected; increased yields of 22%; pest downgraded to minor economic importance	Neuenschwander *et al.* (2003)
Citrus black fly	Several parasitic wasps	Caribbean and Central America	Reduced losses: significant	DeBach and Rosen (1991); White *et al.* (2005)

CASSAVA MEALYBUG IN AFRICA One of the most notable examples of the use of parasitoids in classical biological control to reduce crop losses for food security is for cassava mealybug in Africa (Herren and Neuenschwander, 1991; Neuenschwander, 2001; Nweke, 2009). Cassava is central to the food security and incomes of some of Africa's poorest farmers, especially during droughts (Spielman and Pandya-Lorch, 2009). The cassava mealybug (*Phenacoccus manihoti*) was accidentally introduced into Africa from South America in the early 1970s. In just 10 years, the mealybug threatened to wipe out cassava in Africa (Norgaard, 1988). Yield losses in infested plants were up to 60% in roots and 100% in leaves.

After a systematic search for potential natural enemies in South America, several candidate parasitoids and predators were introduced both on the ground and by aerial release into Africa in the 1980s (Neuenschwander, 2001). The most successful and dominant species was the parasitic wasp *Anagyrus lopezi*, which dispersed up to 100 km/year. Continuous field monitoring in Nigeria and Ghana over 7 years post-release found that the mealybug was effectively under control, remaining at about 10% of outbreak levels, and yield losses were significantly reduced (Nweke, 2009). This self-spreading innovation was highly sustainable and there was no resurgence of the pest as has occurred with other biological control programmes (Neuenschwander, 2001).

The mealybug control programme is estimated to have reduced losses from infestations by an estimated 2.5 t/ha. The control programme cost US$47 million and brought returns of US$9.4 billion over a 40 year period in Africa, yielding a benefit cost ratio of 200:1 (Zeddies *et al.*, 2001). Further study in Nigeria found that the major economic benefits from the control of cassava mealybug and cassava mosaic virus accrued to consumers as the price of cassava was reduced by 40% (Nweke, 2009). In addition, among producers, small-scale farmers benefited more than large farmers.

Without question, the biological control of cassava mealybug is one of the most important scientific success stories in African history (Nweke, 2009). Extensive research was the driving force in West Africa between 1971 and 1989 together with international collaboration especially between researchers in South America, Africa and Europe (Neuenschwander, 2004). Strong leadership, ongoing donor support and political will were also important ingredients in the successful programme, which helped to increase food production and reduce food prices and poverty (Nweke, 2009).

CEREAL STEMBORERS IN AFRICA Lepidopteran stemborers are a major constraint to maize and sorghum production in East and Southern Africa (Omwega *et al.*, 2006). The key stemborer pests are *Busseola fusca*, common at higher altitudes, and *Chilo partellus*, common at lower altitudes. The Trichogrammatid egg parasitoid *Cotesia flavipes* was released in coastal Kenya in 1993. Average annual parasitism of *C. partellus* increased linearly from the time of introduction to reach 20% by 2004 (Kipkoech *et al.*, 2006). The net reduction in total stemborer density over the past 10 years was 34%, thus preventing 47% yield loss. It is estimated that the region will accumulate a net present value of US$183 million in economic benefits during 20 years post-release of the parasitoid. Further study has shown that farmers could improve maize yields by up to 42% by also improving efficiency of labour and fertilizer use (Kipkoech *et al.*, 2008). Future yield improvement efforts should promote both biological control and improved management as an holistic strategy to improve maize yields.

ADVANTAGES AND LIMITATIONS Several thousand introductions of parasitoids for biological control of pests have been made over the past 100 years. Of these, approximately 60% have completely, substantially or partially controlled the target pest (Van Driesche *et al.*, 2008). Although it may take time to realize the full economic benefits, they are often substantial (La Salle, 1999) as has been shown above for cassava mealybug and cereal stemborers. Furthermore, estimated savings from seven major biological control successes in California over a 50-year period were about US$250 million (van den Bosch *et al.*, 1982).

Successful manipulation of parasitoids to control insect pests, whether through classical, augmentative or conservation biological control, is ecologically and economically advantageous and often more feasible, more efficient and less environmentally damaging than using pesticides (Van Driesche *et al.*, 2008). Pesticides usually need to be reapplied several times per season each year to achieve pest control; parasitoid biological control has the unique advantage of being self-regulatory (Neuenschwander, 2004). Successful biological control can therefore solve pest problems permanently as shown for the cassava mealybug in Africa. In addition, parasitoid biological control is specifically targeted at the pest while pesticides not only kill pests but also natural enemies. Importantly, for small-scale farmers, successful manipulation and use of parasitoids substantially reduces the costs associated with control. The greatest limitations to the use of parasitoids are: (i) reduced efficacy due to unpredictable environmental conditions; and (ii) potential parasitism of non-target pests. However, the advantages far outweigh the limitations.

Predators

Unlike parasitoids, predators tend to be more polyphagous, often nocturnal and are usually natural indigenous fauna in agroecosystems (La Salle, 1999; Van Driesche *et al.*, 2008). They readily persist in agroecosystems, especially when unaffected by insecticides, and can rapidly colonize newly cultivated fields. Several groups are important, including Hemiptera bugs, Carabid, Staphylinid, and Coccinellid (including ladybirds) beetles, Chrysopid lacewings, Syrphid flies as well as mites and spiders (La Salle, 1999; Van Driesche *et al*., 2008).

Spiders play an especially important role as predators of insect pests in rice systems (Barrion and Litsinger, 1995). Over 340 species of spiders have been identified in rice production systems in South and South-east Asia (La Salle, 1999). They are some of the most ubiquitous predaceous organisms; they feed almost exclusively on insects and are very important in the control of several important rice pests (La Salle, 1999; Chen, 2008). Mites are another important group of predators which have been important components of biological control programmes, especially to control other mites such as the red-spider mite (La Salle, 1999). They have also been used successfully in biological control programmes against nematodes, grasshoppers, locusts and scale insects (Hoy *et al*., 1983; Gerson and Smiley, 1990; Van Driesche *et al*., 2008).

PEST MANAGEMENT IN IRRIGATED RICE SYSTEMS The cultivation of tropical Asian rice represents an agricultural ecosystem of unrivalled ecological complexity. It has a rich invertebrate biodiversity if pesticides are avoided early in the crop cycle (Settle *et al*., 1996; Schoenly *et al*., 1998). High populations of generalist predators, especially spiders, are likely to be supported, in the early season, by feeding on abundant populations of detritus-feeding and plankton-feeding insects. This abundance of alternative prey gives the predator populations a 'head start' on later-developing pest populations such as plant hoppers (e.g. devastating brown plant hopper) and leaf hoppers. This process strongly suppresses pest populations and gives stability to rice ecosystems by decoupling predator populations from a strict dependence on herbivore populations. Management of tropical rice insect pests in irrigated rice fields through the conservation of generalist predators such as spiders is one of the most widely successful and best understood agroecosystems (Settle *et al*., 1996; Chen, 2008).

Research on the impact of IPM has documented the declining levels of pesticide use in irrigated rice systems and validated that natural control through zero-pesticide use was the most profitable option for farmers in South-east Asia when health costs were taken into account (Rola and Pingali, 1993; Pingali and Roger, 1995; Pingali, 2001). The value of private health savings at this time was estimated at US$117 million (Templeton and Jamora, 2007). The benefit–cost ratio was 98:1 with an IRR of 65%. Surprisingly, although there has been continued investment in promoting the IPM approach in rice, there has not been a recent wide-scale study of its impact. The stability of monoculture irrigated rice is one of the best examples of the durability of monoculture agriculture with respect to insect pest attack. However, there remains an ongoing need for IPM extension education to discourage unnecessary insecticide use that upsets this natural balance (Matteson, 2000) as has happened recently in Vietnam, Indonesia and China (Chen, 2008).

CASSAVA GREEN MITE IN AFRICA A particularly noteworthy example of predator-based biological control is the use of the South American phytoseiid mite *Typhlodromalus aripo* to successfully control the neotropical spider mite cassava green mite (*Mononychellus tanajoa*), which causes up to 80% reduction in cassava root yield in sub-Saharan Africa (Yaninek and Hanna, 2003). This was the first example of classical biological control of a phytophagous mite by a phytoseiid predator on a continental scale (Neuenschwader *et al*., 2003). From 1984 until 2001, over 400,000 *T. aripo* predator mites were released at 220 sites in 16 countries (Yaninek and Hanna, 2003). It rapidly spread beyond the release sites and established in 20 sub-Saharan African countries (West, Eastern and Southern Africa) covering more than 3.8 million km^2 by 2000. Success is also complemented by interactions between *T. aripo* and the cassava varieties in the system. Some varieties, including cassava mosaic virus-resistant varieties, have hairy tips where the green mite vector finds refuge (Neuenschwander, 2004; Nweke, 2009). Within 2 years, cassava green mite populations were reduced by more than 40% in countries in West and Southern Africa (where surveys were done) (Yaninek and Hanna, 2003). Reduction in root yield loss was 80–95% with estimated savings of US$2157 million (Neuenschwander, 2004). The economic impact in four West African countries alone was estimated at more than US$200 million per season (Yaninek and Hanna, 2003).

ADVANTAGES AND LIMITATIONS Successful manipulation of predators to control insect and arthropod pests whether through classical, augmentative or conservation biological control is ecologically and economically advantageous and often more feasible, more efficient and less environmentally damaging than using pesticides (Van Driesche *et al.*, 2008). Successful biological control can solve pest problems permanently as shown for the cassava green mite in Africa (Yaninek and Hanna, 2003). However, if the ability of generalist predators to colonize the agroecosystem early is compromised, management of pests such as the brown plant hopper in irrigated rice systems in South-east Asia can be challenging (Chen, 2008). Importantly, for small-scale farmers, the successful manipulation of generalist predators reduces the costs associated with control through pesticides.

Insect pathogens and biopesticides

Naturally occurring entomopathogens are important regulatory factors in insect populations (Lacey *et al.*, 2001). Pathogens, including bacteria, fungi and viruses of specific insect pests, are also increasingly being used in biological control and IPM programmes in field and glasshouse crops, orchards, ornamentals, rangeland, turf and stored products (Lacey *et al.*, 2001; Van Driesche *et al.*, 2008; see Table 8.3 for examples).

Bacteria, particularly *Bacillus thuringiensis*, have been the most successfully commercialized and widely used biopesticides against Lepidoptera, Coleoptera and Diptera (Lacey *et al.*, 2001; Federici, 2007). Application of *B. thuringiensis* in agroecosystems allows survival of beneficial insects and natural enemies of targeted pests, making it an ideal component of IPM systems. Cry1 proteins, which are primarily active against Lepidopteran larvae, and the genes responsible have been extensively studied. And, as discussed in Chapter 5, *Bt* genes have been successfully

Table 8.3. Successful examples of biological control of major crop pests by pathogens.

Crop/Pest	Control agent	Geographical location	Impact	Key references
Locusts and grasshoppers[a]	*Metarhizium anisopliae* var. *acridum* (fungus)	At least 11 African countries	Successful treatment of red locust outbreak in Malawi, Mozambique and Tanzania	Neuenschwander *et al.* (2003); Moore (2008)
Army worm[a]	SpexNPV (nucleo-polyhedrosisvirus)	Tanzania	90% kill of army worm in wide-scale tests; 70% reduced costs of control	Mushobozi *et al.* (2005); Grzywacz *et al.* (2008)
Palm rhinoceros beetle (oil and coconut palms)	Non-occluded virus	Pacific Islands	Reduction of beetles below economic thresholds	Lacey *et al.* (2001)
Lepidopteran pests[a]	*Bacillus thuringiensis* (bacterium; numerous commercial products available worldwide)	Global – 80% biopesticides market	Many successful examples of control of pests of major food crops, e.g. maize	Lacey *et al.* (2001); Van Driesche *et al.* (2008)
Soyabean velvet bean caterpillar	AgMNPV (baculovirus)	Brazil	Reduced pesticide application by 17 million l over 20 years	Lacey et al. (2001); Rohrman (2008)

[a] Crop non-specific

used for insect resistance in important GM crops such as maize, soybean and cotton.

Fungal pathogens successfully used as biological control agents include species of *Metarrhizium*, *Beauveria* and *Entomophthora* (Lacey *et al.*, 2001; Bateman, 2004). The mycopathogens *Beauveria bassiana* and *Metarhizium anisopliae* infect many insects and mites over a wide range of environmental conditions. Most research has focused on efforts to develop them as biopesticides, for example Green Muscle® discussed below.

Baculoviruses are the most important viral pathogens of insects (Lacey *et al.*, 2001; Van Driesche *et al.*, 2008). The two most important groups are nucleopolyhedroviruses (NPV) and the granuloviruses (GV). Some baculoviruses have been used successfully as introduced biological control agents (Fuxa, 1990) although some require reintroduction and management for continued efficacy (Van Driesche *et al.*, 2008). Others have been developed as biopesticides, for example SpexNPV discussed below.

GREEN MUSCLE® FOR LOCUSTS AND GRASSHOPPERS IN AFRICA Throughout Africa, swarms of locusts and grasshoppers periodically devastate food crops with millions of hectares having to be treated with chemical pesticides costing hundreds of millions of US$ for each outbreak (Neuenschwander, 2004). Although the development of an appropriate biopesticide has been frustrated by practical problems, 15 years of international collaboration through the LUBILOSA (Lutte Biologique contre les Locustes et les Sateriaux) programme, involving CAB International, IITA and the DFPV (Department of Crop Protection Training) of the AGRHYMET Regional Centre, Niger with funding from the Dutch and German governments, successfully produced Green Muscle®. This is a formulation of *Metarhizium anisopliae* var. *acridum*, indigenous and highly selective against transboundary locusts and grasshoppers with no adverse effects observed on non-target organisms (Langewald *et al.*, 2003; Neuenschwander, 2004; Moore, 2008; New Agriculturalist, 2009). The programme developed high quality formulations, thus improving the efficiency of delivery; mass production methods to maximize yield of fungal spores; and drying and packaging to achieve a shelf life of 18 months at 30°C. The end result is an effective product which is persistent and environmentally safe. Green Muscle® has been tested successfully in a number of African countries and permanent collaboration has been established with the crop protection agencies of Niger, Benin, Burkina Faso, Chad, Mali, Senegal and the Gambia. It is manufactured commercially in South Africa and Senegal. In 2009, Green Muscle® was used to successfully treat significant outbreaks of red locusts in Tanzania, Malawi and Mozambique (New Agriculturalist, 2009). The successful development and deployment of Green Muscle® to control locust plagues in Africa combined 15 years of international collaboration, strong leadership, ongoing donor support, political will and permanent collaboration with governments and policy makers in a number of African countries (Neuenschwander, 2004). It was a remarkable achievement.

SPEXNPV FOR ARMY WORM IN TANZANIA The African army worm *Spodoptera exempta* is a major episodic, migratory pest of cereals and rangeland over much of Eastern and Southern Africa (Grzywacz *et al.*, 2008). Outbreaks may extend over many square kilometres. Control has been reliant on the use of environmentally dangerous chemical pesticides with associated environmental and health risks. The native nucleopolyhedrosisvirus SpexNPV is an alternative control. Field trials have demonstrated that both ground and large-scale aerial application of SpexNPV to army worm outbreaks can cause NPV disease and population collapse (Mushobozi *et al.*, 2005; Grzywacz *et al.*, 2008; also see www.lancs.ac.uk/staff/wilsonk4/armyweb). SpexNPV is as effective as currently used chemical insecticides (e.g. Diazanon), achieving >90% kill rate. Field-based production of SpexNPV in Tanzania is both feasible and affordable, costing approximately US$3 per ha – much less than the current cost of chemical insecticides of around US$10 per ha (Mushobozi *et al.*, 2005). As for

Green Muscle®, the successful development and use of SpexNPV involved extensive research, over 10 years of collaboration between the UK and Tanzania, strong leadership, donor support and national policy support in Tanzania (Mushobozi *et al.*, 2005).

ADVANTAGES AND LIMITATIONS The most important advantages of the successful deployment of entomopathogens and biopesticides over chemical pesticides are efficacy and low cost (Lacey *et al.*, 2001). In addition, they are safe for humans, non-target organisms and the environment, preserve other natural enemies in agroecosystems and reduce pesticide residues in food. However, for entomopathogens and biopesticides to replace chemical insecticides, further attention is needed to: (i) increase pathogen virulence and speed of kill; (ii) improve pathogen performance under challenging and unpredictable environmental conditions; (iii) greater efficiency in production; (iv) improvements in formulation to ease application, increase environmental persistence and prolong shelf life; (v) better understanding of integration with other control systems; and, above all, (vi) acceptance by farmers and the general public which will foster market growth (Lacey *et al.*, 2001). One of the main problems with biopesticides is the lack of commercial interest due to small market size and high cost of mass production (Lacey *et al.*, 2001; Van Driesche *et al.*, 2008).

Area-wide management for invasive and migratory pests such as locusts and army worm is an ideal target for biopesticides. However, currently, they are required to pass through regulatory processes in each country. International support and political will is needed to develop a uniform regulatory framework that could operate regionally (Neuenschwander, 2004). It is probable that entomopathogens and biopesticides will be used more widely in the field in future in synergistic combinations with other management strategies to enhance the effectiveness and sustainability of integrated strategies. In greenhouse crops, especially in Europe and the USA, their use has increased substantially in the past 20 years.

Biological control of arthropod pests in greenhouse crops

It has been estimated that, globally, the area of protected or greenhouse (glasshouses, plastic houses and tunnels) crops is more than 300,000 ha, with vegetables occupying 65% of this area (Ferguson and Murphy, 2002). The trend is for continued growth in greenhouse production. Growing conditions of year-round warmth and high humidity within the protected environment are highly favourable to arthropod pests including thrips (*Frankliniella occidentalis, Thrips tabaci*), whiteflies (*Trialeurodes vaporariorum, Bemisia* spp.), spider mites (*Tetranychus urticae*), aphids (e.g. *Aphis gossypii, Myzus persicae*) and leaf miners (*Liriomyza* spp.) (Van Driesche *et al.*, 2008). The damage inflicted by such pests on greenhouse crops can be substantial and varies with the pest, geographic region and season.

As a result of international and government policy and consumer demand for safe, high quality food, there has been a global move to reduce pesticide use in greenhouse crops (Van Steekelenberg, 2006). The current trend is to use IPM, incorporating monitoring for pests with a range of control strategies, with an increasing use of biological control (De Buck and Beerling, 2006), for example in Europe (see www.koppert.com/pest-control) and in developing countries (see Dudutech, 2009; Real IPM, 2009). Good sanitation practices and physical controls including coloured sticky traps, light traps and insect barriers are widely used. An increasing range of predators, parasitoids and entomopathogenic fungi are available for many of the major greenhouse pests to minimize the use of pesticides. These include ladybirds, predatory mites and bugs, gall midge, parasitic wasps, lacewings, hoverflies and the fungi *Beauveria bassiana* and *Verticillium lecanii* (Van Driesche *et al.*, 2008). For example, in the Netherlands by 2000, 90% of all tomatoes, cucumbers and sweet peppers were produced in greenhouses under IPM conditions (De Buck and Beerling, 2006) while in Almeria, Spain, 8000 ha of sweet pepper production is cultivated in IPM greenhouses (Markus Knapp, Koppert, the Netherlands, 2009, personal communication).

Biological control of weeds and invasive plant species

Insect enemies of weeds

Up until 1999, over 350 insect natural enemies have been introduced into 75 countries targeting over 130 terrestrial and aquatic weed and invasive species (Julien and Griffiths, 1998). Although the long-term economic and social outcomes from weed biological control programmes are often not well documented, a meta-review of the success of biological control programmes on weeds found that 66% were completely, substantially or partially successful (Van Driesche *et al.*, 2008). Furthermore, Page and Lacey (2006) conducted an economic analysis of over 100 years of weed biological control projects in Australia finding that the annual return over the period was AUS$95.3 million for an annual investment of AUS$4.3 million. The total return was estimated at AUS$10 billion, making it one of the most successful scientific programmes in Australia's history. In addition, successful programmes to control some major weeds will no doubt have had measurable human health benefits, through reducing allergenic pollen and water-borne diseases such as malaria and schistosomiasis (Morin *et al.*, 2009). Insect enemies of weeds have therefore been responsible for some striking successes with massive benefits both to pasture and rangeland and natural terrestrial and aquatic ecosystems (La Salle, 1999; Van Driesche *et al.*, 2008) and, as a result, to livestock and fish production. Table 8.4 lists some successful examples, including prickly pear cactus and St John's Wort in farmland and water hyacinth in aquatic systems.

Fungal control of invasive weeds

Exotic and indigenous pathogens have also been successfully used to control weeds either through biological control or as biological herbicides (TeBeest, 1996; Barton, 2004; Hallett, 2005). From 1971 until 2004, 26 species of fungi, originating from 15 different countries, have been used as classical biological control agents against over 26 species of weeds in seven countries (Barton, 2004). Table 8.4 lists some successful examples, including rush skeleton weed, *Mikana*, strangler vine, northern joint vetch and *Striga* spp.

Advantages and limitations of biological controls of weeds

Successful examples of the use of natural enemies of invasive weeds in pastures, rangeland, natural vegetation and aquatic ecosystems are cost-effective, environmentally safe and contribute to health benefits (Page and Lacey, 2006; Morin *et al.*, 2009). The main concern is the potential for entomopathogens to move to non-target species (Morin *et al.*, 2009). Extensive reviews of almost 400 cases of classical biological control have identified only 7.25% of cases where natural enemies moved to non-target plant species (McFadyen, 1998; Barton, 2004). Rigorous host-range testing remains a priority before the introduction of natural enemies.

Similarly, successful examples of fungal control of invasive weeds have highlighted their cost effectiveness, efficacy and safety compared to alternative methods of control, e.g. herbicides, cultural control, manual weeding etc. In contrast to some entomopathogenic fungi, no examples of non-target problems with exotic pathogens have been identified to date (McFadyen, 1998) – they are extremely safe (Barton, 2004). Risk assessments based on rigorous host-range testing, combined with a good understanding of the taxonomy, biology and ecology of the agent, the target weed and non-target species, can ensure that the introduction of exotic pathogens is a safe and environmentally benign method of weed control.

It is estimated that over 200 plant pathogens have been or are under evaluation for their potential as bioherbicides (Hallett, 2005). However, with the exceptions of Collego® and Devine® commercialized in the early 1980s, few have been registered, commercially produced and are available for use. The main problems are the lack of commercial interest due to small market size and high cost of mass production as well as efficacy and

Table 8.4. Successful examples of biological control of weeds by insect enemies and fungal pathogens.

Weed	Control agent	Geographical location	Impact	Key references
Prickly pear cactus	*Cactoblastis cactorum* (beetle)	Australia (also Hawaii, South Africa and the Caribbean)	Successful control over 25 million ha in Australia alone	DeBach and Rosen (1991); La Salle (1999)
St John's Wort	*Chrysolina quadrigemina* (beetle)	North-western USA	Successful control over 2 million ha; estimated savings: US$3.5 million per year	DeBach and Rosen (1991); La Salle (1999)
Water hyacinth	*Neochetina* spp. (weevils)	Africa (Benin)	Reduced losses: 36%; estimated savings: US$260 million; B:C = 124:1 (over 20 years)	Neuenschwander (2004); De Groote *et al.* (2003)
Rush skeleton weed	*Puccinia chrondrillina*	Australia, USA	Successful control	Julien and Griffiths (1998); TeBeest (1996)
Mikania micrantha	*Puccinia spegazzinia*	India, Pacific		Ellison *et al.* (2005); Orapa *et al.* (2008)
Strangler vine	*Phythophthora palmivora* (Devine®)	Florida	Successful control but limited market	TeBeest (1996)
Northern joint vetch	*Colletotrichum gloeosporioides* f. sp. *aeschynomene* (Collego®)	Arkansas	Successful control but limited market	TeBeest (1996)
Striga	*Fusarium oxysporum* f. sp. *strigae* (pesta granules)	Africa	Reduced Striga emergence: >90%; Potential application: millions ha	Venne *et al.* (2009)

resistance problems. Continued success with bioherbicides is likely to depend on their development for weeds and invasive species of national and worldwide importance, such as *Striga* spp. Future efforts should be targeted at developing techniques for the cultural and genetic enhancement of potential bioherbicidal candidates (Hallett, 2005).

Importance of Pollinators

Any treatment of crop-associated biodiversity in farming systems is not complete without some consideration of the importance of pollinators. Animal pollinators include many different species of bees, flies, butterflies, moths, bats and birds. Although most of the world's important staple food crops (such as rice, wheat and maize) are wind pollinated or self pollinated, about 30% of human food is derived from animal-pollinated plants, bees being responsible for 80% of all insect pollination of crops (La Salle, 1999). Contributing to food crop production is only one benefit of animal pollinators (Free, 1993). The timely presence of pollinators can result in earlier, more uniform crops and higher quality fruits and seeds. Pollinators provide key ecosystem services. Although the importance of pollinators in supporting production of food crops that add nutritional richness to human diets (particularly vegetables, nuts and fruits) is clear, their direct importance to back-stopping basic global food security is often over-estimated (Aizen *et al.*, 2008; Aizen and Harder, 2009). The International Initiative for the Conservation and Sustainable Use of Pollinators claims that two-thirds of the world's 3000 species of agricultural crops require animals for pollination (FAO, 2009). In reality, few of these crops depend solely on animal pollination, owing largely to their capacity for self pollination (Aizen *et al.*, 2008). The food production potential of very few staple food crops would suffer from an absence of effective pollinators.

Although some studies indicate strong evidence of pollinator decline in Europe and North America (Potts *et al.*, 2010), other studies on long-term global trends in crop yield, production and cultivated areas of pollinator-dependent and non-dependent crops in both developed and developing countries have revealed no overall global pollinator shortage as well as ongoing crop yield increases of about 1.5% per year since 1961 (Aizen *et al.*, 2008). These findings do not support the commonly held view that regional pollinator shortages are affecting crop yield globally and hence, food security. And, if in the highly unlikely scenario of bees disappearing altogether, global agricultural production would decrease by only 4–6% (Aizen and Harder, 2009). Although the current pollin ation crisis is largely mythology, agriculture is becoming more pollinator dependent because of the increases in the area cultivated with crops benefiting from – but not solely dependent on – pollinators, e.g. soybean, rape/canola, sunflower etc. This may have important economic and ecological consequences in terms of growing pollinator demands in future and needs ongoing research, especially on monitoring and fostering pollinators (Aizen *et al.*, 2008; Potts *et al.*, 2010).

Effects of Genetically Modified Crops on Non-target Insects and Non-GM Crops

The rapid, recent expansion of insect-resistant GM crops with genes from *Bacillus thuringiensis* expressing Cry proteins adds further to vegetational diversity in agroecosystems (McIntyre *et al.*, 2009; Morin *et al.*, 2009). However, concerns about their potential adverse effects on non-target insects, especially natural enemies of crop pests, have been raised (Sanvido *et al.*, 2007). Potentially, such insect-resistant crops could affect the quantity and quality of prey/hosts for non-target predators and parasitoids, key natural enemies of major crop pests and the mainstay of biological control in many IPM programmes.

The risks from insect-resistant GM crops for the environment and for biodiversity are extensively assessed before and during their development in the laboratory, greenhouse and the field and a substantial database has been amassed during the past 10 years (Sanvido *et al.*, 2007; Romeis *et al.*, 2008a). Review of this extensive scientific knowledge from worldwide experimental field research and commercial cultivation so far provides no sound scientific evidence that the presently commercialized insect-resistant GM crops have caused any environmental harm at either field or landscape levels (Romeis *et al.*, 2008b, 2009; Storer *et al.*, 2008). Specific studies with *Bt* maize and the web-building spider *Theridion impressum* (Meissle and Romeis, 2009), *Bt* maize pollen and the adult green lacewing *Chrysoperla carnea* (Li *et al.*, 2008), Indian *Bt* cotton and cotton aphids (Lawo *et al.*, 2009) and high concentrations of Cry1Ab feed on bumble bees (Babendreier *et al.*, 2008) clearly showed no effects on fecundity, fertility, survival, mortality, or performance, according to the various traits studied. The risks on such non-target insects from currently available insect-resistant GM crops are therefore negligible due to their narrow spectrum of activity which targets only insects eating plant tissue. Furthermore, past and recent claims of negative impacts of insect-resistant GM crops on natural enemies (e.g. Schmidt *et al.*, 2009 for Cry1Ab protein on larvae of the two-spotted ladybird) have been shown to be scientifically flawed (Rauschen, 2010; Ricroch *et al.*, 2010).

The millions of hectares of insect-resistant GM crops under commercial cultivation worldwide are proving to be safe, effective and easy to use insect suppression tools that are compatible with other IPM tools, including the conservation of natural enemies as important agents of biological control (Kennedy, 2008). In fact, in cropping systems where *Bt* crops have been deployed with an associated decline in insecticide use, biological control organisms have benefited significantly (Romeis *et al.*, 2008b). Furthermore, a recent survey in five US states has shown that insect pest control by *Bt* maize has spilled-over to non-*Bt* varieties grown nearby (Black, 2010). Over 14 years, use of *Bt* varieties improved farmers' profits by US$3.2 billion with most of the benefit – US$2.4 billion – accrued on non-*Bt* fields. Similar findings are emerging from China for *Bt* cotton (Wu *et al.*, 2008). Therefore there is great potential to sustainably and profitably improve food production by fostering insect-resistant GM crop-based IPM systems (Kennedy, 2008; Romeis *et al.*, 2008b; Morin *et al.*, 2009).

Role of Associated Vegetation in Managing Harmful Biodiversity in Farming Systems

In addition to harmful pests and beneficial natural enemies, C-AB includes other plants, particularly natural vegetation and associated crops in mixed farming systems (Lenné and Wood, 1999). In developing countries, natural vegetation includes uncultivated, semi-wild and wild bush land surrounding cultivated fields. In some tropical regions, it may be part of a shifting cultivation or bush fallow system. In developed countries, natural vegetation surrounding cultivated fields includes fallowed and bush land as well as agri-environmental schemes (Jordan, 1999; Smith *et al.*, 2007). Mixed cropping agroecosystems allow partitioning of crop diversity in space, e.g. intercrops and polycultures and in time, e.g. rotation, relay cropping and seasonal plantings, e.g. *kharif* and *rabi* crops in South Asia. Associated vegetation potentially provides options for managing harmful pests in farming systems.

Associated natural vegetation

Associated, natural vegetation surrounding crops can be a source of beneficial predators and parasitoids of the pests of the nearby crop (Polaszek and Khan, 1998; Altieri, 1999; La Salle, 1999; Landis *et al.*, 2000). For example, a comprehensive survey of species of cereal stemborer parasitoids conducted in four agroecological zones of Kenya from 2005 to 2007 found that natural vegetation surrounding cereal crops served as refugia for sustaining stemborer parasitoids (Mailafiya *et al.*, 2009). On the other hand, it may harbour damaging insect pests, pathogens (Thresh, 1981) and invasive weeds of the crop. Locust and grasshopper plagues and army worm infestations provide striking examples of the way in which serious problems can arise in agriculture when insect pests invade crops from natural surrounding vegetation (Dobson and Magor, 1999; Mushobozi *et al.*, 2005).

Because of the complexity of the interrelationships between the crop and its associated natural vegetation, the outcomes for the crop are likely to be unpredictable (Andow, 1991), that is, a 'mixed blessing' (Lenné and Wood, 1999). Positive results from the manipulation of surrounding vegetation will depend on the composition of the vegetation and its ability to host and increase the abundance of predators and parasitoids and to deter pests (Polaszek and Khan, 1998). This implies the need for site-specific research to understand and manage local problems – a substantial challenge for researchers and today's funding. The enormity of this challenge helps to explain why very limited new research has been done in the past 20 years on the beneficial role of natural vegetation associated with crops in managing harmful biodiversity (Marshall, 2002; Neueschwander *et al.*, 2003).

It is interesting to note that during the past 40 years, much of the published literature on crop-associated vegetation has been somewhat romantically and narrowly focused on its potential *beneficial properties* as a refuge for birds and a source of predators and parasitoids to manage crop pests (Altieri and Nicholls, 2004; see www.defra.gov.uk), rather than on its potential *detrimental role* in harbouring damaging crop pests such as locusts, grasshoppers and *Qualea* (Olson and Wackers, 2007) and human pests such as tsetse fly and mosquitoes. In spite of the lack of research in the past 20 years (Neueschwander *et al.*, 2003), the use of associated, natural vegetational diversity to manage crop pests continues to be recommended both in developed and developing countries (Landis *et al.*, 2000; Altieri and Nicholls, 2004).

In the past 20 years, deliberately managed, natural or semi-natural vegetation in field margins, hedgerows and headlands through agri-environmental schemes has become a common component of temperate farming systems (Jordan, 1999; Marshall, 2002; Smith *et al.*, 2007). Farmers are financially encouraged to establish and maintain such C-AB. The main aim of these schemes is to conserve biodiversity, maintain and enhance landscape quality and character, and protect natural resources. Just as for unmanaged natural vegetation, these plant communities are also expected to contain populations of natural enemies (predators, parasitoids and insect pathogens) potentially available to manage pests of the nearby crop (Jordan, 1999). However, their success will depend on seasonal synchrony of the natural enemies and host prey and the ability of the both to reproduce and migrate. Studies on the effectiveness of field margins to enhance cereal aphid control in the UK have shown variable results (Holland *et al.*, 2008), which has not helped to persuade farmers to adopt such approaches. If field margins, hedgerows and headlands in associated crops are to be effectively used to manage crop pests, sound understanding of their ecology is needed for informed management (Marshall, 2002). In most cases, the required research has not been done in sharp contrast to the extensive knowledge base available on classical and augmentative biological control and biopesticides.

Associated crops

There are many opportunities for partitioning diversity spatially in the crop itself, which may affect pests, natural enemies, pest management and subsequent crop yields. The most commonly practised systems are intercrops and polycultures. We mainly look at intercrops as almost no studies have been done to understand the multiplicity of interactions occurring between crops, weeds, insect pests, pathogens, parasites and beneficial organisms in polyculture systems. The one exception is the 'push-pull' strategy involving intercrops and grass borders to manage stemborers and the weed striga (Khan *et al.*, 2000). As we pointed out in Chapter 5, the spatial and temporal associations exploited by farmers in polycultures are driven by socio-economic factors, unrelated to pest management. Chapter 5 dealt with issues related specifically to crop diversity in such systems. Here we focus on using intercrops to manage pests.

Many studies of pathogens and insect pests in intercrops have focused on cereal–legume associations (Risch *et al.*, 1983; Allen, 1990; Cardona, 1990; Thurston, 1992; Francis and Adipala, 1994). The most commonly reported effect of intercropping is reduced incidence and severity of diseases and reduced populations of insect pests (Risch, 1983; Trenbath, 1993; Smith and McSorley, 2000). For example, in maize/bean intercrops, reduction of anthracnose, rust, haloblight and bean common mosaic virus ranged from 28% to 47% (Allen, 1990) and leafhoppers were also significantly reduced (Cardona, 1990). In contrast, angular leaf spot severity increased by 37% and populations of a highly polyphagous chysomelid beetle were also higher (Allen, 1990; Cardona, 1990). In the majority of studies, however, the effect of reduced pest incidence and damage on crop yield has not been measured (Cardona, 1990).

A comprehensive review by Risch *et al.* (1983) found that only 19 of 153 studies reported yield data and Trenbath (1993) noted that few published studies had linked pest problems in intercrops with yield loss. More recent studies have looked at both pest damage and yield. For example, studies conducted in Cameroon of stemborer damage on sole maize and intercrops with cassava, cowpea and soybean found as much as three times more stemborer damage and yield loss in sole maize crops (Chabi-Olaye *et al.*, 2005). However, intercropping maize with other cereals had limited advantage in reducing yield losses due to stemborers in Kenya (Songa *et al.*, 2007). Under some conditions, it appears that intercropping can usefully contribute to the control of pest and disease populations and the reduction of yield loss (Allen, 1990; Trenbath, 1993). But without a comprehensive understanding of the effects of intercropping on both pests and yield, there are limited grounds to recommend it as a strategy to support increased food security, especially as there are additional management problems associated with intercrops, as highlighted in Chapter 5, this volume. More research is needed to understand the mechanisms that underlie the observed effects on management of insect pests and pathogens and the yield benefits of intercropping.

A review of literature on weeds in intercrops found less weed biomass in the intercrop compared to the sole crop in 47 of 54 studies (Liebman and Dyck, 1993). Increased crop density in intercrops results in increased competition and possibly more pronounced allelopathic effects, making the intercrop less hospitable for weeds. At the same time, however, intercropping often increases labour requirements for weeding and reduces the choice of suitable herbicides for managing weeds (Ransom, 1990). Though intercropping has some drawbacks for large-scale, mechanized farming, it has benefits for weed control in small-scale farming systems in developing countries. However, this potential remains largely untapped because of the slow uptake of this farming system (Francis and Adipala, 1994). Better understanding of intercropping and improved intercropping systems could lead to the increased adoption of intercropping in the smallholder agricultural sector.

The presence of weeds in any cropping system adds a further level to spatial diversity in crops and, potentially, to the complexity of interactions between the crop and its pests (Polaszek *et al.*, 1999). However, the consequences of these multi-species associations are difficult to predict (Norris and Kogan, 2000, 2005). On the one hand, weeds may be alternative hosts to pests (Thresh, 1981; Terry, 1991; Hillocks *et al.*, 1996). This is especially important when weedy relatives of crops are represented in the crop and provide sources of inoculum of pathogens and additional food and shelter for insect pests, for example: rice yellow mottle virus and its vectors on wild *Oryza* species (Terry, 1991); many sorghum diseases on *Sorghum halepense* (Warwick and Black, 1983); and many potato diseases on wild solanaceous species (Thresh, 1981). The inclusion of cover crops in apple orchards in Australia resulted in increased insect pest and disease problems as well as detrimental effects of production in some cases (Bone *et al.*, 2009). Thresh (1981) provides numerous examples of weeds in crops hosting viruses, fungal and bacterial pathogens, nematodes and insect pests of the crop.

On the other hand, the presence of weeds in a crop can also increase the activity of pest predators and parasites by providing shelter, modifying crop microclimates and altering crop background to enhance predator colonization (Andow, 1991; Altieri, 1994, 1999). There is increasing interest in managing the weed diversity available in a crop to enhance predation of pest species (Altieri, 1999), however, until our understanding of the combined impact of the negative and positive effects of weed diversity within crops is greatly improved, there will always be risks of crop loss associated with this practice. More research is needed to understand the mechanisms that underlie the observed effects on the yield benefits of intercropping and how this can be applied more widely.

It is somewhat paradoxical that spatial diversity in crops is more common in those agroecosystems where limited research has been done to understand its functional role in pest management, i.e. in developing countries, and least common where considerable research effort has been carried out, i.e. in developed countries (Smithson and Lenné, 1996; Kiaer *et al.*, 2009 for crop mixtures). However, very little research has been done in either system in the past 15 years, probably due to lack of funding for such research. The degree of the relationship between increased diversity and increased food (versus biomass) production also merits much more study in agroecosystems (Frankel *et al.*, 1995; Wood and Lenné, 1999) before increasing diversity in agroecosystems can be promoted as a sound pest management strategy and/or for increasing food security in developing countries.

'Push-pull' for insect pest and weed management

Lepidopteran stemborers and the weed *Striga hermonthica* are major constraints to maize production in East Africa, causing combined yield losses of up to 40–50% (Khan *et al.*, 2000; Omwega *et al.*, 2006; Amudavi *et al.*, 2009). In smallholder systems, chemical control of both pests is uneconomical and impractical while cultural control for *Striga* is labour-intensive and ineffective. Push-pull was developed through a partnership between ICIPE in Kenya and Rothamsted in the UK as an alternative strategy for smallholders to manage both the insect pest and the weed.

Push-pull uses a combination of intercropped repellent plants to deter the stemborers (*Busseola fusca*) from the maize crop ('push') and trap crops to attract the repelled pest ('pull') (Khan *et al.*, 2000; Cook *et al.*, 2007; Amudavi *et al.*, 2009). Silverleaf desmodium (*Desmodium uncinatum*) is commonly used as the repellent while Napier grass (*Pennisetum purpureum*) and Sudan grass (*Sorghum sudanense*) are common trap crops. Silverleaf desmodium produces several semiochemical volatiles that repel stemborers, including ocimene, nonatriene and other sesquiterpenes (Khan *et al.*, 2000). Research has also shown that Napier grass may produce chemical substances, some of which attract female stemborers to oviposit (Amudavi *et al.*, 2009). In response to the stemborer larvae, Napier grass produces a gummy substance that reduces larval survival, thus trapping the pest. Alternatively, van den Berg *et al.* (2006) suggest that reduced larval survival on Napier grass may be related to its dense trichome covering.

However, research in Cameroon and Uganda found no consistent evidence that stemborers preferred Napier grass to maize (Matama-Kauma *et al.*, 2006, 2008; Ndemah *et al.*, 2006). Similarly, in two-choice tests with Napier grass and maize, *B. fusca* moths had no marked difference in oviposition preference (van den Berg *et al.*, 2006). Furthermore, studies in wind tunnels revealed that volatiles produced by maize and Napier grass did not appear to influence female stemborer behaviour (Calatayud *et al.*, 2008). It was concluded that female moths recognized their preferred host – maize in preference to Napier grass – only after landing, indicating that tactile and contact-chemoreception stimuli from plants played the major role in oviposition decisions of the stemborer. Some questions may therefore be raised about the role of Napier grass as a trap crop in the push-pull system.

Push-pull also suppresses *Striga* through shading, nitrogen fixation and allelopathy. *Desmodium* roots produce isoschaftoside, a di-C-glycosylflavone, which has been found to interfere with *in vitro* germination of *Striga* seed (Hooper *et al.*, 2008). Whether this alone is responsible for the suppressing *Striga* parasitism on maize is still under investigation. This discovery of *Striga* suppression is serendipitous, as the choice of *Desmodium* as an intercrop was based on its ability to act as a repellent of maize stemborers.

Push-pull provides several benefits to smallholder farmers. These include: increased maize yields from 30% to 100% depending on the level of control of both stemborers and *Striga*; reduced soil erosion through improved ground cover and increased soil fertility through nitrogen fixation by *Desmodium*; increased fodder availability for livestock from *Desmodium* and Napier grass; and increased farm incomes from surplus maize, fodder and *Desmodium* seed (Amudavi *et al.*, 2007). The push-pull strategy has been adopted by more than 10,000 farmers in Kenya, Uganda and Tanzania and increased maize yields have been the key incentive for adoption.

Push-pull provides a potential model for diverse combinations of crops and other useful plants which contribute to mutually beneficial pest management provided the plants included in the system have food or fodder value for farmers. But the approach is complex and knowledge-intensive, requiring monitoring and decision systems and currently incurs higher operating costs than simple pest management options (Cook *et al.*, 2007). Such complex, multiple cropping systems are not likely to be adopted unless each component in the system has some benefit to the farmers growing them which, overall, is valued above the limitations.

Conclusions

We have shown that the manipulation and utilization of beneficial C-AB to manage harmful C-AB in biological control programmes can dramatically and successfully reduce food crop losses and contribute to food security in both developed and developing countries. Where economic analyses have been undertaken, the benefit:cost ratios are notably high, for example 200:1 for control of cassava mealybug in Africa. Biological control is a key ecosystem service. Several studies have analysed and identified the basic ingredients of successful initiatives (Neuenschwander, 2004; Nweke, 2009). These include: sound, comprehensive research input over many years; international collaboration; leadership; and government/policy support. With the declining support for basic agricultural research, it is worrying that today's donors would be unlikely to support the long-term research that was necessary for successful programmes such as cassava mealybug control and Green Muscle® in Africa.

Moreover, the potential to rely on successful control strategies in the long term is not a foregone conclusion due to changing conditions, e.g. climate change and changes in pest biotypes and their effects on beneficial C-AB. Resurgence of the brown plant hopper in rice systems in South-east Asia is a good example of the need for further research and manipulation of past successful strategies. Continued successful implementation of biological control programmes will require often long-term scientific research, donor support and government and international commitment (Neuenschwander, 2004).

In contrast to biological control, crop-associated vegetational diversity can be a mixed blessing for farmers (Lenné and Wood, 1999). Planned vegetational diversity based on scientific understanding of crop–pest interactions can, in many cases, make a valuable contribution to improved pest management. In contrast, the impact of natural vegetational diversity cannot be predicted or relied on for pest management. Each agricultural situation must be assessed separately, since pest–crop interactions will vary depending on the pest, crop, associated vegetation, associated beneficial biodiversity, location and size of field, climate and cultural practices. In the majority of cases, lack of resources to support such complex studies suggests that, in the foreseeable future, use of vegetational diversity to manage pests should be restricted to planned diversity supplemented by biological control, genetic diversity in the crop and IPM. Much more research is needed before vegetational diversity can be recommended as a sound pest management strategy for enhancing food security in agroecosystems.

References

Aizen, M.A. and Harder, L.D. (2009) The global stock of domesticated honey bees is growing slower than agricultural demand for pollination. *Current Biology* 19, 915–918.

Aizen, M.A., Garibaldi, L.A., Cunningham, S.A. and Klein, A.M. (2008) Long-term global trends in crop yield and production reveal no current pollination shortage but increasing pollinator dependency. *Current Biology* 18, 1572–1575.

Allen, D.J. (1990) The influence of intercropping with cereals on disease development in legumes. In: Waddington, S.R., Palmer, A.F.E. and Edje, O.T. (eds) Workshop on Research Methods for Cereal/Legume Intercropping in Eastern and Southern Africa, *CIMMYT Eastern and Southern Africa On-Farm Research Report* No. 17, pp. 62–67.

Allen, D.J., Lenné, J.M. and Waller, J.M. (1999) Pathogen biodiversity: its nature, characterization and consequences. In: Wood, D. and Lenné, J.M. (eds) *Agrobiodiversity: Characterization, Utilization and Management.* CAB International, Wallingford, UK, pp. 123–153.

Altieri, M.A. (1994) *Biodiversity and Pest Management in Agroecosystems.* Food Products Press, New York.

Altieri, M.A. (1999) The ecological role of biodiversity in agroecosystems. *Agriculture, Ecosystems and Environment* 74, 19–31.

Altieri, M.A. and Nicholls, C.I. (2004) *Biodiversity and Pest Management in Agroecosystems,* 2nd edn. Food Products Press, New York.

Amudavi, D., Khan, Z. and Pickett, J. (2007) Enhancing the push-pull strategy. *Leisa Magazine* 23.4, December 2007.

Amudavi, D.M., Khan, Z.R., Wanyama, J.M., Midega, C.A.O., Pittchar, J., Nyangau, I.M., Hassanali, A. and Pickett, J.A. (2009) Assessment of technical efficiency of farmer teachers in the uptake and dissemination of push–pull technology in Western Kenya. *Crop Protection* 28, 987–996.

Andow, D.A. (1991) Vegetational diversity and arthropod population response. *Annual Review of Entomology* 36, 561–586.

Babendreier, D., Reichhart, B., Romeis, J. and Bigler, F. (2008) Impact of insecticidal proteins expressed in transgenic plants on bumblebee microcolonies. *Entomologia Experimentalis et Applicata* 126, 148–157.

Barrion, A.T. and Litsinger, J.A. (1995) *Riceland Spiders of South and South-east Asia.* CAB International, Wallingford, UK.

Barton, J. (2004) How good are we at predicting the field host-range of fungal pathogens used for classical biological control of weeds? *Biological Control* 31, 99–122.

Bateman, R. (2004) Constraints and enabling technologies for mycopesticide development. *Outlook on Pest Management* 15, 64–69.

Black, R. (2010) GM crops bring cash harvest to non-GM varieties. *BBC News: Science and Environment,* 8 October 2010. Available at: www.bbc.co.uk/news/science-environment-11496710 (accessed 18 October 2010).

Bone, N.J., Thompson, L.J., Ridland, P.M., Cole, P. and Hoffman, A.A. (2009) Cover crops in Victorian apple orchards: effects on production, natural enemies and pests across a season. *Crop Protection* 28, 675–683.

Calatayud, P.A., Guenego, H., Ahuya, P., Wanjoya, A., Le Ru, B., Silvain, J.F. and Frerot, B. (2008) Flight and oviposition behaviour of the African stem borer *Busseola fusca,* on various host plant species. *Entomologia Experimentalis et Applicata* 129, 348–355.

Cardona, C. (1990) Effect of intercropping on insect populations: the case of beans. In: Waddington, S.R., Palmer, A.F.E. and Edje, O.T. (eds) Workshop on Research Methods for Cereal/Legume Intercropping in Eastern and Southern Africa. *CIMMYT Eastern and Southern Africa On-Farm Research Report* No. 17, pp. 56–61.

Carpenter, J.E. (2010) Peer-reviewed surveys indicate positive impact of commercialised GM crops. *Nature Biotechnology* 28, 319–321.

Chabi-Olaye, A., Nolte, C., Schulthess, F. and Borgemeister, C. (2005) Effects of grain legumes and cover crops on maize yield and plant damage by *Busseola fusca* (Fuller) (Lepidoptera: Noctuidae) in the humid forest of southern Cameroon. *Agriculture, Ecosystems and Environment* 108, 17–28.

Chen, A. (2008) The unsung heroes of the rice field. *Rice Today* 7, 30–31.

Cook, S.M., Khan, Z.R. and Pickett, J.A. (2007) The use of push-pull strategies in integrated pest management. *Annual Review of Entomology* 52, 375–400.

Cramer, H.H. (1967) *Plant Protection and World Crop Production.* Bayer, Leverkusen, Germany.

DeBach, P. and Rosen, D. (1991) *Biological Control by Natural Enemies.* Cambridge University Press, Cambridge.

De Buck, A.J. and Beerling, A.M. (2006) Implementation of biocontrol and IPM in Dutch horticulture. In: Eilenberg, J. and Hokkanen, H.M.T. (eds) *Progress in Biological Control,* Vol. 2. *An Ecological and Societal Approach to Biological Control.* Springer, Dordrecht, the Netherlands, pp. 73–90.

De Groote, H., Ajuonu, O., Attignon, S., Djessou, R. and Neuenschwander, P. (2003) Economic impact of biological control of water hyacinth in Southern Benin. *Ecological Economics* 45, 105–117.

Dobson, H.M. and Magor, J.I. (1999) Ancient plagues and modern solutions: locust management in the new millennium. In: Terry, P.J. (ed.) *International Crop Protection: Achievements and Ambitions.* British Crop Protection Council Proceedings No. 73, pp. 3–32.

Dudutech (2009) Dudutech Crop Protection in Kenya. Available at: www.flamingoholdings.com/dudutech.asp?bandwidth=big (accessed 14 December 2009).

Ellison, C.A., Murphy, S.T. and Rabindra, R.J. (2005) Facilitating access for developing countries to invasive alien plant classical biocontrol technologies: the Indian experience. In: Harris, D., Richards, J.I., Silverside, P., Ward, A.F. and Witcombe, J.R. (eds) Pathways Out of Poverty. *Aspects of Applied Biology* 75, 71–80.

Evans, E.W., Karren, J.B. and Israelsen, C.E. (2006) Interactions over time between cereal leaf beetle (Coleoptera: Chrysomelidae) and larval parasitoid *Tetrastichus julius* (Hymenoptera: Eulophidae) in Utah. *Journal of Economic Entomology* 99, 1967–1973.

Evans, L.T. (2003) Agricultural intensification and sustainability. *Outlook on Agriculture* 32, 83–89.

FAO (2009) International Initiative for the Conservation and Sustainable Use of Pollinators. Available at: www.internationalpollinatorsinitiative.org/jsp/intpollinitiative.jsp (accessed 18 January 2010).

Federici, B.A. (2007) Insecticidal bacteria: an overwhelming success for invertebrate pathology. *Journal of Invertebrate Pathology* 89, 30–38.

Ferguson, G. and Murphy, G. (2002) Glasshouse crop pest management: insects and mites. In: Pimentel, D. (ed.) *Encyclopedia of Pest Management*. Taylor Francis CRC Press, Boca Raton, Florida, pp. 342–344.

Francis, C.A. and Adipala, E. (1994) Tropical inter-cropping systems: what is their future? *African Crop Science Journal* 2, 131–133.

Frankel, O.H., Brown, A.H.D. and Burdon, J.J. (1995) *The Conservation of Plant Biodiversity*. Cambridge University Press, Cambridge.

Free, J.B. (1993) *Insect Pollination of Crops*, 2nd edn. Academic Press, London.

Fuxa, J.R. (1990) New directions for insect control with baculoviruses. In: Baker, R.R. and Dunn, P.E. (eds) *New Directions in Biological Control: Alternatives for Suppressing Agricultural Pests and Diseases*. Alan R. Liss, New York, pp. 97–113.

Gerson, U. and Smiley, R.L. (1990) *Acerine Biocontrol Agents: an Illustrated Key and Manual*. Chapman & Hall, London.

Greathead, D.J. (1986) Parasitoids in classical biological control. In: Waage, J. and Greathead, D.J. (eds) *Insect Parasitoids*. Academic Press, London, pp. 289–318.

Grzywacz, D., Mushobozi, W.L., Parnell, M., Jolliffe, F. and Wilson, K. (2008) Evaluation of *Spodoptera exempta* nucleopolyhedrosisvirus (SpexNPV) for the field control of African armyworm (*Spodoptera exempta*) in Tanzania. *Crop Protection* 27, 17–24.

Hallett, S.G. (2005) Where are the bioherbicides? *Weed Science* 53, 404–415.

Herren, H.R. and Neuenschwander, P. (1991) Biological control of cassava pests in Africa. *Annual Review of Entomology* 36, 257–283.

Hillocks, R.J., Logan, J.W.M., Riches, C., Russell-Smith, A. and Shaxson, L.J. (1996) Soil pests in traditional farming systems in sub-Saharan Africa – a review. Part 1. Problems. *Tropical Pest Management* 42, 241–251.

Holland, J.M., Oaten, H., Southway, S. and Moreby, S. (2008) The effectiveness of field margin enhancement for cereal aphid control by different natural enemy guilds. *Biological Control* 47, 71–76.

Hooper, A.M., Hassanali, A., Chamberlain, K., Khan, Z. and Pickett, J.A. (2008) New genetic opportunities from legume intercrops for controlling *Striga* spp. parasitic weeds. *Pest Management Science* 65, 546–552.

Hoy, M.A., Cunningham, G.L. and Knutson, L. (1983) *Biological Control of Pests by Mites*. Publication 3304, Division of Agricultural Sciences, University of California, Berkeley.

Jordan, V.W.L. (1999) The role of integrated production and integrated pest management for crop and environmental protection. In: Terry, P.J. (ed) *International Crop Protection: Achievements and Ambitions*. British Crop Protection Council Proceedings No. 73, pp. 75–98.

Julien, M.H. and Griffiths, M.W. (eds) (1998) *Biological Control of Weeds: a World Catalogue of Agents and their Target Weeds*, 4th edn. CAB International, Wallingford, UK.

Kennedy, G.G. (2008) Integration of insect-resistant genetically modified crops within IPM programs. In: Romeis, J., Shelton, A.M. and Kennedy, G.G. (eds) *Integration of Insect-Resistant Genetically Modified Crops within IPM Programs*. Springer Science + Business Media B.V., pp. 1–26.

Khan, Z.R., Pickett, J.A., Van Den Berg, J., Wadhams, L. and Woodcock, C.M. (2000) Exploiting chemical ecology and species diversity: stem borer and striga control for maize and sorghum in Africa. *Pest Management Science* 56, 957–962.

Kiaer, L.P., Skovgaard, I.M. and Ostergard, H. (2009) Grain yield increase in cereal variety mixtures: a meta-analysis of field trials. *Field Crops Research* 114, 361–373.

Kipkoech, A.K., Schulthess, F., Yabann, W.K., Maritim, H.K. and Mithoefer, D. (2006) Biological control of cereal stemborers in Kenya: a cost benefit approach. *Annales de la Société Entomologique de France* 42, 519–528.

Kipkoech, A.K., Mithoefer, D. and Yabann, W.K. (2008) Assessing yield and efficiency implications of relying on parasitoids for control of cereal stemborers: the case of small-scale maize farmers in Kenya. *Crop Protection* 27, 1318–1326.

Lacey, L.A., Frutos, R., Kaya, H.K. and Vail, P. (2001) Insect pathogens as biological control agents: do they have a future? *Biological Control* 21, 230–248.

Landis, D.A., Wratten, S.D. and Gurr, G.M. (2000) Habitat management to conserve natural enemies of arthropod pests in agriculture. *Annual Review of Entomology* 45, 175–201.

Langewald, J., Stolz, I., Everts, J. and Peveling, R. (2003) Towards the registration of microbial insecticides in Africa: non-target arthropod testing on Green Muscle, a grasshopper and locust control product based on the fungus *Metarhizium anisopliae* var. *acridum*. In: Neuenschwander, P., Borgemeister, C. and Langewald, J. (eds) *Biological Control in IPM Systems in Africa*. CAB International, Wallingford, UK, pp. 207–226.

La Salle, J. (1999) Insect biodiversity in agroecosystems: function, value and optimization. In: Wood, D. and Lenné, J.M. (eds) *Agrobiodiversity: Characterization, Utilization and Management*. CAB International, Wallingford, UK, pp. 155–182.

Lawo, N.C., Wachers, F.L. and Romeis, J. (2009) Indian *Bt* cotton varieties do not affect the performance of cotton aphids. *PLoS* 4, e4804, 1–9.

Lenné, J.M. and Wood, D. (1999) Vegetational diversity in agroecosystems: a mixed blessing for successful pest management? In: Terry, P.J. (ed.) *International Crop Protection: Achievements and Ambitions*. British Crop Protection Council Proceedings No. 73, pp. 75–98.

Li, Y., Meissle, M. and Romeis, J. (2008) Consumption of *Bt* maize pollen expressing Cry1Ab or Cry3Bb1 does not harm adult green lacewings, *Chrysoperla carnea* (Neuroptera: Chrysopidae). *PLoS* 3, e2909, 1–8.

Liebman, M. and Dyck, E. (1993) Crop rotation and intercropping strategies for weed management. *Ecological Applications* 3, 92–122.

Mailafiya, D.M., le Ru, B.P., Kairu, E.W., Calatayud, P.A. and Dupas, S. (2009) Species diversity of lepidopteran stem borer parasitoids in cultivated and natural habitats in Kenya. *Journal of Applied Entomology* 133, 416–429.

Marshall, E.J.P. (2002) Introducing field margin ecology in Europe. *Agriculture, Ecosystems and Environment* 89, 1–4.

Matama-Kauma, T., Schulthess, F., Mueke, J.M., Omwega, C.O. and Ogwang, J.A. (2006) Effect of wild grasses planted as border rows on stemborer infestations in Maize in Uganda. *Annales de la Société Entomologique de France* 42, 455–460.

Matama-Kauma, T., Schulthess, F., le Ru, B.P., Ogwang, J.A. and Omwega, C.O. (2008) Abundance and diversity of lepidopteran stemborers and their parasitoids on selected wild grasses in Uganda. *Crop Protection* 27, 505–513.

Matteson, P.C. (2000) Insect pest management in tropical Asian irrigated rice. *Annual Review of Entomology* 45, 549–574.

McFadyen, R.E. (1998) Biological control of weeds. *Annual Review of Entomolology* 43, 369–393.

McIntyre, B.D., Herren, H.R., Wakhungu, J. and Watson, R.T. (eds) (2009) *Agriculture at the Crossroads*. The global report of the International Assessment of Agricultural Knowledge, Science and Technology, Island Press, Washington, DC.

Meissle, M. and Romeis, J. (2009) The web-building spider *Theridion impressum* (Araneae: Theridiidae) is not adversely affected by *Bt* maize resistant to corn rootworms. *Plant Biotechnology* 7, 645–656.

Moore, D. (2008) A plague on locusts – the LUBILOSA story. *Outlooks on Pest Management* 19, 14–17.

Morin, L., Reid, A.M., Sims-Chilton, N.M., Buckley, Y.M., Dhileepan, K., Hastwell, G.T., Nordblom, T.L. and Raghu, S. (2009) Review of approaches to evaluate the effectiveness of weed biological control agents. *Biological Control* 51, 1–15.

Mushobozi, W.L., Grzywacz, D., Musebe, R., Kimani, M. and Wilson, K. (2005) New Approaches to improve the livelihoods of poor farmers and pastoralists in Tanzania through monitoring and control of African armyworm, *Spodoptera exempta*. *Aspects of Applied Biology: Pathways out of Poverty* 75, 37–46.

Ndemah, R., Schulthess, F. and Nolte, C. (2006) The effect of grassy field margins on soil water, plant nutrient levels, stem borer attacks and yield of maize in the humid forest zone of Cameroon. *Annales de la Société Entomologique de France* 42, 461–470.

Neuenschwander, P. (2001) Biological control of the cassava mealybug in Africa: a review. *Biological Control* 21, 214–229.

Neuenschwander, P. (2004) Harnessing nature in Africa: biological pest control can benefit the pocket, health and the environment. *Nature* 432, 801–802.

Neuenschwander, P., Borgemeister, C. and Langewald, J. (2003) *Biological Control in IPM Systems in Africa.* CAB International, Wallingford, UK.

New Agriculturalist (2009) Tanzania flexes its green Muscle. *New Agriculturalist* September 2009. Available at: www.new-ag.info/developments/devItem.php?a=943 (accessed 30 November 2009).

Norgaard, R.B. (1988) The biological control of cassava mealybug in Africa. *American Journal of Agricultural Economics* 70, 366–371.

Norris, R.F. and Kogan, M. (2000) Interactions between weeds, arthropod pests, and their natural enemies. *Weed Science* 48, 94–158.

Norris, R.F. and Kogan, M. (2005) Ecology of interactions between weeds and arthropods. *Annual Review of Entomology* 50, 479–503.

Nweke, F. (2009) Controlling cassava mosaic virus and cassava mealybug in Sub-Saharan Africa. *IFPRI Discussion Paper* 00912.

Oerke, E.C. and Dehne, H.W. (2004). Safeguarding production-losses in major crops and the role of crop protection. *Crop Protection* 23, 275–285.

Oerke, E.C., Dehne, H.W., Schonbeck, F. and Weber, A. (1994) *Crop Production and Crop Protection: estimated losses in major food and cash crops.* Elsevier, New York.

Olson, D.M. and Wackers, F.L. (2007) Management of field margins to maximize multiple ecological services. *Journal of Applied Ecology* 44, 13–21.

Omwega, C.O., Muchugu, E., Overholt, W.A. and Schulthess, F. (2006) Release and establishment of *Cotesia flavipes* Cameron (Hymenoptera: Braconidae) an exotic parasitoid of *Chilo partellus* (Swinhoe) (Lepidoptera: Crambidae) in East and Southern Africa. *Annales de la Société Entomologique de France* 42, 511–517.

Orapa, W., Day, M. and Ellison, C. (2008) New efforts at biological control of *Mikania micrantha* H.B.K. (Asteraceae) in Papua New Guinea and Fiji. In: *Proceedings of the Australia and New Zealand IOBC Biological Control Conference*, Sydney, p. 45.

Page, A.R. and Lacey, K.L. (2006) Economic impact assessment of Australian weed biological control. *CRC for Australian Weed Management Technical Series* No. 10., Glen Osmond, Australia. Available at: www.weedscrc.org.au/publications/technical_series.html (accessed 16 October 2009).

Pingali, P.L. (2001) *Milestones in Impact Assessment Research in the CGIAR, 1970–1999, with an Annotated Bibliography of Impact Assessment Studies Conducted in the CGIAR, 1970–1999.* Standing Panel on Impact Assessment, Technical Advisory Committee (TAC) of the CGIAR, World Bank, Washington, DC.

Pingali, P. and Roger, P. (eds) (1995) *Impact of Pesticides on the Rice Environment and Human Health.* Kluwer Academic Publishers, Boston, Massachusetts.

Polaszek, A. and Khan, Z.R. (1998) Host plants. In: Polaszek, A. (ed.) *African Cereal Stem Borers: economic importance, taxonomy, natural enemies and control.* CAB International, Wallingford, UK, pp. 3–10.

Polaszek, A., Riches, C. and Lenné, J.M. (1999) The effects of pest management strategies on biodiversity in agroecosystems. In: Wood, D. and Lenné, J.M. (eds) *Agrobiodiversity: Characterization, Utilization and Management.* CAB International, Wallingford, UK, pp. 273–303.

Potts, S.G., Biesmeijer, J.C., Kremen, C., Neumann, P., Schweiger, O. and Kunin, W.E. (2010) Global pollinator declines: trends, impacts and drivers. *Trends in Ecology and Evolution* 25, 345–353.

Ransom, J.K. (1990) Weed control in maize/legume intercrops. In: Waddington, S.R., Palmer, A.F.E. and Edje, O.T. (eds) Workshop on Research Methods for Cereal/Legume Intercropping in Eastern and Southern Africa. *CIMMYT Eastern and Southern Africa On-Farm Research Report* No. 17, pp. 41–44.

Rauschen, S. (2010) A case of 'pseudo-science'? A study claiming effects of the Cry1Ab protein on larvae of the two-spotted ladybird is reminiscent of the case of the green lacewing. *Transgenic Research* 19, 13–16.

Real IPM (2009) Holistic IPM Programmes in Kenya. Available at: www.realipm.com (accessed 14 December, 2009).

Ricroch, A., Bergé, J.B. and Kuntz, M. (2010) Is the German suspension of MON810 maize cultivation scientifically justified? *Transgenic Research* 19, 1–12.

Risch, S.J. (1983) Intercropping as cultural pest control: prospects and limitations. *Environmental Management* 7, 9–14.

Risch, S.J., Andow, D. and Altieri, M.A. (1983) Agroecosystem diversity and pest control: data, tentative conclusions and new research directions. *Environmental Entomology* 12, 625–629.

Rohrman, G. (2008) *Baculovirus Molecular Biology.* National Center for Biotechnological Information, Bethesda, Maryland.

Rola, A. and Pingali, P. (1993) *Pesticides, Rice Productivity and Farmers' Health.* Jointly published by the

International Rice Research Institute, Manila, the Philippines and the World Resources Institute, Washington, DC.

Romeis, J., Bartsch, D., Bigler, F., Candolfi, M.P., Gielkens, M.M.C., Hartley, S.E., Hellmich, R.L., Huesing, J.E., Jepson, P.C., Layton, R., Quemada, H., Raybould, A., Rose, R.I., Schiemann, J., Sears, M.K., Shelton, A.M., Sweet, J., Vaituzis, Z. and Wolt, J.D. (2008a) Assessment of risk of insect-resistant transgenic crops to nontarget arthropods. *Nature Biotechnology* 26, 203–208.

Romeis, J., Van Driesche, R.G., Barratt, B.I.P. and Bigler, F. (2008b) Insect-resistant transgenic crops and biological control. In: Romeis, J., Shelton, A.M. and Kennedy, G.G. (eds) *Integration of Insect-Resistant Genetically Modified Crops within IPM Programs.* Springer Science + Business Media B.V., pp. 87–117.

Romeis, J., Meissle, M., Raybould, A. and Hellmich, R.L. (2009) Impact of insect-resistant transgenic crops on above-ground non-target arthropods. In: Ferry, N. and Gatehouse, A.M.R. (eds) *Environmental Impact of Genetically Modified Crops.* CAB International, Wallingford, UK, pp. 165–198.

Royal Society (2009) *Science and the Sustainable Intensification of Global Agriculture.* Royal Society Policy Document 11/09.

Sanvido, O., Romeis, J. and Bigler, F. (2007) Ecological impacts of genetically modified crops: ten years of field research and commercial cultivation. *Advances in Biochemical Engineering and Biotechnology* 107, 235–278.

Schmidt, J.E.U., Braun, C.U., Whitehouse, L.P. and Hilbeck, A. (2009) Effects of activated Bt transgene products (Cry1Ab, Cry3Bb) on immature stages of the ladybird *Adalia bipunctata* in laboratory ecotoxicity testing. *Archives of Environmental Contamination Toxicology* 56, 221–228.

Schoenly, K.G., Justo, H.D., Barrion, A.T., Harris, M. and Bottrell, D.G. (1998) Analysis of invertebrate biodiversity in a Philippines farmer's irrigated rice field. *Environmental Entomology* 27, 1125–1136.

Settle, W.H., Ariawan, H., Astuti, E.T., Cahyana, W., Hakim, A.L., Hindayana, D., Lestari, A.S. and Pajarningsih (1996) Managing tropical rice pests through conservation of generalist natural enemies and alternative prey. *Ecology* 77, 1975–1988.

Shiva, V.S. and Jafri, A.H. (2004) Failure of GMOs in India. *Synthesis/Regeneration* 33. Available at: www.greens.org/s-r/33/33-04.html (accessed 18 October 2010).

Smith, H.A. and McSorley, R. (2000) Intercropping and pest management: a review of major concepts. *American Entomologist* 46, 154–161.

Smith, J., Potts, S.G., Woodcock, B.A. and Eggleton, P. (2007) Can arable field margins be managed to enhance their biodiversity, conservation and functional value for soil macrofauna? *Journal of Applied Ecology* 45, 269–278.

Smithson, J.B. and Lenné, J.M. (1996) Varietal mixtures: a viable strategy for sustainable productivity in subsistence agriculture. *Annals of Applied Biology* 128, 127–158.

Songa, J.M., Jiang, N., Schulthess, F. and Omwega, C. (2007) The role of intercropping different cereal species in controlling lepidopteran stemborers on maize in Kenya. *Journal of Applied Entomology* 131, 40–49.

Spielman, D.J. and Pandya-Lorch, R. (2009) *Millions Fed: proven success in agricultural development.* International Food Policy Institute, Washington, DC.

Storer, N.P., Dively, G.P. and Herman, R.A. (2008) Landscape effects of insect-resistant genetically-modified crops. In: Romeis, J., Shelton, A.M. and Kennedy, G.G. (eds) *Integration of Insect-Resistant Genetically Modified Crops within IPM Programs.* Springer Science + Business Media B.V., pp. 273–302.

TeBeest, D.O. (1996) Issues and prospects confronting biological control of weeds with plant pathogens in the future. *Phytoparasitica* 24, 91–95.

Templeton, D. and Jamora, N. (2007) *Economic Assessment of the Policy-oriented Research on the Private Health Costs of Pesticide use in the Philippines.* PORIA Case Study, CGIAR Science Council, Rome.

Teng, P.S. (1999) Current and future importance of biotechnology to crop protection. In: Terry, P.J. (ed.) *International Crop Protection: Achievements and Ambitions.* British Crop Protection Council Proceedings No. 73, pp. 101–122.

Terry, P.J. (1991) Grassy weeds- a general overview. In: Baker, F.W.G. and Terry, P.J. (eds) *Tropical Grassy Weeds.* CAB International, Wallingford, UK, pp. 5–38.

Thresh, J.M. (ed.) (1981) *Pests, Pathogens and Vegetation.* Pitman Press, Bath, UK.

Thurston, H.D. (1992) *Sustainable Practices for Plant Disease Management in Traditional Systems.* Westview Press, Boulder, Colorado.

Trenbath, B.R. (1993) Intercropping for the management of pests and diseases. *Field Crops Research* 34, 381–405.

van den Berg, J., de Bruyn, A.J.M. and van Hamburg, H. (2006) Oviposition preference and survival of the maize stem borer, *Busseola fusca* (Lepidoptera: Noctuidae), on Napier grasses, and maize. *African Entomology* 14, 211–218.

Van den Bosch, R., Messenger, P.S. and Gutierrez, A.P. (1982) *An Introduction to Biological Control.* Plenum Press, New York.

Van Driesche, R., Hoddle, M. and Center, T. (2008) *Control of Pests and Weeds by Natural Enemies: An Introduction to Biological Control.* Blackwell Publishing, Oxford.

Van Steekelenberg, N.A.M. (2006) Novel approaches to integrated pest and disease control in glasshouse vegetables in the Netherlands. *Pest Management Science* 36, 359–362.

Venne, J., Beed, F., Avocanh, A. and Watson, A. (2009) Integrating *Fusarium oxysporum* f. sp. *strigae* into cereal cropping systems in Africa. *Pest Management Science* 65, 572–580.

Warwick, S.I. and Black, L.D. (1983) The biology of Canadian weeds. 61. *Sorghum halepense* (L.) Pers. *Canadian Journal of Plant Science* 62, 997–1014.

White, G.L., Kairo, M.T.K. and Lopez, V. (2005) Classical biological control of the citrus blackfly *Aleurocanthus woglumi* by *Amitus hesperidum* in Trinidad. *Biocontrol* 50, 751–759.

Wood, D. and Lenné, J.M. (eds) (1999) *Agrobiodiversity: Characterization, Utilization and Management.* CAB International, Wallingford, UK.

Wu, K.-M., Lu, Y.-H., Feng, H.-Q., Jiang, Y.-Y. and Zhao, J.-Z. (2008) Suppression of cotton bollworm in multiple crops in China in areas with *Bt* toxin containing cotton. *Science* 321, No. 5896, 1676–1678.

Yaninek, S. and Hanna, R. (2003) Cassava Green Mite in Africa – a unique example of successful classical biological control of a mite pest on a continental scale. In: Neuenschwander, P., Borgemeister, C. and Langewald, J. (eds) *Biological Control in IPM Systems in Africa.* CAB International, Wallingford, UK.

Zeddies, J., Schaab, R.P., Neuenschwander, P. and Herren, H.R. (2001) Economics of biological control of cassava mealybug in Africa. *Agricultural Economics* 24, 209–219.

9 Biodiversity and Ecosystem Functioning Below-ground

T.W. Kuyper and K.E. Giller

Introduction

Soils contain more (known and unknown) species diversity than other terrestrial habitat. Biodiversity is therefore strongly an underground phenomenon. This observation is even more valid for agroecosystems, where human planned food production from crops often reduces above-ground biodiversity, without concomitant reductions in associated, below-ground biodiversity. However, soil biodiversity is cryptic for at least four related reasons: (i) the soil is an opaque and extremely heterogeneous medium, making observations difficult (Crawford *et al.*, 2005); (ii) a large part of that diversity consists of organisms with (very) small body sizes and these organisms cannot be directly observed by the naked eye; (iii) not all bacterial and fungal species are active and a large part of biodiversity is dormant (Lavelle *et al.*, 1995) unless awakened by soil disturbance such as tillage; and (iv) while the number of described soil-dwelling species is already huge, that number is dwarfed by expert estimates of unknown species richness (Table 9.1).

Molecular methods demonstrate that even the expert judgements could be at the lower end of species diversity estimates. Soil metagenomic approaches have exploded old estimates of bacterial 'species' richness, which were in the order of 10^3 or 10^4 per g of soil (Torsvik *et al.*, 2002). New methods increase that number by one or two orders of magnitude (Gans *et al.*, 2005). But phylogenies (most often based on ribosomal RNA) do not necessarily reflect ecological function and these huge 'species' numbers do not capture functional diversity (Jaspers and Overmann, 2004; Gamper *et al.*, 2010). Molecular phylogeny of rhizobia based on the genes required for nodulation not surprisingly is more strongly related to their legume host-range than the phylogeny based on 16S rRNA (Young and Haukka, 1996).

It is not surprising that the debate on the relation between soil species diversity and soil ecosystem functioning has not progressed as much as its above-ground counterpart. A further consequence for farmers of its invisibility is that the agroecosystem consequences of soil biodiversity are much less visible and tangible than above-ground biodiversity. However, it is virtually certain that a substantial number of species of this cryptic soil biodiversity are already extinct and that these unnoticed extinctions were not manifested in reduced ecosystem functioning.

The aim of this chapter is to reflect on new developments in the decade after the first book (Wood and Lenné, 1999), in which the second author contributed to a similar paper on the functional significance of soil biodiversity in agroecosystems (Wardle *et al.*,

Table 9.1. Known and unknown diversity of species groups that are mainly soil dwelling (data from Coleman (2008) and Turbé *et al.* (2010)).

Species group	Species described	Species estimates
Bacteria	5,000	1,000,000–1,000,000,000
Fungi	70,000	1,500,000–7,000,000
Protozoa	40,000	20,000–200,000
Nematodes	25,000	50,000–1,900,000
Oligochaeta	3,650	7,000–8,000
Acari	45,000	80,000–1,000,000
Collembola	7,500	15,000–50,000
Isoptera	2,600	10,000

1999). We address how the field has developed in that decade, and to what extent the above-ground consensus about the relation between biodiversity and ecosystem function (Hooper *et al.*, 2005; Loreau, 2010) can be extended to the more specific relation between soil biodiversity, agroecosystem functioning and sustainable food production. We thus address one of the hundred most important questions for the conservation of biodiversity listed by Sutherland *et al.* (2009).

Soil Biodiversity Loss and Agricultural Intensification

Perfecto and Vandermeer (2008) provided a conceptual scheme that links agricultural intensification to biodiversity, both planned and associated biodiversity, including soil biodiversity (Fig. 9.1).

Despite the lack of a measurable parameter (or a proxy that supports the claim that intensification is a one-dimensional concept) on the *x*-axis, conceptual schemes are highly important for the biodiversity discourse. The main message of such graphs is that agricultural intensification inevitably leads to biodiversity loss; however, Giller *et al.* (1997) have argued that intensification could first lead to an increase in soil biodiversity before it collapses. Such non-quantitative graphs can easily become mantras, expressing generalized belief rather than factual knowledge that agricultural intensification results in biodiversity losses and that such losses impair agricultural sustainability (Wood and Lenné, 2005; McIntyre *et al.*, 2009; Phelan, 2009). Biodiversity could then easily be transformed from a descriptive towards a normative concept (biodiversity is intrinsically good). We return to the biodiversity discourse in the final section of this chapter.

Agricultural intensification has resulted in a take-over of several soil ecosystem functions and services by human management (fertilizer instead of N-fixation; tillage instead of soil faunal activity, etc.). There is evidence that some of the food production increases under intensification are not sustainable, and history shows how flourishing civilizations have perished through inadequate soil management (Hillel, 1991). Agricultural intensification also often had a negative impact on soil biodiversity through: (i) non-selective use of pesticides (fungicides, insecticides, herbicides); (ii) overuse of fertilizers; (iii) increased monoculture (reduced above-ground diversity); (iv) inadequate practices, resulting in soil compaction or erosion; and (v) inadequate management of organic matter. Consequently, McIntyre *et al.* (2009) claimed that business-as-usual is no longer an option. A movement towards less intensified, more *natural* or *eco-efficient* agricultural practices emerged (Brussaard *et al.*, 2010).

The Power of Metaphor

Soil life, the 'unseen majority of biodiversity' (Van der Heijden *et al.*, 2008), is responsible for crucial life-support functions or ecosystem services. In agroecosystems the first and

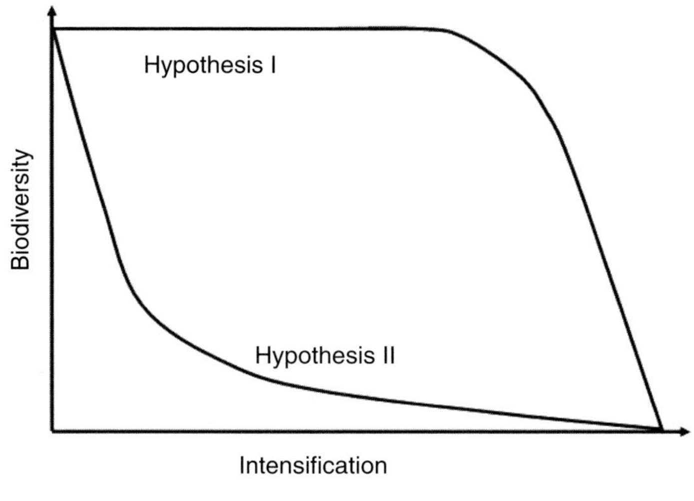

Fig. 9.1. Hypothesized relationship between agricultural intensification and biodiversity (After Perfecto and Vandermeer, 2008).

foremost ecosystem service is the production (or provisioning) service. Other ecosystem services provided by soils and soil biota include supporting and regulating functions. From a non-utilitarian viewpoint cultural 'services' can be added as a fourth category (Millennium Ecosystem Assessment, 2005; Kibblewhite *et al.*, 2008; Turbé *et al.*, 2010).

Giller (1996) took up a metaphor, earlier expressed by Usher *et al.* (1979), that soil biodiversity is the *poor* man's tropical rainforest. That metaphor contributed to a narrative that links the importance of biodiversity to resource-poor people. Díaz *et al.* (2006) argued that biodiversity change is inextricably linked to poverty, because resource-poor farmers rely more directly on ecosystem services than farmers in the developed countries. Therefore subsistence farmers (who are organic-by-default) face the greatest risk from biodiversity losses. Consequently, biodiversity loss could aggravate global inequality and further marginalize resource-poor farmers. McIntyre *et al.* (2009) also suggested that maintenance and careful management of biodiversity could provide an escape route from poverty.

The metaphor is strengthened by attempts to translate ecosystem services in monetary terms and suggests mechanisms (payment for ecosystem services) that could simultaneously contribute to biodiversity conservation, poverty alleviation and food security. Attempts to put monetary values on ecosystem services demonstrated that the value of soil processes surpasses that of other domains. Pimentel *et al.* (1997) calculated the economic value of biodiversity and claimed the benefits for the global economy to be US$2.9 trillion per year. Costanza *et al.* (1997) calculated the monetary value of all ecosystem services to be at least US$33 trillion per year (range US$16–54 trillion). With that huge number in mind, the claim that the consequences of soil biodiversity mismanagement are in excess of US$1 trillion per year (Turbé *et al.*, 2010) seems moderate. Little progress has been made in methodology to estimate the value of these services, because studies still produce results that differ by one or two orders of magnitude. Porter *et al.* (2009) estimated the monetary value of ecosystem services of natural ecosystems at US$2000/ha/year, and those of agroecosystems around US$1000/ha/year, with 50–80% of the value attributable to supporting and regulating services (production services were estimated at US$200–500/ha/year). In a comparison between conventional and organic agriculture in New Zealand, Sandhu *et al.* (2008)

estimated the production service to be worth US$3000–4000/ha/year, and the value from supporting and regulating services to be only 20–30% of that value. If production services outweigh the supporting, regulating and cultural services a situation may arise where human well-being (including food security) increases globally, while the other services decline (Raudsepp-Hearne *et al.*, 2010).

A relevant question for soil biodiversity is whether we can link monetary values for ecosystem services to specific soil organisms or soil biodiversity. Several processes are uniquely ascribed to certain taxonomic and/or functional groups. The economic value of N_2-fixation through rhizobia and other symbiotic N_2-fixing bacteria has been estimated at US$90 billion (Pimentel *et al.*, 1997). No estimates have yet been made for phosphorus savings through the arbuscular mycorrhizal (AM) symbiosis. Assuming that AM fungi increase phosphorus uptake by crops with 10%, we can calculate that the ecosystem service provided by AM fungi amounts to US$2 billion. As N_2-fixation of legumes is often limited by availability of P (Giller and Cadisch, 1995), the value of ecosystem services provided by AM fungi is arguably substantially higher. Another ecosystem service is soil formation, for which US$25 billion was attributed to earthworms (Pimentel *et al.*, 1997; Porter *et al.*, 2009). But this attribution is simplistic at best, because other organisms (fungi, bacteria) also play a role in weathering and soil formation. Other services by earthworms are probably more important. Barrios (2007) reported that earthworms increased tea production by US$500/ha/year, almost two orders of magnitude larger than their role in soil formation. However, again attribution of this value to earthworms only (and not to organic matter and the primary decomposers of that organic matter) is not evident, because the correlation between earthworm activity and production increases is not a demonstration of a causal relation. But the general message is clear: we should be careful in assigning monetary values to specific soil biota.

However impressive, these calculations contain implications that are often not addressed. We mention four. While there is no doubt that below-ground biota are essential for the functioning of agro-ecosystems, the key question of how many different species of soil organisms are crucial is not addressed by such calculations. Replacement of indigenous earthworms by exotic species, which results in global biodiversity loss, may still increase the value of the soil biota. Second, considering the value of services related to organic matter and nutrient mineralization, agricultural intensification would often be the preferred option for resource-poor farmers. Zingore *et al.* (2005) compared productivity of smallholder and commercial farmers in Zimbabwe. Next to increased direct value of the crops, the much larger organic matter contents of the soil on commercial farms would more than double the monetary value of the supporting ecosystem service. Third, depending on the economic value of the production services compared to the other services, such arguments could lead both to the conclusion that an eco-efficient agriculture (*eco-agriculture*) is the preferred solution and to the conclusion that maximally intensive agricultural practices and biodiversity conservation should be separated. Finally, the dazzling numbers of the economic value of ecosystem services confront us with the fact that no real economy is willing to contemplate payment for such services. Rather, conversely, the short-term benefits by unsustainable production increases for the poor outweigh long-term benefits of ecosystem services through biodiversity conservation – except at unrealistically high prices for C sequestration. These examples suggest that putting monetary value on ecosystem services, and linking these services to soil biodiversity, will not suffice as a defence for maintaining soil biodiversity.

The Consensus View on Biodiversity and Ecosystem Function

The consensus view (Hooper *et al.*, 2005; Loreau, 2010) contains two classes of mechanisms that explain the positive relationship between biodiversity and ecosystem functioning. These are *functional complementarity* and *selection effects* (which include sampling effects). The *portfolio effect* (increased diversity results in reduced variance rather than in increased ecosystem functioning) has also been mentioned, but it is debatable whether this can be called a mechanism.

A class of relationships that is particularly relevant for the biodiversity–ecosystem function relationship in the context of agroecosystems is the situation where biodiversity and ecosystem function co-vary due to a common driving factor, often human management. Most studies that link soil biodiversity to ecosystem functioning are of a correlative nature (Reed and Martiny, 2007). Agricultural intensification with indiscriminate use of fertilizers and pesticides will reduce both biodiversity and ecosystem services. Soil disturbance (ploughing) results in increased carbon turnover (as a consequence of increased microbial activity) and also reduces microbial species richness (through a homogenizing effect). It would be erroneous to link soil biodiversity loss and enhanced microbial activity (and therefore increased supporting services) in a mechanistic framework.

Overemphasis on soil species diversity at the neglect of abiotic driving variables (e.g. soil organic matter or nutrient availability) may result in failure to restore agroecosystems. Plant species richness and composition may have a small impact on the composition of soil microbial assemblages (Kielak *et al.*, 2008), in part because abiotic factors (the legacy of previous soil use, or the *ghost of intensive agriculture past*) override biotic effects. Buckley and Schmidt (2003) found that 9 years after stopping agricultural management soil abiotic factors were still dominant influences on microbial diversity; and that fields required more than 45 years to erase the historical effect of tillage. Kulmatiski and Beard (2008) confirmed that the legacy of past land use could persist for fifty years. Such legacies seem to be larger under agricultural extensification than under agricultural intensification (Postma-Blaauw *et al.*, 2010).

The consensus view has been criticized by Hillebrand and Matthiessen (2009), who highlighted two shortcomings. One critique is that biodiversity studies were executed under conditions that lacked ecological realism. For instance, the study by Hanson *et al.* (2008) that claimed niche partitioning through specialized resource use was based on additions of single carbon compounds. A study by Orwin *et al.* (2006) also added single carbon compounds, resulting in both changes in microbial community structure and reduced plant growth as the added C sources resulted in N immobilization. It is unlikely that such studies mimic conditions that test for effects of litter diversity. Their other criticism refers to the inadequate definition and operationalization of functional groups (see below).

Soil Biodiversity and Soil Ecosystem Functioning

Hooper *et al.* (2005) were rather brief on the issue whether the above-ground consensus is also valid for the underground, although they noted that the huge soil biodiversity in combination with limited niche differentiation implies that the relationship saturates at low species numbers. There is no *a priori* reason why the consensus view should not be valid below-ground. The study by Van der Heijden *et al.* (1998) that demonstrated that increased species richness of AM fungi resulted in increased plant species diversity, plant productivity and resource use, is a prime example.

Bell *et al.* (2005) manipulated bacterial species richness and investigated the relation between species richness and carbon respiration. Their systems contained 1–72 species (a fraction of what occurs in 1 mg of soil or water) and showed a linear increase in respiration with the natural logarithm of species richness. The effect was largely due to a positive selection effect, as the differences between the best 2-species or 4-species and 36-species or 72-species treatments were small. The clearest example for a relationship between saprotrophic fungal species richness and decomposition rate was observed by Setälä and McLean (2004), but their graphs indicate a huge effect of having at least one species (the zero-fungal species treatment had a significantly lower decomposition rate) and a relationship that saturates at low diversity (5–10 species of a maximum of 43 species). No evidence for niche partitioning in the case of complex substrates with lignin-like compounds was obtained. Studies of potential niche differentiation of soil fauna along a food axis showed that most soil animals are generalists rather than specialists, making it unlikely that enhanced biodiversity would enhance litter decomposition (Hättenschwiler *et al.*, 2005). The studies that did show a positive relation between diversity and decomposition rates were based on species with large functional dissimilarity (Heemsbergen *et al.*, 2004). In contrast to studies on plant species diversity, negative selection effects, where dominant species do contribute significantly less to that ecosystem function (Jiang *et al.*, 2008), seem to be more common among saprotrophic bacteria (Jiang, 2007) and fungi (Gessner *et al.*, 2010).

A major concern in many of these experimental studies is that it is not the number of species inoculated or added that counts when explaining diversity effects, but the number of species that survive. In several of these experiments the species richness actually realized was much less than the number of species inoculated.

A meta-analysis by Balvanera *et al.* (2006) concluded that diversity effects on ecosystem properties were weak and only slightly positive. More specifically they noted that plant diversity enhanced soil biodiversity, but that it was unclear whether plant or soil biodiversity had a positive effect on soil nutrient supply. Caution is clearly needed. Srivastava *et al.* (2009) reported significant diversity effects on decomposition but no significant effects on total detrital standing stocks. While this difference might be due to the fact that the analytical methods are more sensitive to capture differences in disappearance rates than in standing stocks, an alternative explanation could be that saprotrophic species diversity enhances decomposition only in the initial stages.

For real soil ecosystems the consensus is that a reduction in soil microbial diversity does not have a negative impact on 'generalist' functions such as decomposition or nitrogen mineralization (Giller *et al.*, 1998; Nannipieri *et al.*, 2003). The claim that reduced microbial diversity leads to a decreased capacity of more specialist functions is still contested, although a study by Wertz *et al.* (2007) yielded strong evidence that a very drastic reduction of soil microbial diversity did not impair two narrow ecological functions, denitrification and nitrite oxidation.

Functional Biodiversity – Competing Claims on a Concept

The consensus view has remained controversial. A major reason for the continued controversy and the underlying criticism that species richness is an inadequate parameter for establishing biodiversity–ecosystem function relationships is that many studies showed saturating relationships at low diversity (Díaz and Cabido, 2001). There are more reasons why taxonomic diversity has been gradually replaced in the scientific debate by functional diversity. Scientists and policy makers have become aware that unprecedented losses in species diversity could negatively impact ecosystem processes and services delivered by species. As stated by Gardi and Jeffery (2009): 'It is the diversity of processes, the functional diversity, carried out by the soil biota which gives soil biodiversity such high value.'

However, this (paradigm) shift from species diversity to functional diversity is not without problems, as it necessitates a theory that disconnects both diversities and explains how different species with similar or even identical ecosystem effects (functions) can coexist (Fitter, 2005). Or alternatively: under what conditions is species diversity a good surrogate for functional diversity? Díaz and Cabido (2001) showed that both forms of diversity are largely congruent if each species occupies its own niche, and if niche overlap is equal to or less than is expected by random models. If different species show niche convergence, both diversity parameters are disconnected, implying functional redundancy.

Functional diversity is difficult to operationalize. Which functional traits are useful for an assessment of functional diversity? It is difficult to explain situations where there is no obvious link between functional diversity and ecosystem functioning. Petchey and Gaston (2006) suggested that inappropriate classification of functional diversity is often used as an argument to interpret (or explain away) the absence of significant links. The authors listed three other explanations, including incorrect measure of functional diversity, other ecological factors that override functional diversity and lack of statistical power, before they proposed as a fifth alternative that functional diversity has no effect. The first explanation, that the lack of a significant relationship is due to incorrect classification of functional traits, is particularly dangerous – it could lead to iterations of alternative classifications until a significant relationship appears. Then functional diversity becomes an unfalsifiable concept.

Despite the fact that the concept of functional biodiversity was introduced recently, the literature lacks clarity on its definition and operationalization. Tilman *et al.* (1997) defined plant functional diversity as the number of functional types or groups. Turbé *et al.* (2010) used a similar aggregate grouping for soil biota (see below). However, other authors used the term functional diversity for variation within a specified function, often even within one species. This usage is widespread among mycorrhizal researchers (Van der Heijden and Scheublin, 2007). Munkvold *et al.* (2004) described intraspecific variation in hyphal length for two AM fungal species, and demonstrated a good correlation between hyphal length and P-uptake. But variation within a function, which could serve as an insurance mechanism, is not too dissimilar from functional redundancy.

Functional classifications of soil biota are in their infancy – although one would *a priori* expect much more functional diversity among microbes than among primary producers. Turbé *et al.* (2010) proposed three functional groups:

- Chemical engineers, including saprotrophic fungi and bacteria, arbuscular mycorrhizal fungi and N_2-fixing rhizobia;
- Biological regulators, including nematodes, mites and springtails; and
- Ecosystem engineers, including earthworms, termites and isopods.

However, these functional groups are too crude and general to be useful for the debate between biodiversity and soil ecosystem function. First, one may wonder whether including saprotrophic and mutualistic, biotrophic microorganisms in a single group is useful. Second, it is debatable whether including saprotrophic fungi and saprotrophic bacteria in one group is effective. A link has repeatedly been proposed between litter quality and the relative contribution that fungi and bacteria make to decomposition of organic matter. But contrary to received wisdom, Joergensen and Wichern (2008) showed that shifts within the fungal assemblage (i.e. mycorrhizal fungi versus saprotrophic fungi) have a much larger impact on soil ecosystem function than a shift within the saprotrophs between fungi and bacteria.

Ecological or functional classifications of bacteria are still in their infancy. Fierer *et al.* (2007) collected soil samples across North America and tested whether a classification of bacterial phyla in copiotrophs (r-strategists) and oligotrophs (K-strategists) was meaningful. While their data fitted in general this dichotomy, the authors also noted that such a dichotomy is very crude – too crude to contribute to the current biodiversity debate.

Early classifications of rhizobia separated them into two groups: the slow-growing 'cowpea miscellany' that were more promiscuous in their host range, and the fast growing, more host-specific '*Rhizobium*'. Although these different groups were later classified into the slow-growing *Bradyrhizobium* and the fast-growing *Rhizobium*, the growth rate of the bacteria has no effect on the speed of nodulation in soil. Molecular phylogeny has revealed an increasing diversity of root-nodulating bacteria with a large number of genera of α-proteobacteria and β-proteobacteria that can nodulate legumes (Rivas *et al.*, 2009). A biogeographical analysis of the global distribution of new species of N_2-fixing bacteria tells us more about the distribution of scientists interested in this topic, and their itinerant wanderings, than of the distribution of the bacteria (Giller *et al.*, 2005).

If we consider mycorrhizal fungi as a functional group, should we discriminate between ectomycorrhizal fungi, AM fungi and dark septate endophytic fungi (DSE)? Are within AM fungi further functional subdivisions useful, separating root colonizers from soil colonizers (Hart and Reader 2002)? Are generalists (i.e. the species that associate with almost all species and that are the species that can be grown in culture and sold as commercial inoculum) functionally different from the specialists (with about ten times as many species, to judge from environmental DNA sequences)? And how should functional traits that are expressed in the interaction between plants and fungi be dealt with, e.g. situations where the primary service of AM fungi is either supporting (enhanced nutrient uptake) or regulating (protection against root pathogens) (Newsham *et al.*, 1995)?

Similar questions are pertinent for saprotrophic fungi, where we could arrive at further subdivisions of fast-growing r-selected versus slow-growing K-selected species; or cellulolytic versus ligninolytic fungi; or white-rot versus brown-rot fungi.

It is usual to recognize three groups of earthworms: epigeics, endogeics and anecics. But is such a functional classification preferable over a functional classification of endogeics that separates decompacting species with small-body sizes from compacting species with larger body sizes (Blanchart *et al.*, 2004)?

These questions imply that at present functional classifications possess a degree of arbitrariness that makes their application highly problematic. But of course this arbitrariness is useful to explain away the lack of significant relationships between functional diversity and ecosystem function.

A Neutral View on Soil Biodiversity and Redundancy

Niche theories have been dominant in explaining the vastness of soil biodiversity. Giller (1996) explained tremendous soil biodiversity through the Hutchinsonian niche, where every species occupies its own niche and competitive interactions between species (including the ghost of competition past) result in resource partitioning. Other authors explained the huge biodiversity through the extremely heterogeneous nature of soil, which provides almost infinite potential for niche differentiation (Young *et al.*, 2008). But with every species occupying its own niche, it becomes a riddle why soil biodiversity–ecosystem function relationships saturate at low diversity.

These observations paved the way to conclude that apparently soil assemblages show a large degree of redundancy (Giller *et al.*, 1997; Swift *et al.*, 2004). Bardgett (2002) and Wardle (2006) also subscribe to the view that there is no predictable relationship between species diversity and soil ecosystem functioning, that there is redundancy in soil communities and that traits of dominant organisms play a much larger role on ecosystem process rates. Cases where the relationship between taxonomic diversity and ecosystem functioning saturated at low diversity levels were then explained as caused by *functional redundancy*.

A major reason why ecologists have felt unease with the concept of redundancy relates to the colloquial use of the term redundant as equivalent to superfluous. For scientific and policy reasons the saying that certain species are superfluous (and even that almost all soil biota are superfluous) is difficult to digest; as is the suggestion that redundancy implies that there is 'excess' biodiversity (Welbaum *et al.*, 2004).

Several attempts to 'save' biodiversity from redundancy have been undertaken. First, Díaz and Cabido (2001) mentioned the distinction between functional effect and functional response. Species with similar functional effects (hence showing functional redundancy) could still have differential functional responses (and hence not show redundancy). Second, Hector and Bagchi (2007) and Gamfeldt *et al.* (2008) argued that it is risky to posit redundancy from one ecosystem function or service only and that ecosystem multifunctionality should rather be the focus. They claim that for this reason a larger biodiversity is needed. Their argument results in a larger number of functional groups and a narrowing of the gap between functional diversity and species diversity. It remains doubtful, however, how these views relate to functional diversity of soil biota. For short-term decomposition apparently three functional groups (with in total two to six, exceptionally ten species) seemed sufficient, and for long-term decomposition again at most three functional groups and four species were deemed essential (Hector and Bagchi, 2007). So for these soil processes the data equally support the claim that ecosystem functioning saturates at very low levels of both species and functional diversity. Ultimately, different functional traits would allow different independent classifications and including multiple traits would increase the number of functional groups. Eviner and Chapin (2003) even proposed that each species could have a unique suite of functional traits – which ultimately explodes the distinction between functional diversity and species diversity.

Another escape route is that to some authors redundancy is a relative concept, and degrees of redundancy are recognized, depending on the number of organisms that can fulfil that function. Some functions (decomposition) can be carried out by many bacteria and fungi, whereas N_2-fixing bacteria and AM fungi belong to less species-rich (and therefore less redundant) groups. Also nitrification and denitrification are executed by more limited numbers of species. However, Wertz *et al.* (2007) noted that for two ecosystem functions, nitrification and denitrification, a decrease in diversity did not affect the resilience and resistance of both microbial groups. Gardi and Jeffery (2009) suggested that for breakdown of some highly recalcitrant or xenobiotic compounds, no functional redundancy exists at all. McGuire and Treseder (2010) also suggested that decomposition of recalcitrant carbon was a narrow process – but we are unaware of data to substantiate these claims. Rather, their degradation is not limited by enzymatic capability but by the supply of easily degradable carbon compounds that are essential for co-metabolism of recalcitrant carbon.

Because soil microbial assemblages show very large redundancy, many authors still feel at ease with ecosystem models where microbial communities are treated as kinetic constants and response functions (Allison and Martiny, 2008; but cf. Strickland *et al.*, 2009). In such models there is no need to better understand soil microbial diversity in order to improve predictions of decomposition and nutrient transformation. One way to test under what conditions species identity and diversity would matter for such models would be to operationalize the concept of a Minimum Workable Decomposer Community (as proposed by Ekschmitt and Griffiths, 1998), which up to now has remained elusive.

The apparent failure of niche theories to explain the huge soil biodiversity has given rise to alternative theories that are more compatible with the idea of widespread functional redundancy. Neutral models (Hubbell, 2005) are one such class. Neutral communities are characterized by a very long tail of rare species – much longer than in a log-normal distribution. When species–area curves do not saturate, such distributions probably follow the predictions of the neutral model. In such cases our ability to describe the microbial assemblage remains inadequate. A major element of neutral theory is the assumption of dispersal limitation. At first sight, soil biota violate this assumption because every species seems to be everywhere (but the environment selects). Recently for bacteria, Zhou *et al.* (2002) and Martiny *et al.* (2006) disputed the idea that everything is everywhere, and noted that the distribution of bacterial assemblages over small distances supports dispersal limitation. A study by Noguez *et al.* (2005) showed dispersal limitation even over very short spatial scales. Such dispersal limitation could explain why disturbance such as tillage has a large impact on species diversity – but without functional consequences. Dispersal limitation, in combination with severe P-limitation that limits horizontal gene transfer, has been invoked for high bacterial diversity – which also explains why agricultural intensification leads to reduced bacterial diversity (Souza *et al.*, 2008). Also for AM fungi suggestions have been made that neutral theory provides an attractive alternative to explain species richness (Lekberg *et al.*, 2007; Dumbrell *et al.*, 2010).

Cases of cosmopolitan species do of

course occur, but sometimes reflect human-aided dispersal. The AM fungus *Glomus mosseae* is now globally distributed but a genetic analysis indicated that this is due to recent area expansion, driven by agriculture (Rosendahl *et al.*, 2009). Similarly, cosmopolitan occurrence of certain N_2-fixing *Burkholderia* species with invasive species of *Mimosa* is due to spreading of plants with their root symbiont (Bontemps *et al.*, 2010). Human-aided dispersal of earthworms has been reviewed by Hendrix *et al.* (2006).

While there is a natural tendency to juxtapose niche-based with neutral theories and to treat them as mutually exclusive, this is not necessarily the case. Both neutral (chance) and niche (deterministic) processes are responsible for shaping soil communities. Edaphic habitat specialization can still occur, as shown for bacteria and AM fungi with regard to pH (Fierer and Jackson, 2006; Helgason and Fitter, 2009).

The Concept of Functional Dissimilarity

In view of the diversity of studies that did and did not report relations between diversity and ecosystem functioning, ecologists have tried to understand under what specific conditions such positive relationships hold. Heemsbergen *et al.* (2004) introduced the concept of *functional dissimilarity* to explain why in certain combinations ecosystem properties did scale with diversity, whereas in others it did not. Inspection of these cases suggests that functional dissimilarity is especially large if species are phylogenetically divergent; and functional dissimilarity is small in cases of conservative functional traits and niche convergence. This relation has been shown for both plants and in the case of soil biota for AM fungi and soil animals. Maherali and Klironomos (2007) tested coexistence of different species of AM fungi. They produced different treatments of eight AM fungi, consisting of experimental units where all eight species belonged to the same family, units where the eight species belonged to two families or to three families. The realized species richness was in all cases lower than the number of species that were inoculated. The realized species richness was especially small (around three species) when all eight species belonged to the same family. Apparently, species co-existence of closely related species was unlikely. One major consequence of this study is that earlier studies that investigated the relation between species richness and ecosystem properties, but where the realized species richness was not investigated (as in the study by Van der Heijden *et al.*, 1998), may have yielded unreliable results.

Soil Biodiversity in Brown Worlds

One class of biodiversity–ecosystem functioning studies that used experimental approaches is based on effects of litter mixtures (diversity) on biodiversity of saprotroph species and on process rates of carbon decomposition and nitrogen mineralization. This topic was discussed by Wardle *et al.* (1999), who concluded that very few generalities had emerged and that effects of litter diversity on ecosystem processes were idiosyncratic. A reconsideration of the published literature suggests that this conclusion is still valid. Gartner and Cardon (2004) indicated the importance of litter interactions, as a majority of the published studies found a significant response where the behaviour of litter mixtures was different from the predicted values based on the behaviour of the litter decomposing singly. They also noted more responses where decomposition was enhanced (synergistic responses) than reduced (antagonistic responses), and the opposite for nitrogen mineralization. The average response of positive and negative responses was, however, similar, suggesting that despite such interactions a simple additive approach is often sufficient. Later analyses by Wardle *et al.* (2006) and Srivastava *et al.* (2009) supported the conclusion of Gartner and Cardon (2004) that plant litter diversity did not have a positive effect on decomposition. With increasingly diverse plant litters, effects converge towards a simple additive model and antagonistic and synergistic effects seem to cancel out. Unsurprisingly, the diversity of outcomes challenges the usefulness of traditional classifications of plant functional types but also their replacement by classifications based on chemical diversity or dissimilarity.

Of relevance for the biodiversity–ecosystem functioning debate is the observation by Tiunov and Scheu (2005), who showed that a rate-enhancing effect of biodiversity on decomposition was larger with a single, well-defined substratum than with a natural, multi-resource substratum, suggesting that niche differentiation according to different carbon sources contributed only a minor part of the effect.

Outlook – Biodiversity as Model and Metaphor

This paper has (deliberately) sketched a paradoxical situation. Despite a plethora of theory (and publications!) that support positive correlations (and are consistent with or even suggest causal relationships) between (species and functional) diversity and ecosystem functioning, there is equally widespread acceptance of redundancy hypotheses. The available evidence supports the conclusion that the general theory may only pertain in situations at the lower end of biodiversity (unrealistic in real-world soil ecosystems, including agroecosystems) and with specific combinations that exhibit functional dissimilarity (also unlikely in nature because functional trait conservatism is more important than trait dissimilarity for related species). Therefore *evidence* for a relation between soil biodiversity and sustainable agroecosystem functioning is at best anecdotal and scattered (Brussaard *et al.*, 2007). However, support for a link between soil biodiversity and ecosystem function has also received wide support in circles outside science because it makes intuitive sense that having more species is advantageous. This paradox (a credible theory that shows a poor match with empirical findings) has found clear expression in the review prepared for the EU by Turbé *et al.* (2010). In their report one reads the claim that '*soil biodiversity is the driving force behind regulation of ecosystem services*'. But they also noted that '*no consistent relation between soil species diversity and soil function*' has been found. Similarly, the review wavers somewhat uneasily between suggestions that '*policies aimed at above-ground biodiversity may not do much for the protection of soil biodiversity*' and that '*soil biodiversity plays an important role for the conservation of above-ground diversity*'.

Apparently the word biodiversity has multiple meanings as concept and as a metaphor. For many soil biodiversity is not a technical concept with a precise definition and operationalization, but rather a metaphor. As a metaphor, soil biodiversity simply represents soil life or living soil. The term soil biodiversity is used to raise awareness with farmers and policy makers, and the public at large, of the importance of soil biota. We do not disagree with the conclusion that soil biota are essential for the crucial ecosystem services provided by soil. We are convinced that an attempt to completely replace soil biota by external inputs cannot lead to sustainable soil management. However, we think that the case for a *causal link* between soil biodiversity and ecosystem functioning has been overstated. While agricultural soils on Anthropogenic Dark Earths (Terra Preta) harbour a much higher microbial diversity than their neighbouring oxisols (Naverrete *et al.*, 2010), while the System of Rice Intensification (SRI) leads to increases in soil microbial biomass (Zhao *et al.*, 2010), and while organic agriculture leads (or does not lead) to increases in AM fungal diversity (Oehl *et al.*, 2004; Galván *et al.*, 2009) we should not conclude that the higher microbial diversity or biomass drives or controls the increased productivity. Considering the present state of soil biodiversity experiments (where controlled soil biodiversity manipulation turns out to be quite complicated) we should be cautious in suggesting specific forms of soil agrobiodiversity management. But there are forms of judicious soil and crop management that take the importance of soil life and living soil into account.

The metaphorical use of the term soil biodiversity puts biodiversity science (and soil biodiversity scientists) in a difficult position. There is societal support for a theoretical link between soil biodiversity and ecosystem function because it seems credible, but the empirical base is weak to say the least. The use of metaphors is ultimately not without risk. Overstating a positive relationship between soil biodiversity and agroecosystem functioning could erode support for biodiversity conservation and soil biodiversity science in the longer term.

References

Allison, S.D. and Martiny, J.B.H. (2008) Resistance, resilience, and redundancy in microbial communities. *Proceedings of the National Academy of Sciences of the USA* 105, 11512–11519.

Balvanera, P., Pfisterer, A.B., Buchmann, N., He, J.-S., Nakashizuka, T., Raffaelli, D. and Schmid, B. (2006) Quantifying the evidence for biodiversity effects on ecosystem functioning and services. *Ecology Letters* 9, 1146–1156.

Bardgett, R.D. (2002) Causes and consequences of biological diversity in soil. *Zoology* 105, 367–374.

Barrios, E. (2007) Soil biota, ecosystem services and land productivity. *Ecological Economics* 64, 269–285.

Bell, T., Newman, J.A., Silverman, B.W., Turner, S.L. and Lilley, A.K. (2005) The contribution of species richness and composition to bacterial services. *Nature* 436, 1157–1160.

Blanchart, E., Albrecht, A., Brown, G., Decaens, T., Duboisset, A., Lavelle, P., Mariani, L. and Roose, E. (2004) Effects of tropical endogeic earthworms on soil erosion. *Agriculture, Ecosystems and Environment* 104, 303–315.

Bontemps, C., Elliott, G.N., Simon, M.F., Dos Reis Júnior, F.B., Gross, E., Lawton, R.C., Neto, N.E., de Fátima Loureiro, M., de Faria, S.M., Sprent, J.I., James, E.K. and Young, J.P.W. (2010) *Burkholderia* species are ancient symbionts of legumes. *Molecular Ecology* 19, 44–52.

Brussaard, L., de Ruier, P.C. and Brown, G.G. (2007) Soil biodiversity for agricultural sustainability. *Agriculture, Ecosystems and Environment* 121, 233–244.

Brussaard, L., Caron, P., Campbell, B., Lipper, L., Mainka, S., Rabbinge, R., Babin, D. and Pulleman, M. (2010) Reconciling biodiversity conservation and food security: scientific challenges for a new agriculture. *Current Opinion in Environmental Sustainability* 2, 34–42.

Buckley, D.H. and Schmidt, T.M. (2003) Diversity and dynamics of microbial communities in soils from agroecosystems. *Environmental Microbiology* 5, 441–452.

Coleman, D.C. (2008) From peds to paradoxes: linkages between soil biota and their influences on ecological processes. *Soil Biology and Biochemistry* 40, 271–289.

Costanza, R., d'Arge, R., Farber, R.S., Grasso, M., Hannon, B., Limburg, K., Naeem, S., O'Neill, R.V., *et al.* (1997) The value of the world's ecosystem services and natural capital. *Nature* 387, 253–260.

Crawford, J.W., Harris, J.A., Ritz, K. and Young, I.M. (2005) Towards an evolutionary ecology of life in soil. *Trends in Ecology and Evolution* 20, 81–87.

Díaz, S. and Cabido, M. (2001) Vive la différence, plant functional diversity matters to ecosystem processes. *Trends in Ecology and Evolution* 16, 646–655.

Díaz, S., Fargione, J., Chapin III, F.S. and Tilman, D. (2006) Biodiversity loss threatens human well-being. *PLoS Biology* 4(8), e277.

Dumbrell, A.J., Nelson, M., Helgason, T., Dytham, C. and Fitter, A.H. (2010) Relative roles of niche and neutral processes in structuring a soil microbial community. *ISME Journal* 4, 337–345.

Ekschmitt, K. and Griffiths, B.S. (1998) Soil biodiversity and its implications for ecosystem functioning in a heterogeneous and variable environment. *Applied Soil Ecology* 10, 201–215.

Eviner, V.T. and Chapin III, F.S. (2003) Functional matrix: a conceptual framework for predicting multiple plant effects on ecosystem processes. *Annual Review of Ecology, Evolution and Systematics* 34, 455–485.

Fierer, N. and Jackson, R.B. (2006) The diversity and biogeography of soil bacterial communities. *Proceedings of the National Academy of Sciences USA* 103, 626–631.

Fierer, N., Bradford, M.A. and Jackson, R.B. (2007) Toward an ecological classification of soil bacteria. *Ecology* 88, 1354–1364.

Fitter, A.H. (2005) Darkness visible: reflections on underground ecology. *Journal of Ecology* 93, 231–243.

Galván, G., Parádi, I., Burger, K., Baar, J., Kuyper, T.W., Scholten, O.E. and Kik, C. (2009) Molecular diversity of arbuscular mycorrhizal fungi in onion roots from organic and conventional farming systems in the Netherlands. *Mycorrhiza* 19, 317–328.

Gamfeldt, L., Hillebrand, H. and Jonsson, P.R. (2008) Multiple functions increase the importance of biodiversity for overall ecosystem functioning. *Ecology* 89, 1223–1231.
Gamper, H.A., van der Heijden, M.G.A. and Kowalchuk, G.A. (2010) Molecular trait indicators: moving beyond phylogeny in arbuscular mycorrhizal ecology. *New Phytologist* 185, 67–82.
Gans, J., Wolinsky, M. and Dunbar, J. (2005) Computational improvements reveal great bacterial diversity and high metal toxicity in soil. *Science* 309, 1387–1390.
Gardi, C. and Jeffery, S. (2009) Soil biodiversity. *JRC Scientific and Technical Reports*, Office for Official Publications of the European Communities, Luxembourg, 24 pp.
Gartner, T.B. and Cardon, Z.G. (2004) Decomposition dynamics in mixed-species litter. *Oikos* 104, 230–246.
Gessner, M.O., Swan, C.M., Dang, C.K., McKie, B.G., Bardgett, R.D., Wall, D.H. and Hättenschwiler, S. (2010) Diversity meets decomposition. *Trends in Ecology and Evolution* 25, 372–380.
Giller, K.E. and Cadisch, G. (1995) Future benefits from biological nitrogen fixation: an ecological approach to agriculture. *Plant and Soil* 174, 255–277.
Giller, K.E., Beare, M.H., Lavelle, P., Izac, A.-M.N. and Swift, M.J. (1997) Agricultural intensification, soil biodiversity and ecosystem function. *Applied Soil Ecology* 6, 3–16.
Giller, K.E., Witter, E. and McGrath, S.P. (1998) Toxicity of heavy metals to microorganisms and microbial processes in agricultural soils – a review. *Soil Biology and Biochemistry* 30, 1389–1414.
Giller, K.E., Bignell, D.E., Lavelle, P., Swift, M.J., Barrios, E., Moreira, F., van Noordwijk, M., Barois, I., Karanja, N. and Huising, J. (2005) Soil biodiversity in rapidly changing tropical landscapes: scaling down and scaling up. In: Usher, M.B., Bardgett, R. and Hopkins, D.W. (eds) *Biological Diversity and Function in Soils*. Cambridge University Press, Cambridge, pp. 295–318.
Giller, P.S. (1996) The diversity of soil communities, the 'poor man's tropical rainforest'. *Biodiversity and Conservation* 5, 135–168.
Hanson, C.A., Allison, S.D., Bradford, M.A., Wallenstein, M.D. and Treseder, K.K. (2008) Fungal taxa target different carbon sources in forest soil. *Ecosystems* 11, 1157–1167.
Hart, M.M. and Reader, R.J. (2002) Taxonomic basis for variation in the colonization strategy of arbuscular mycorrhizal fungi. *New Phytologist* 153, 335–344.
Hättenschwiler, S., Tiunov, A.V. and Scheu, S. (2005) Biodiversity and litter decomposition in terrestrial ecosystems. *Annual Review of Ecology, Evolution and Systematics* 36, 191–218.
Hector, A. and Bagchi, R. (2007) Biodiversity and ecosystem multifunctionality. *Nature* 448, 188–191.
Heemsbergen, D.A., Berg, M.P., Loreau, M., Van Hal, J.R., Faber, J.H. and Verhoef, H.A. (2004) Biodiversity effects on soil processes explained by interspecific functional dissimilarity. *Science* 306, 1019–1020.
Helgason, T. and Fitter, A.H. (2009) Natural selection and the evolutionary ecology of the arbuscular mycorrhizal fungi (Phylum Glomeromycota). *Journal of Experimental Botany* 60, 2465–2480.
Hendrix, P.F., Baker, G.H., Callaham Jr, M.A., Damoff, G.A., Fragoso, C., Gonzalez, G., James, S.W., Lachnicht, S.L., Winsome, T. *et al.* (2006) Invasion of exotic earthworms into ecosystems inhabited by native earthworms. *Biological Invasions* 8, 1287–1300.
Hillebrand, H. and Matthiessen, B. (2009) Biodiversity in a complex world: consolidation and progress in functional biodiversity research. *Ecology Letters* 12, 1405–1419.
Hillel, D. (1991) *Out of the Earth. Civilization and the Life of the Soil.* University of California Press, Berkeley, California.
Hooper, D.U., Chapin III, F.S., Ewel, J.J., Hector, A., Inchausti, P., Lavorel, S., Lawton, J.H., Lodge, D.M. *et al.* (2005) Effects of biodiversity on ecosystem functioning: a consensus of current knowledge. *Ecological Monographs* 75, 3–35.
Hubbell, S.P. (2005) Neutral theory in community ecology and the hypothesis of functional equivalence. *Functional Ecology* 19, 166–172.
Jaspers, E. and Overmann, J. (2004) Ecological significance of microdiversity: identical 16S rRNA sequences can be found in bacteria with highly divergent genomes and ecophysiologies. *Applied and Environmental Microbiology* 70, 4831–4839.
Jiang, L. (2007) Negative selection effects suppress relationships between bacterial diversity and ecosystem functioning. *Ecology* 88, 1075–1085.
Jiang, L., Pu, Z. and Nemergut, D.R. (2008) On the importance of the negative selection effect for the relationship between biodiversity and ecosystem functioning. *Oikos* 117, 488–493.
Joergensen, R.G. and Wichern, F. (2008) Quantitative assessment of the fungal contribution to microbial tissue in soil. *Soil Biology and Biochemistry* 40, 2977–2991.
Kibblewhite, M.G., Ritz, K. and Swift, M.J. (2008) Soil health in agricultural systems. *Philosophical Transactions of the Royal Society B* 363, 685–701.

Kielak, A., Pijl, A.S., van Veen, J.A. and Kowalchuk, G.A. (2008) Differences in vegetation composition and plant species identity lead to only minor changes in soil-borne microbial communities in a former arable field. *FEMS Microbiology Ecology* 63, 372–382.

Kulmatiski, A. and Beard, K.H. (2008) Decoupling plant-growth from land-use legacies in soil microbial communities. *Soil Biology and Biochemistry* 40, 1059–1068.

Lavelle, P., Lattaud, C.., Trigo, D. and Barois, I. (1995) Mutualism and biodiversity in soils. *Plant and Soil* 170, 23–33.

Lekberg, Y., Koide, R.T., Rohr, J.R., Aldrich-Wolfe, L. and Morton, J.B. (2007) Role of niche restrictions and dispersal in the composition of arbuscular mycorrhizal fungal communities. *Journal of Ecology* 95, 95–105.

Loreau, M. (2010) Linking biodiversity and ecosystems: towards a unifying ecological theory. *Proceedings of the Royal Society B* 365, 49–60.

Maherali, H. and Klironomos, J.N. (2007) Influence of phylogeny on fungal community assembly and ecosystem functioning. *Science* 316, 1746–1748.

Martiny, J.B.H., Bohannan, B.J.M., Brown, J.H., Colwell, R.K., Fuhrman, J.A., Green, J.L., Horner-Devine, M.C., Kane, M. *et al.* (2006) Microbial biogeography: putting microorganisms on the map. *Nature Reviews Microbiology* 4, 102–112.

McGuire, K.L. and Treseder, K.K. (2010) Microbial communities and their relevance for ecosystem models: decomposition as a case study. *Soil Biology and Biochemistry* 42, 529–535.

McIntyre, B.D., Herren, H.R., Wakhungu, J. and Watson, R.T. (eds) (2009) *Agriculture at a Crossroads: International Assessment of Agricultural Knowledge, Science and Technology for Development.* Island Press, Washington, DC.

Millennium Ecosystem Assessment (2005) *Ecosystems and Human Well-being: Biodiversity Synthesis.* World Resources Institute, Washington, DC.

Munkvold, L., Kjøller, R., Vestburg, M., Rosendahl, S. and Jakobsen, I. (2004) High functional diversity within species of arbuscular mycorrhizal fungi. *New Phytologist* 164, 357–364.

Nannipieri, P., Ascher, J., Ceccherini, M.T., Landi, L., Pietramellara, G. and Renella, G. (2003) Microbial diversity and soil functions. *European Journal of Soil Science* 54, 655–670.

Navarrete, A.A., Cannavan, F.S., Taketani, R.G. and Tsai, S.M. (2010) A molecular survey of the diversity of microbial communities in different Amazonian agricultural model systems. *Diversity* 2, 787–809.

Newsham, K.K., Fitter, A.H. and Watkinson, A.R. (1995) Multi-functionality and biodiversity in arbuscular mycorrhizas. *Trends in Ecology and Evolution* 10, 407–411.

Noguez, A.M., Arita, H.T., Escalante, A.E., Forney, L.J., García-Oliva, F. and Souza, V. (2005) Microbial macroecology: highly structured prokaryotic soil assemblages in a tropical deciduous forest. *Global Ecology and Biogeography* 14, 241–248.

Oehl, F., Sieverding, E., Mäder, P., Dubois, D., Ineichen, K., Boller, T. and Wiemken, A. (2004) Impact of long-term conventional and organic farming on the diversity of arbuscular mycorrhizal fungi. *Oecologia* 138, 574–583.

Orwin, K.H., Wardle, D.A. and Greenfield, L.G. (2006) Ecological consequence of carbon substrate identity and diversity in a laboratory study. *Ecology* 87, 580–593.

Perfecto, I. and Vandermeer, J. (2008) Biodiversity conservation in tropical agroecosystems – a new conservation paradigm. *Annals of the New York Academy of Sciences* 1134, 173–200.

Petchey, O.L. and Gaston, K.J. (2006) Functional diversity: back to basics and looking forward. *Ecology Letters* 9, 741–758.

Phelan, P.L. (2009) Ecology-based agriculture and the next green revolution – is modern agriculture exempt from the laws of ecology? In: Bohlen, P.J. and House, G. (eds) *Sustainable Agroecosystem Management – Integrating Ecology, Economics and Society.* CRC Press, Boca Raton, Florida, pp. 97–135.

Pimentel, D., Wilson, C., McCullum, C., Hung, R., Dwen, P., Flack, J., Tran, Q., Saltman, T. and Cliff, B. (1997) Economic and environmental benefits of biodiversity. *BioScience* 47, 747–757.

Porter, J., Costanza, R., Sdhu, H., Sigsgard, L. and Wratten, S. (2009) The value of producing food, energy, and ecosystem services within an agro-ecosystem. *Ambio* 38, 186–193.

Postma-Blaauw, M.B., de Goede, R.G.M., Bloem, J., Faber, J.H. and Brussaard, L. (2010) Soil biota community structure and abundance under agricultural intensification and extensification. *Ecology* 91, 460–473.

Raudsepp-Hearne, C., Peterson, G.D., Tengo, M., Bennett, E.M., Holland, T., Benessaiah, K., MacDonald, G.K. and Pfeifer, L. (2010) Untangling the environmentalists paradox: why is human well-being increasing as ecosystem services degrade? *BioScience* 60, 576–589.

Reed, H.E. and Martiny, J.B.H. (2007) Testing the functional significance of microbial composition in natural communities. *FEMS Microbiology Ecology* 62, 161–170.

Rivas, R., Garcia-Fraile, P. and Velázquez, E. (2009) Taxonomy of bacteria nodulating legumes. *Microbiology Insights* 2, 51–69.
Rosendahl, S., McGee, P. and Morton, J.B. (2009) Lack of global population genetic differentiation in the arbuscular mycorrhizal fungus *Glomus mosseae* suggests a recent range expansion which may have coincided with the spread of agriculture. *Molecular Ecology* 18, 4316–4329.
Sandhu, H.S., Wratten, S.D., Culen, R. and Case, B. (2008) The future of farming: the value of ecosystem services in conventional and organic arable land – an experimental approach. *Ecological Economics* 64, 838–848.
Setälä, H. and McLean, M.A. (2004) Decomposition rate of organic substrates in relation to the species diversity of soil saprophytic fungi. *Oecologia* 139, 98–107.
Souza, V., Eguiarte, L.E., Seifert, J. and Elser, J.J. (2008) Microbial endemism: does phosphorus limitation enhance speciation? *Nature Reviews Microbiology* 6, 559–564.
Srivastava, D.S., Cardinale, B.J., Downing, A.L., Duffy, J.E., Jousseau, C., Sankaran, M. and Wright, J.P. (2009) Diversity has stronger-top than bottom-up effects on decomposition. *Ecology* 90, 1073–1083.
Strickland, M.S., Lauber, C., Fierer, N. and Bradford, M.A. (2009) Testing the functional significance of microbial community composition. *Ecology* 90, 441–451.
Sutherland, W.J., Adams, W.M., Aronson, R.B., Aveling, R., Blackburg, T.M., Broad, S., Ceballos, G., Côté, I.M., *et al.* (2009) One hundred questions of importance to the conservation of global biological diversity. *Conservation Biology* 23, 557–567.
Swift, M.J., Izak, A.M. and Van Noordwijk, M. (2004) Biodiversity and ecosystem services in agricultural landscapes – are we asking the right questions? *Agriculture, Ecosystems and Environment* 104, 113–124.
Tilman, D., Knops, J., Wedin, D., Reich, P., Ritchie, M. and Siemann, E. (1997) The influence of functional diversity and composition on ecosystem processes. *Science* 277, 1300–1302.
Tiunov, A.V. and Scheu, S. (2005) Facilitative interactions rather than resource partitioning drive diversity-functioning relationships in laboratory fungal communities. *Ecology Letters* 8, 618–625.
Torsvik, V., Øvreås, L. and Thingstadt, T.F. (2002) Prokaryotic diversity – magnitude, dynamics, and controlling factors. *Science* 296, 1064–1066.
Turbé, A., De Toni, A., Benito, P., Lavelle, P., Ruiz, N., van der Putten, W.H., Labouze, E. *et al.* (2010) *Soil Biodiversity: Functions, Threats and Tools for Policy Makers.* Bio Intelligence Service, IRD, and NIOO, Report for European Commission (DG Environment), 249 pp.
Usher, M.B., Davis, P., Harris, J. and Longstaff, B. (1979) A profusion of species? Approaches towards understanding the dynamics of the populations of microarthropods in decomposer communities. In: Anderson, R.M., Turner, B.D. and Taylor, L.R. (eds) *Population Dynamics.* Blackwell Scientific, Oxford, pp. 359–384.
Van der Heijden, M.G.A. and Scheublin, T. (2007) Functional traits in mycorrhizal ecology: their use for predicting the impact of arbuscular mycorrhizal fungal communities on plant growth and ecosystem functioning. *New Phytologist* 174, 244–250.
Van der Heijden, M.G.A., Klironomos, J.N., Ursic, M., Moutoglis, P., Streitwolf-Engel, R., Boller, T., Wiemken, A. and Sanders, I.R. (1998) Mycorrhizal fungal diversity determines plant biodiversity, ecosystem variability and productivity. *Nature* 396, 69–72.
Van der Heijden, M.G.A., Bardgett, R.D. and van Straalen, N.M. (2008) The unseen majority: soil microbes as drivers of plant diversity and productivity in terrestrial ecosystems. *Ecology Letters* 11, 296–310.
Wardle, D.A. (2006) The influence of biotic interactions on soil biodiversity. *Ecolology Letters* 9, 870–886.
Wardle, D.A., Giller, K.E. and Barker, G.M. (1999) The regulation and functional significance of soil biodiversity in agroecosystems. In: Wood, D. and Lenné, J.M. (eds) *Agrobiodiversity: Characterization, Utilization and Management.* CAB International, Wallingford, UK, pp. 87–121.
Wardle, D.A., Yeates, G.W., Barker, G.M. and Bonner, K.I. (2006) The influence of litter diversity on decomposer abundance and diversity. *Soil Biology and Biochemistry* 38, 1052–1062.
Welbaum, G.E., Sturtz, A.V., Dong, Z. and Nowak, J. (2004) Managing soil microorganisms to improve productivity of agro-ecosystems. *Critical Reviews in Plant Sciences* 23, 175–193.
Wertz, S., Degrange, V., Prosser, J.I., Poly, F., Commeaux, C., Guillaumaud, N. and Le Roux, X. (2007) Decline of soil microbial diversity does not influence the resistance and resilience of key soil microbial functional groups following a model disturbance. *Environmental Microbiology* 9, 2211–2219.
Wood, D. and Lenné, J.M. (eds) (1999) *Agrobiodiversity: Characterization, Utilization and Management.* CAB International, Wallingford, UK.
Wood, D. and Lenné, J.M. (2005) 'Received wisdom' in agricultural land use policy: 10 years from Rio. *Land Use Policy* 22, 75–93.

Young, I.M., Crawford, J.W., Nunan, N., Otten, N. and Speirs, A. (2008) Microbial distribution in soils: physics and scaling. *Advances in Agronomy* 100, 81–121.
Young, J.P.W. and Haukka, K.E. (1996) Diversity and phylogeny of rhizobia. *New Phytologist* 133, 87–94.
Zhao, L., Wu, L., Li, Y., Animesh, S., Zhu, D. and Uphoff, N. (2010) Comparisons of yield, water use efficiency, and soil microbial biomass as affected by the system of rice intensification. *Communications in Soil Science and Plant Analysis* 41, 1–12.
Zhou, J.Z., Xia, B.C., Treves, D.S., Wu, L.Y., Marsh, T.L., O'Neill, R.V., Palumbo, A.V. and Tiedje, J.M. (2002) Spatial and resource factors influencing high microbial diversity in soil. *Applied and Environmental Microbiology* 68, 326–334.
Zingore, S., Manyame, C., Nyamugafata, P. and Giller, K.E. (2005) Long-term changes in organic matter of woodland soil cleared for arable cropping in Zimbabwe. *European Journal of Soil Science* 56, 727–736.

10
Agrobiodiversity Conservation Policy: a 'Tragedy of Errors'

D. Wood and J.M. Lenné

Few of the 850 varieties of pear, for example, that were listed by T.W. Field in 1858, could now, I suppose, be found anywhere in the world. It is the fate of varieties to come and go.

Fairchild (1938)

Introduction

For agrobiodiversity conservation, by far the greatest focus has been on the conservation technology and international policy of crop genetic resources. After thousands of years of informal moving of crops and domestic animals around for use in traditional farming – and *de facto* conservation through use on farm – around 250 years ago more formal systems of conservation began to be used, first in tropical botanic gardens for plantation crops, and then in developed countries to store safely and then to provide raw materials for plant breeding. In the past 50 years the 'Green Revolution' institutes of the Consultative Group for International Agricultural Research (CGIAR) began targeted collecting of their mandate crops to feed into active breeding programmes focused on developing countries. It is only fairly recently that developing countries – always a major source of genetic resources – have actually conserved them nationally. Introduction of genetic resources by developed countries, and then storage, has been the usual pattern. Chapter 4 (this volume) on crop introduction was a precursor to this chapter, as there is an indistinct line between introduction for direct use by farmers and introduction for the purpose of storage and further use in plant breeding.

Ex situ Conservation

Colonial botanic gardens, economic botany and the 'acclimatization' movement

Following prehistoric (or certainly not recorded) trans-Pacific and -Indian Ocean crop introduction of banana, sugarcane and sweet potato the 'Columbian Exchange' between the Americas and the rest of the world post-1492 opened a cornucopia of crops for wider use. Clonally-propagated crops, such as the three above, and, increasingly, seed crops, were widely distributed. Early movement of seed and cuttings was mainly unrecorded and introductions went to the end-user and were not specifically conserved. However, a combination of the increasing role of nation-states servicing colonial agriculture and the difficulties and cost of long-distance movement of living plants in sailing ships placed a premium on the long-term maintenance of introduced samples, often in a wide range of botanic gardens, some of which persist today, still with living specimens of the early introductions of economic plants. These are the earliest records of what is known as '*ex situ*' conservation, i.e. away from the location of the original collections, in contrast to *in situ* conservation, addressed later in this chapter.

These botanic gardens were usually and specifically to service the needs of colonies for production from introduced crops and domestic animals. Around the 1850s a belief flourished that introduced plants could be 'acclimatized' (or 'seasoned') by a period of growth in a local botanic garden before commercial cultivation. While the genetic basis of this is questionable, its belief underpinned crop introduction and, importantly, conservation, for half a century or more. For example, in Soviet Russia, Lysenko continued to promote the inheritance of environmentally acquired characters until the 1950s, much to the detriment of Soviet agricultural science and the great plant collector Vavilov, whose ideas were discredited and who died in prison in 1943. It is interesting that the now all-important conservation role associated with plant introduction is the by-product of discredited science: plants could not be 'acclimatized' except that they were maintained, that is, conserved, over the period of time needed for 'acclimatization'. The 'acclimatization' movement was worldwide: for example, of Australia, which had no native crop plants (*Macadamia* came later), Weigl (2003) writes:

> The whole of colonization is a vast act of acclimatization.... Australian science of the 1860s and beyond was closely associated with the acclimatization movement [as were] all the directors of the Botanical Gardens in Australia. In addition 'Acclimatization Societies' were founded in all colonies, supported by unusually high government subsidies.

Whatever the merits of 'acclimatization', botanic gardens, and the associated conservation of perennial economic crops, flourished. The Indian Botanic garden in Calcutta was established as a garden of 'acclimatization' in 1787. Mahogany was introduced from the West Indies, and 17,000 tea plants brought back from China establishing the Indian tea industry. The British had gardens in Mauritius (1735), the West Indies (St Vincent, 1764; Jamaica, 1774 and Trinidad, 1819), Ceylon (1812), Singapore (1822), Sydney (1816), Melbourne (1854) and Cape Town (1848). The Portuguese established a botanic garden at Rio de Janeiro in 1808, with some of the early samples stolen from a French colonial acclimatization garden in the West Indies (Brockway, 1979; Alexander and Alexander, 2008). The Dutch had a major botanic garden in Bogor – now in Indonesia – which still maintains the original oil palm introduction that established a major industry in South-east Asia.

Ex situ perennial crop research collections

Over the past century the role of these botanic gardens changed. Crop introduction for direct use became less important as most suitable crops were, over time, introduced and moved into wider cultivation (or failed). The gardens themselves took on the dual role of public parks and plant identification by their associated herbaria, as with Singapore, Peradeniya and Bogor, with a notable reduction in their agricultural work. In contrast with introduction for direct use (as with the great introduction programme of the late 1800s by the USDA for seed plants) the newer trend through the 1900s was to introduce more varieties of proven crops for crop improvement through plant breeding at a network of research stations. At first, varieties were introduced, used in breeding, and the introductions discarded or lost through poor maintenance. However, as a result of several trends, this wasteful practice began to be replaced by a formal system of conservation. These trends included: the use of simple cold-storage technology, which dramatically extends the life of stored seed at an economic cost; the rise of a large number of nation states no longer under colonial control; increasing knowledge that the undoubted success of plant breeding actually replaced the older varieties used in breeding newer varieties ('genetic erosion'); and the increasing value of older varieties and wild relatives in breeding for disease and pest resistance, and for yield characters in new varieties.

Conservation changed from a major emphasis on collections grown in botanic gardens to seed collections maintained at research stations for the primary use of plant breeders. However, as with the early colonial voyages of plant exploration, there were many especially tropical species for which seed could not be dried and stored in cold stores. Many temperate fruit, nut and tuber crops were clonally propagated, either not producing seed at all, or with seed that was genetically different from the parent (and therefore usually less adapted). These were maintained as growing plants in *ex situ* orchard collections very similar in style (but not purpose) to the 'acclimatization' gardens, and usually associated with research stations where samples could be evaluated either for direct use or through plant breeding. For example, the orchard collections maintained at the research institute CATIE in Costa Rica contained global collections of cocoa (*Theobroma*) and its wild relatives; coffee (*Coffea*), with several crop species and wild relatives; *Bactris gasipaes* (peach palm, pejibaye); and multiple collections of important Central American tree crops, including *Bixa orellana* (achiote), *Byrsonima crassifolia* (nance), *Pouteria* spp. (sapotes), *Annona* spp. (chirimoyas, guanábanas) and many more. These field collections of major plantation crops are used to rapidly access samples for evaluation, breeding and direct use by farmers. For example, the CATIE collection holds the local species *Elaeis oleifera* (American oil palm), used for breeding with the main commercial oil palm, *Elaeis guineensis*, from West Africa, now a major oil crop of South-east Asia. In contrast to CATIE, a similar research garden at Lancetilla in Honduras, while meeting the requirements for conservation, remains more of a 'crop introduction' garden: for example, it has a 3.4 ha collection of introduced *Garcinia mangostana* (mangosteen), the largest collection in the western hemisphere (Dickson, 1977).

Seed and tissue-culture *ex situ* collections

The largest existing *ex situ* collections of genetic resources are those of annual seed crops. Such seed is easy to store under cold, dry, conditions; easy to use for field evaluation; easy to multiply to provide duplicate samples; and easy to move between countries for evaluation, trials and duplicate storage (but still subject to quarantine). However, the present system of conservation took time to evolve. Early collections in the now-model USDA system, which was formalized in 1898, went directly to researchers, breeders, or farmers with no requirement for long-term maintenance. Until 1948 samples were not maintained by the USDA – no facilities existed at that time. By 1948 only 33% of accessions received were placed in the GRIN database. Unfortunately: 'most germplasm accessions obtained before 1948 are no longer available' (Committee on Managing Global Genetic Resources, 1991). Given the vast range of genetic resources needed for US agriculture, the Plant Introduction Office both ensured adequate storage through the advanced technology of the National Seed Storage laboratory and also distributed samples for local storage. Importantly, it also promoted evaluation and assessment for breeding in a wide range of regional plant introduction stations and State Agricultural Experiment Stations covering the wide range of crops from subtropical crops in Hawaii, Florida and Puerto Rica to clonal crops in Oregon, woody ornamentals in the National Arboretum in Washington, DC, to more temperate crops to the four regional plant introduction stations in Washington State, Georgia, New York State and Iowa. Each of these research stations had a responsibility to store samples received from the Plant Introduction Office (Committee on Managing Global Genetic Resources, 1991). While most of the samples are stored as seed in cold storage, all storage technology is used, from tissue culture, through orchard and arboretum collections, to storage in liquid nitrogen at −196°C. Tissue culture is used for perennial plants that either do not produce or do not breed true from seed, for example, cassava. This is a form of micro-vegetative propagation in test tubes stored under controlled cool conditions.

Breeding and genetic erosion

As plant breeding advanced in developed countries it was realized that there was a need for more, and better, crop genetic resources. For example, potato breeders in Britain, faced with diseases such as late blight and a cytologically complex origin of cultivated potato, needed access to genes from wild species from Andean countries. Over a period of years expeditions were mounted and, rather than being discarded after use, as before, the collected samples carefully preserved as true seed in cold stores for immediate and future use (Bradshaw and Ramsey, 2005). The argument then, as now, was that we do not know what future constraints on crop production will be and it is easier to store collections than to re-collect. This pattern of collecting in developing countries, storing in developed countries and feeding samples into breeding programmes was repeated numerous times. Examples include Japanese collections of wild wheat from Persia, US collections of wild tomatoes in Chile, Australian collecting of pasture genetic resources in many different countries and Canadian collections of barley.

As these stored collections built up to service the needs of breeders in developed countries, there was little emphasis on effective storage in many of the smaller developing countries (although Brazil, India and China were early adopters of storage technology and associated data management). However, there were excellent, large, well-managed collections directly servicing global plant breeding in the eight international agricultural research centres of the CGIAR located in developing countries. These were mainly crop-specific collections, built up over up to 50 years. For example, there were maize and wheat collections stored at CIMMYT in Mexico, sorghum and chickpea at ICRISAT in India, rice at IRRI in the Philippines, faba bean, lentils, wheat and barley at ICARDA in Syria, potato and sweet potato in CIP in Peru, cassava, banana and soybean at IITA in Nigeria, and beans and tropical pastures at CIAT in Colombia. These were the over 600,000 samples that underpinned the 'Green Revolution', sourced worldwide, well documented for features of value to breeders, widely duplicated for safety and freely available worldwide. There are more details of these international activities in Engels and Wood (1999).

However, there were clouds on the horizon for this valuable *ex situ* system of storage, use of samples in crop improvement and free availability of samples. First, use of seed samples in advanced breeding produced new varieties that at least partly replaced older varieties in the field. This loss of older varieties is called 'genetic erosion' (the degree and importance of this has at times both been exaggerated and also disputed). But the possibility of genetic erosion underpinned the drive for urgent 'rescue' collecting, nationally and internationally. The international CGIAR institutes were very active in seed collecting over decades, partly driven by the need for conservation for future use and partly to directly service their own plant breeding work. For most years between 1972 and 1998 accessions to CGIAR genebanks exceeded 10,000 seed samples: the number peaked in 1977 at 32,000 samples taken into CGIAR stores in that year.

Second, most of these samples of crop seed were sourced from developing countries, which still had traditional agriculture based on multitudes of traditional varieties ('landraces'). As most of the funding for collection came from developed countries (often routed through CGIAR institutes) most of the samples ended up in the genebanks of developed countries with, again, the motive of long-term conservation and also direct use in breeding programmes. As it is technically easy to grow out and multiply samples of most important crops, a common result was, for each collection, a sample stored in the country of collection, a duplicate sample in the CGIAR genebank that specialized in the crop, and a further duplicate sample in a designated international 'base' collection, usually in a developed country. Each of these three genebanks could further duplicate elsewhere for security and use in plant breeding. Importantly, until 1993, when the Convention on Biological Diversity became operational, this pattern of duplication had resulted in hundreds of thousands of *ex situ* samples stored outside of the country of origin and therefore not covered by the CBD. Brush (2004) calculates that: 'of the 6,159,248 accessions inventoried among all gene banks in 1996, slightly more than half (3,447,469) were held by gene banks in Europe, North America, Japan and international agricultural research centers of [CGIAR]': significantly, a 'large portion of the total diversity of the world's major crops has been captured and stored in gene banks of major industrial countries and agricultural research centers.'

Biopiracy

Subsequent events – still ongoing – depend on the perception by many countries that their genetic resources were being exploited by others (as indeed they were but the benefits were reciprocal). The two reasonable technical developments – the high volume of rescue collecting of crop genetic resources and the associated sample storage *ex situ* in developed countries – provided the opportunity for a highly misguided campaign by the Canadian NGO, RAFI (now ETC Group). This campaign, termed 'seed wars' by Paul (1984), was originally targeted at developing countries with the slogan: 'No patents on life'. The argument was that multinational corporations were patenting samples derived from developing countries and selling the patented products back to developing countries that had provided the original samples, a target made more believable by associating all plant breeding with restrictive (and exploitative) intellectual property protection. In fact the 'Green Revolution' was nothing whatever to do with multinational corporations or monopolistic control of seed – quite the opposite – and RAFI was told this repeatedly. An associated idea was that genetic diversity in itself was more important for food security than plant breeding (criticized by Duvick, 1986). The NGO belief seemed to be that traditional crops and their diversity somehow got better (see below for our refutation of this belief). But ETC Group (no date): 'believes that intellectual property is predatory on the rights and knowledge of farming communities and indigenous peoples,' ignoring both the public nature of the highly important 'Green Revolution' and also the great benefits which farmers in developing countries had gained from the cultivation of introduced crops (Chapter 4, this volume), that is, farmers benefiting from the 'rights and knowledge' of farmers elsewhere. In a retrospect, Charles (2001) writes: 'In the 1990s, the system of free exchange began to unravel. Ironically, the roots of its downfall lie in a campaign begun by people who wanted to preserve it.' Charles (2001) then describes what went wrong: activists, specifically RAFI, fighting against the exploitation of: 'cashpoor but gene-rich developing nations by gene-hungry multinational corporations.' Unfortunately for global food security, depending as it does on the free movement of crop genetic resources, this activist campaign, searching for a slogan, came up with the word 'biopiracy'. Things became worse for food security: the Indian activist Vandana Shiva joined the debate (Shiva, 1996).

This high-profile and continual focus on the slogan of biopiracy and the apparent exploitation of plant varieties from developing countries by multinational countries produced an atmosphere of distrust in those countries that had hitherto freely provided samples. Developing countries were led to believe they were sitting on a genetic goldmine and in Charles's (2001) words: 'many decided to claim those treasures for themselves.' No mention was made by NGOs (even if they understood the facts) of the massive interdependence of developing countries on past crop introduction for their present food security. The result of this campaign of sowing the seeds of distrust was inevitable: the former free movement of crop genetic resources was compromised and began to slow to a trickle (with the exception of the CGIAR institutes, who, although closely targeted by seed activists, carried on their essential work scarcely hampered by NGO activism). In a wide-ranging review of just what went wrong, Aoki (2007) writes that a 'lack of consensus relates to perceptions of the increased economic value of PGRs [plant genetic resources] and fears of the theft of such resources.' In retrospect, we believe that this campaign, which spread like a virus through the international NGO community, is a major long-term danger to global food security.

Convention on Biological Diversity (CBD)

At this stage – leading into the early 1990s – United Nations agencies became involved to try to resolve the 'lack of consensus' on the global movement of genetic resources (looking forward, this led to two international agreements, neither of which has solved the problem caused by the 'seed wars' and 'biopiracy' campaigns – see below). Towards the end of 1988 the United Nations Environmental Programme (UNEP) began work on what became the Convention on Biological Diversity (CBD), which entered into force at the end of December 1993. This covered all biological diversity, including crops. However, the 'fears of theft of genetic resources' which had resulted from the NGO 'biopiracy' campaign prompted the CBD to reinforce a regime of 'national sovereignty' over biological resources. Months later, RAFI (1994) claimed the CBD was condoning 'biopiracy'. In the nature of international meetings, CBD talks on 'access and benefit-sharing' over genetic resources are still ongoing and unresolved. However, as late as 2004, ETC Group (formerly RAFI) was compounding the damage of the biopiracy campaign by criticizing the CBD for its attempt to formulate 'Access and Benefit-Sharing' guidelines, needed to counter the damage done by NGOs to the international movement of genetic resources (ETC, 2004).

More importantly for agricultural biodiversity, the very high numbers of existing *ex situ* collections (Brush, 2004, above, identifies almost 3.5 million samples of these) could not be covered retroactively by the CBD. At a late stage in negotiations for the CBD one of the authors (D.W.) began promoting the word 'agrobiodiversity' (Wood, 1992). This was intended to emphasize that agricultural biodiversity was a legitimate – indeed, very important – concern for the CBD. This was of little avail: in what now seems to have been a major error by the CBD process, when the CBD was being adopted in May 1992 the issue of 'Access to ex-situ collections not acquired in accordance with this Convention' was designated by the CBD to the UN Food and Agriculture Organization (FAO) (CBD, 1992).

The International Treaty on Plant Genetic Resources (ITPGR)

Before the CBD, FAO had already moved into the international management and policy control of plant genetic resources – interestingly, promoted by the 'seed wars' NGOs, who perhaps (mistakenly as it turned out) thought that internationalizing genetic resources would prevent their use by the dreaded multinational seed companies. FAO initiatives were the International Undertaking on Plant Genetic Resources (IUPGR) and the FAO Commission on Plant Genetic Resources (CPGR), both from 1983. Significantly, as noted by Aoki (2007): 'The IUPGR and the CPGR were spearheaded by a group of developing countries and were supported by an array of NGOs allied with the International Coalition for Development Action.' ICDA was a major 'seed wars' NGO, and its staff later became part of RAFI, now ETC Group.

The declaration of national sovereignty over genetic resources by the CBD (which, with the required number of ratifications, came into force in December 1993) rang alarm bells within FAO and many OECD countries, used as they were to centuries of free access to genetic resources. Policy control over the highly important genetic resources collections in the legally independent CGIAR Centres was quickly transferred to the World Bank in 1994 in exchange for US$24 million funding to the Centres. Almost immediately, also in 1994, FAO concluded an Agreement with all the individual CGIAR Centres that were holding very large *ex situ* collections, mainly from developing countries. Most of these samples would be designated by FAO as germplasm 'in trust for the benefit of the international community' – a clear side-stepping of national sovereignty. One of us (D.W.) was in Rome, working on a review of CGIAR genetic resources practice and policy, and was presented with a late draft of the FAO-CGIAR Agreement. It was seriously inadequate in what genetic resource managers considered a point of honour and standard practice: the unconditional return of duplicate samples to the country or institution that had provided samples to the CGIAR genebanks. This error was corrected then but later reversed. But the FAO-CGIAR Agreement, with its mention of 'in trust', provided a platform for the FAO's next move.

Instead of working within the UNEP-sponsored CBD process – a full International Convention covering all biodiversity – FAO decided to press ahead with its own Treaty. After a long process of negotiation the International Treaty on Plant Genetic Resources for Food and Agriculture (ITPGR) was adopted in November 2001 and, after the required ratifications, entered into force in June 2004. Prior to the adoption of the ITPGR in the decade following the CBD, lobbying against the CBD continued (RAFI, 1994) in an apparent attempt to undermine the CBD and to promote the ITPGR. Also, the CBD was damagingly linked to the fall in annual accessions to CGIAR genebanks (Falcon and Fowler, 2002, p. 210). Yet CGIAR records show this claim to be wrong: the CBD had no discernible impact on accessions. In a continued attempt to promote the adoption of the ITPGR, ETC Group – formerly RAFI, one of the NGO 'godfathers' of the ITPGR, and the originator of the 'biopiracy claim' – urged countries to: 'Ratify, ratify, ratify – and don't mess with the deal.' (ETC Group, 2001). Was this yet more bad advice from northern NGOs to developing countries?

ITPGR expansionism

The ITPGR approach to subsuming national sovereignty into an international system was, on the face of it, reasonable. For a list of basic food crops (Annex I crops) countries would voluntarily place national collections in a 'multilateral system' (the ITPGR does not recognize the concept of 'country of origin'). In addition, unlike the CBD, there would be retroactivity for deposited samples in the multilateral system. If all countries had accepted the ITPGR, and also if countries had placed Annex I crops only from their own countries in the ITPGR there was a chance of equity and even goodwill. But these two provisos were not met: very quickly the operation of the ITPGR began to seriously erode national sovereignty over genetic resources far beyond the Annex I list which countries had agreed to voluntarily place in the ITPGR.

A series of problems now threatens the future of the ITPGR, its relations with the CBD and the hitherto free movement of genetic resources:

- Not all countries accepted the ITPGR. By October 2010 the ITPGR only had 126 parties, 67 fewer than the CBD, which had 193 parties. The 67 countries staying out of the ITPGR included China, Mexico and Nigeria, rich sources of crop genetic resources.
- It was assumed that any country that accepted the ITPGR would automatically place its national collections voluntarily under the ITPGR: this was not to be. A request by FAO to countries to formally place samples in the ITPGR produced an alarming result. With three exceptions, developing countries are not assigning samples to the ITPGR. Only three developing countries (Brazil, Namibia and Zambia) placed part of their collections in the ITPGR, a total of 8449 samples. In contrast, developed countries (France, Germany, the Netherlands, Nordics, Portugal and Switzerland) have placed a total of 221,826 samples in the ITPGR. International Institutes of the CGIAR, under a formal agreement with FAO dating to 2006, placed in excess of 600,000 samples in the ITPGR.

At this stage it was obvious that many developing countries were staying away from the ITPGR and most were not assigning collections. Things were going to become far worse for the 'countries of origin' of plant genetic resources: many developed countries began depositing in the ITPGR samples originating in developing countries.

- Any country or CGIAR institute could place in the ITPGR samples originating from any other country, *without* the knowledge or permission of the country of origin. Very many – we suggest as many as 150,000 of the CGIAR samples – would have come from countries that had not accepted the ITPGR.

- Further, these samples could include crops beyond the Annex I list agreed by the ITPGR negotiators (the only ITPGR agreement was for CGIAR institutes to include a wider range of crops).
- The next major expansion of the ITPGR (still not authorized by parties) was associated with the Svalbard seed deposit under an agreement of the Government of Norway. The Svalbard international seed store was opened in February 2008 on the Norwegian arctic island of Svalbard, sponsored by the Rome-based Global Crop Diversity Trust (located in offices in FAO). The stated intention of Svalbard was to secure seed storage, especially of threatened genetic resources, for future use globally. However, the Svalbard vault did not give any priority to threatened collections: instead it amassed duplicate samples from the well-managed CGIAR seeds stores, the genebanks of developed countries, and the very largest and most secure national store (USDA) in the USA (NordGen, 2010). Svalbard has now become a major mechanism for the vast expansion in coverage of the ITPGR. The culprit was the Svalbard Depositor Agreement, a document of the Royal Norwegian Ministry of Agriculture and Food (2008), which ties originals of deposited samples to the ITPGR. Article 7 of this Depositor Agreement is all-embracing and includes Annex I, non-Annex I crops and, quite remarkably, 'plant genetic resources [that] are not plant genetic resources for food and agriculture', i.e. all other plants. The impact of Article 7 is very large. For example, the USDA, with the largest national genetic resource collection in the world, is depositing in Svalbard, as are many other developed countries. Although the USA has not yet ratified the ITPGR, every single species of the USDA deposit, even medicinal plants (which the ITPGR is not designed to cover) will now be covered by the ITPGR. As any country can deposit in Svalbard samples from any other country, already there are 95,722 samples totaling 25,335,121 seeds of Mexican origin deposited from various sources. Yet Mexico itself has not ratified the ITPGR and has deposited nothing (NordGen, 2010). In contrast, two major countries with vast national seed collections that *have* ratified the ITPGR – Brazil and India – have, as yet, deposited nothing in Svalbard. Yet other countries and collections have deposited 7345 samples of Brazilian origin and 38,967 samples of Indian origin. Ironically, although the ITPGR ignores 'country of origin', the meticulous documentation of Svalbard samples by the Nordic Genebank clearly indicates 'country of origin'.
- There is a further planned expansion of the ITPGR through the work of the Global Crop Diversity Trust. The Trust plans networks of crop evaluation. To take part in this otherwise reasonable programme, countries who are not parties to the ITPGR will have to sign a: 'Solemn Undertaking for ensuring access as interim to ratifying the International Treaty for PGRFA' (GCDT, 2006): that is, they have to place their high-value varieties in the ITPGR.

Finally, there are grey areas between the operation of the CBD, under UNEP, and the ITPGR under FAO. Andersen (2008) calls these clashes of different rules 'regime constellations' and notes that: 'there was confusion with regard to the division of labor between the CBD and the FAO for PGRFA management.' She notes that for more than a decade valuable time was lost: 'we can assume that from 1992 to 2004 regime constellations contributed to postponing crucial efforts necessary for the conservation, sustainable use and utilization of PGRFA'. An obvious grey area can be seen from the figures we present above for Mexican 'orphan' samples (*ex situ* collections removed from the country of origin prior to the CBD). Thus the 95,722 Mexican samples in Svalbard fall under two contrasting regimes: the CBD for the original samples in Mexican national collections which are firmly under Mexican sovereignty; and the ITPGR for duplicates of the same samples from institutes depositing in Svalbard, samples now part of the ITPGR 'multilateral system' of which Mexico is not a part. Remarkably, if Mexico has lost its own original samples of many major crops and wishes to replace them – a not uncommon need – then Mexico will have to sign a SMTA renouncing any rights to samples of Mexican origin it receives from the ITPGR regime. These important rights of repatriation were protected by the FAO-CGIAR Agreements at the insistence of one of us (D.W.), but the ITPGR now ignores them.

Threats to food security

There is now a situation of extreme uncertainty about the limits of 'regime constellations' governing plant genetic resources. The CBD process, after more than 17 years, has still not decided its own requirements for 'access and benefit-sharing'. The ITPGR is not yet accepted by some 67 countries, yet by various mechanisms includes samples from these countries. As a result, the ITPGR is rapidly expanding its coverage well beyond the original intentions of negotiating countries and threatening the CBD access and benefit-sharing proposals. We can predict with some certainty that this 'regime constellation' will cause the movement between nations of the genetic resources of crops vital for world food security to dry up. There is evidence of this from what the ITPGR Secretariat is *not* reporting. But, as noted by Aoki (2007): 'Ironically, the PGR slowdown hurts the poorest countries the most, as the Least Developed Countries are net importers of seed germplasm.' With up to 70% of crops grown by peasant farmers in South America and Africa being introduced, indigenous communities and farmers will, over time, suffer from this slowdown in exchange of crop genetic resources.

Despite the failings of the ITPGR, the food security of developing countries is still protected by the international agricultural research centres of the CGIAR, which continue their valuable work of conserving, evaluating, documenting and breeding important crops and pasture species. But given the confusion and seeds of mistrust engendered by first the NGO 'biopiracy' campaign and then by a Treaty (ITPGR) that conflicts with an existing Convention (CBD), future CGIAR access to samples needed to maintain and enhance agricultural production may be jeopardized.

The ongoing valuable work of the CGIAR is in major contrast to the perverse 'own goal' results of NGO activism in promoting the ITPGR. Given the intense involvement of 'No patents on life' NGOs in the genesis of the ITPGR, it is ironic that the ITPGR will be funded from taxing patents on derivatives of samples covered by the ITPGR, with 'patents on life' becoming the life-blood of the ITPGR. That this is a perverse result of their pro-ITPGR activism ('Ratify, ratify, ratify') seems to be lost on ETC/RAFI, who, in their critique of the CBD write: 'The practice of biopiracy will not ebb as long as genetic resources are a feedstock for industry profits, nor while those resources can be legally monopolized' (ETC, 2004) – this was written the same year as the coming into force of the ITPGR, which now depends both on biopiracy and industry profits for its future success.

There could also be a political falling-out from the ramifications of the ITPGR. Norway, in control of the Svalbard seed vault and advised by the Global Crop Diversity Trust (headed by a former co-founder of RAFI, see Charles, 2001), has linked seed deposit with acceptance of the terms of the ITPGR (rather than the needs of developing countries for secure duplicate storage). As we reported above, 95,722 samples from one single non-ratifier of the ITPGR (Mexico) are now under the control of the ITPGR (and many more samples from other non-ratifiers). We suggest that linking by Norway of Svalbard deposit to the ITPGR is the biggest single act of biopiracy ever, seemingly targeted at the 'countries of origin' of samples – mainly developing countries. Moreover, it has gone totally unremarked by otherwise vociferous 'biopiracy' activists. Whatever quibbles there are over a definition of 'biopiracy', surely taking almost 100,000 samples from the peasant farmers of a single country, making these samples available for patenting, taxing the patents to fund the UN, making the original farmers sign away their rights for access to these samples and returning no benefits to the original farmers must take some beating.

In situ Conservation

One response to the NGO claim that genetic resources were being taken from the country of origin for *ex situ* storage was a new emphasis on *in situ* conservation, where samples remained in developing countries with their farmers or in the wild. However, there were several other reasons for *in situ* conservation. In part it was an attempt by NGOs to obtain funding – for example, the Community Biodiversity Development and Conservation Programme (CBDC) of which RAFI was a partner (this seems to have had little success if judged by formal publications). There was a belief that varieties continued to evolve on-farm to respond to changing environmental forces – notably changing climate, but also pest and disease resistance (we discuss the merits of this belief below). In part it was a political approach – to retain national sovereignty over samples on national territory (a direct response to NGO scaremongering): samples in farmers' fields and seed stores are excluded from the ITPGR.

It is worth emphasizing very strongly indeed that by far the greatest application of *in situ* conservation in developing countries, where the greatest varietal variation is still found, is in farmers' seed stores. Year by year farmers use their own seed to grow their own crops. If farmers are happy with the performance of their suite of varieties and crops, they will continue to grow them: outside support will not be needed. If farmers are not happy with their varieties and crops, they will seek to change them. This fact places a great burden on outsiders who wish to preserve varieties on farm to justify their actions against the normal actions of farmers to change varieties frequently, as and when necessary. A distinction must be made – but often is not made – between the acceptable study of how and why farmers maintain or reject varieties and unacceptable project interventions to induce farmers to continue maintaining obsolescent varieties against their better instincts.

For more than 20 years and, especially since the coming into force of the Convention on Biological Diversity (CBD) (UNEP, 1992), there has been increasing interest in and the generation of considerable literature on *in situ* conservation of plant agrobiodiversity: wild relatives of crops and landraces (Altieri and Merrick, 1987; Hoyt, 1988; Brush, 1995, 2000; Frankel *et al.*, 1995; Maxted *et al.*, 1997, 2002, 2008; Engels and Wood, 1999; Heywood and Dulloo, 2006; Heywood *et al.*, 2007; Maxted and Kell, 2009; Newton *et al.*, 2010). Definitions, contexts, needs, site selection, target species, strategies, priorities, methodologies, management, information systems, policies etc. have been the subject of detailed and ongoing discussion and frequent review as this mountain of literature shows. A considerable amount of funding, mostly through stand-alone projects, has been directed to *in situ* conservation including two Global Environmental Facility (GEF) projects. The international centre Bioversity (formerly IPGRI) has focused much of its research agenda on *in situ* conservation for the past 15 years. In spite of all of this effort, there does not appear to have been: (i) any impact assessment of the contribution of *in situ* conservation to the overall effort on agrobiodiversity conservation for food security; (ii) any analysis of the expected synergy between *in situ* and *ex situ* conservation; and (iii) any attempt to estimate the value or costs of *in situ* versus *ex situ* conservation.

Just as for *ex situ* conservation, the main *functional* purpose of *in situ* conservation of useful food plants – whether wild relatives of crops or landraces on-farm – is for future use in crop breeding programmes to produce more food. The major justification given for conserving useful plants *in situ* rather than *ex situ* in genebanks is that plant populations are maintained in a 'dynamic' state, responsive to environmental change and subject to natural and managed selection in time and space (Maxted *et al.*, 1997). This is especially recommended in the case of diseases against which the host and pathogen have coevolved (Leppik, 1970; Frank, 1993; Frankel *et al.*, 1995), and are expected to continue to interact, hopefully resulting in useful and novel resistances for use in crop breeding (Maxted *et al.*, 1997; Newton *et al.*, 2010). In this next section, we consider the current status of *in situ* conservation of wild crop relatives and landraces on-farm, with some emphasis on diseases.

In situ conservation of wild relatives of crops

It has long been recognized that wild relatives of crop plants are rich sources of valuable traits for crops (Leppik, 1970; Harlan, 1977; Lenné and Wood, 1991; Frankel *et al.*, 1995). The literature is rich with examples of staple food crop breeding programmes benefiting from useful genes from wild relatives. For example, most of the successful and durable resistances to wheat stem rust (*Puccinia graminis* f.sp. *tritici*), the *Sr* genes, have come from wild relatives, especially *Aegilops* spp. (Roelfs, 1988; Ellis *et al.*, 2007). Novel resistance genes to the serious potato disease late blight (*Phytophthora infestans*) are being successfully transferred from wild *Solanum* spp. to cultivated potato (Park *et al.*, 2008). Wild emmer wheat (*Triticum turgidum* ssp. *dicoccoides*) is a rich source of drought resistance genes (Peleg *et al.*, 2007) while wild rice (*Oryza rufipogon*) is providing salinity tolerance to rice (Gregorio *et al.*, 2002). It is likely that wild relatives of crops will continue to be valuable sources of genes for future crop improvement for the foreseeable future, especially as advances in biotechnologies have greatly facilitated their use. Hence, conservation of crop wild relatives will continue to be an important strategy to support crop improvement for food security.

However, in every case where wild relatives have been used in crop breeding to date, the germplasm has been sourced from *ex situ* collections and *not* from *in situ* populations. Clearly, it is more convenient, efficient and probably less costly for the breeder to use *ex situ* material. In many cases, the material may have already been evaluated for useful traits. But why are there *no* examples of the practical use in crop breeding of wild relatives from *in situ* populations which may contain novel traits that have evolved in more recent times? Perhaps this is because there are few established genetic reserves specifically for wild relatives of crops, in spite of the comprehensive attention given to the theory in the literature. One exception is the Erebuni reserve in Armenia, first recommended by Vavilov for wild wheats. This contains *Triticum urartu* and other wild wheat species such as *Triticum boeoticum* and *Triticum araraticum*, which grow in the protected area together with *Aegilops* spp. (Damania, 1994).

Considerable study has also been made of diseases in wild plant populations, including in populations of crop wild relatives (Burdon, 1987; Burdon and Jarosz, 1989; Burdon and Leather, 1990; Frankel *et al.*, 1995; Dinoor and Eshed, 1997). By far the most cited study is the Ammiad Project, implemented in northern Israel from 1984 to 1993, to study the natural dynamics of wild emmer populations (Anikster *et al.*, 1997). Although it was designed to serve as a precursor to conservation of selected wild cereal populations in their native ecosystems, no *in situ* genetic reserves appear to have been established as a result of this project. The study concentrated on biochemical markers as indicators of spatial and temporal variation and variability for disease resistance to several important fungal diseases of wheat. Overall, there was very low incidence of disease in the wild emmer populations studied and none of 1055 accessions showed any resistance to leaf rust (*Puccinia recondita*), all being susceptible even to the least virulent race used (Anikster *et al.*, 1997). The fact that susceptible plants are very common in wild populations casts doubt on the value of *in situ* conservation of wild relatives for disease resistance for future use. Furthermore, after 10 years, one of the main conclusions of the Ammiad Project was that detection of novel traits in wild populations may require very long periods of monitoring (Anikster *et al.*, 1997).

Dinoor and Eshed (1997) highlighted the considerable disagreement in the literature about the implementation of *in situ* conservation of wild relatives, in this case for disease resistance. Some of the key issues include: *how?* and *where?* to designate promising populations and *what?* are the appropriate methodologies for evaluation, sampling, exploitation and monitoring. Furthermore, reservations have been expressed about the value of genetic reserves or parks conserving plant–pathogen relationships (Burdon and Jarosz, 1989). Fundamentally, Harper (1990) has questioned the role of plant diseases in determining the composition of plant communities, since most associations of plants and diseases show very low disease incidence and severity. The occurrence of resistance genes in wild relatives of crops is evidence of powerful selective forces but much of this diversity for disease resistance represents 'the ghosts of diseases past' (Harper, 1990), but how long past (Allen *et al*., 1999)? Due to the complexity of the factors that determine the frequencies of *R* genes and diversity for resistance in the wild relative including pathogen variability, the genetic basis of resistance and virulence, and the breeding systems in both the host plant and the pathogen, inferences as to the recency and severity of the pathogen as a selective force can only be tentative (Frankel *et al*., 1995). No doubt this also holds true for crop landraces on-farm.

Advocacy of plant conservation *in situ* for disease resistance in reserves or 'pathogen parks' (*sensu* Browning, 1974) seems paradoxical: the effective long-term conservation of crop wild relatives for their potential as sources of resistance requires potentially dangerous conservation of the pathogen/s against which the resistances are being sought (Wills, 1996; Lenné, 1998). Both Harper (1990) and Alexander (1992) recommended enhancing the pathogen pressure as this might demonstrate – in the long term – some value for screening *in situ*. However, the problems with this approach outweigh the advantages. Such reserves are sources of diseases for related crops growing nearby, threatening local food security and farmers' livelihoods. This is most probably why the concept has never been put into practice. Furthermore, even after proposing strategies for *in situ* conservation of wild crop relatives for disease resistance, Dinoor and Eshed (1997) highlighted major shortcomings of evaluating disease resistance *in situ*, in particular, the inability to accurately detect resistance. They concluded that optimal utilization of genetic resources for breeding disease-resistant crops should be through random collections and centralized targeted evaluation using relevant pathogen selectors – that is, *ex situ* screening. This is much easier, more convenient, and far safer as screening can be fully controlled and the pathogens are securely contained.

As stated above, the main justification and assumed value of *in situ* conservation of crop wild relatives is that they will evolve with environmental changes, in particular with variable pathogens. But 20 years on, there is *no evidence* of success and no studies appear to have been made on medium- to long-term plant–pathogen coevolution in wild relative populations.

Therefore, it appears that no one can say how far the concept of plant pathogen coevolution is justified in generating useful resistances, since evolutionary changes are slow, and perhaps no obvious changes may be observed in 50 to 100 years (Frankel *et al*., 1995; Maxted *et al*., 1997). In sharp contrast, controlled screening *ex situ* with appropriate pathogen selectors can detect changes immediately. In a comprehensive and detailed review of the status of *in situ* conservation, Maxted *et al*. (1997) observed that there is currently a dearth of practical examples in the field and much remains to be understood about the approach. Ten years on, Heywood *et al*. (2007) concluded that many important scientific and technical issues still need to be addressed. Furthermore, recently Maxted and Kell (2009) highlighted the need for more research and funding, especially under the threat of climate change.

Surely, in the light of 20 years of activity and funding, the complete lack of success in demonstrating a major value of *in situ* conservation for food security means that the concept needs a radical rethink rather than a call for more effort and funding. Viable and practical alternatives exist. In the short term, the most practical and cost-effective approach would be to target those wild relative populations with demonstrable value based on already accessed genes; comprehensively collect them; and screen them under controlled and safe conditions with appropriate selection pressures. Such an approach capitalizes on the postulated synergies between *ex situ* and *in situ* conservation. Furthermore, reliance on useful traits from wild species alone may be less needed in the future as ongoing advances in biotechnologies will allow scientists to access genes from all living organisms.

On-farm conservation of crops

The concept of on-farm conservation, as for *in situ* conservation of wild plants in nature reserves, is based on the perceived value that maintaining plant populations in a 'dynamic' state will result in the evolution of useful traits for the farmer and food production – also referred to as 'local adaptation' (Altieri and Merrick, 1987; Jarvis and Hodgkin, 1998; Brush, 1999; Newton *et al.*, 2010). The main difference between *in situ* conservation of wild plants and on-farm conservation of crops is management by farmers. This is likely to have far greater impact on landrace diversity than environmental pressures. In fact, natural and human-mediated selection may contradict each other. Farmer landrace and seed selection from season to season could quickly reverse any effects of natural selection on landraces on-farm, as noted recently by Mercer and Perales (2010, in the context of climate change).

Also, the normal varietal turnover on-farm, as newer or better varieties catch the farmers' eyes, could completely negate years of adaptation of specific varieties to local conditions. Not just varieties, but whole crops can rapidly disappear from large areas as markets change (a major justification for *ex situ* conservation, capturing the ebb-and-flow of varieties and crops over time). Certainly, most of the varieties stored in *ex situ* genebanks are no longer available *in situ*, a result of the rational decisions of very many farmers over many years to abandon varieties no longer deemed of value.

Landraces are defined as being genetically diverse, although this view is commonly based on visible morphological variation rather than characterized genetic diversity. Few studies have looked in detail at diversity for more than a single useful trait at any one time. Isozyme and molecular studies have indicated 'diversity *per se*' but have rarely further explored the significance of such diversity. Although it is widely believed that the genetic diversity within landraces provides protection against climatic extremes and disease and pest epidemics (Altieri and Merrick, 1987; Maxted *et al.*, 1997; Jarvis and Hodgkin, 1998), there is a substantial lack of evidence for this (Frankel *et al.*, 1995). Disastrous epidemics of wheat rust, rice blast and potato blight – to name a few – have been recorded throughout history on landraces before the time of modern plant breeding (Allen *et al.*, 1999). Furthermore, diversity *per se* will not be an adequate defence unless the diversity includes tolerance/resistance traits to the abiotic and/or biotic factor/s affecting crop production. This critical factor is usually forgotten by proponents of on-farm conservation.

Although landraces have recently been described as 'winning combinations' of genes and traits resulting from the interaction among farmers, the crop and the environment (Bellon, 2009) and considered essential to achieving greater agricultural sustainability (Newton *et al.*, 2010), landraces usually contain many inferior components that limit the productivity of the population as a whole (Frankel *et al.*, 1995). It is generally possible to select within landraces individuals that perform better than the parent landrace. This casts doubt on the current dogma of the *superiority* of landraces over modern varieties. It may well be the *weakness* of selection to which landraces have been exposed, rather than its intensity and direction, which has maintained their *long-term* resilient qualities (Frankel *et al.*, 1995).

It has often been highlighted that remarkably little scientific research, particularly on the genetic basis of the diversity and its evolutionary properties, has been done on landrace populations under on-farm conservation, despite the value of these resources in feeding people, especially poor people (Frankel *et al.*, 1995; Lenné *et al.*, 1997; Brush, 1999; Maxted *et al.*, 2002; Mercer and Perales, 2010). In particular, Brown (1999) stressed the lack of both a scientific basis and optimal procedures for on-farm conservation of landraces. He postulated the following advantages of on-farm conservation of landraces:

- Conservation linked with use which is of value to farmers;
- Conservation of both the crop and its indigenous knowledge;
- Provided the population is very large, allelic richness and genotypic diversity;
- Diversity to meet minor/moderate temporal environmental variation, i.e. resilience and stability but not to extreme environmental variation such as severe drought; and
- Potentially provides scope for ongoing evolution, e.g. for disease resistance, but this will be dependent on the diversity in the landrace, selection pressure and the breeding system of the crop.

Brown (1999) provided a detailed research agenda for addressing all of the above issues and stressed the urgent need to initiate such studies so that both the farmers and the wider community could benefit from on-farm conservation of important food crops. Unfortunately, many of these *postulated* advantages have become a general belief – even a mantra – without most of the necessary scientific studies being made. More than 10 years ago, Bellon *et al.* (1997) noted that the attention that on-farm conservation has attracted and the apparent rush to implement conservation projects seems to be inversely proportional to the research effort being expended. Unfortunately, this continues today for key research areas.

From a review of the recent literature, it appears that most studies have looked at socio-economic, cultural and anthropological issues, diversity *per se* (as numbers of crops and landraces and occasionally isozyme or molecular studies), gene flow, seed management and training (e.g. Sthapit *et al.*, 2005; Smale, 2006; Veteläinen *et al.*, 2009; also see www.bioversityinternational.org). No scientific studies appear to have been done on the genetic basis of the diversity, especially useful traits for increasing food production, so-called 'local adaptation', and the evolutionary potential for future useful genes. There is little information on temporal changes in allele frequencies in landrace populations on-farm (Frankel *et al.*, 1995). Yet, continuing evolution of useful traits is the major rationale for on-farm conservation just as it is for *in situ* conservation of wild relatives. It is difficult to understand why such key studies have been neglected when the necessary areas of research have been identified on several occasions (Lenné *et al.*, 1997; Brown, 1999; Mercer and Perales, 2010; see Box 10.1). The true value of on-farm conservation must be founded on the clear demonstration of progressively valuable changes in the dynamic landrace populations. If projects are not attempting to measure this basic characteristic as a justification for on-farm conservation, their relevance to future food security must be seriously questioned. Also, adding value to varieties maintained on-farm raises economic questions: does the added value, which may be captured in the distant future through formal plant breeding, exceed the ongoing cost to the farmer of foregoing access to newer and better varieties and also the administrative cost of project intervention?

So-called 'local adaptation' is commonly claimed as a property of local landraces as a result of their progressive improvement over time on-farm (Cleveland *et al.*, 1994; de Boef *et al.*, 1995). However, very limited research has been done on characterizing the local adaptation of landraces. Of note, Bunting and Curtis (1968) found precise photoperiod sensitivity in sorghum landraces in Nigeria while Mercer *et al.* (2008) showed that highland landraces of maize were clearly adapted to highland sites, while lowland and midland landraces appear more adapted to the midland site. Although both studies showed local adaptation at one point in time, research was not done on adaptation over time. In fact, we could not find any studies in the literature that demonstrated progressive improvement in landraces over time. Perhaps we should be aware of Darwin's caveat: 'There seems to be no more design in the variability of organic beings, and in the action of natural selection, than in the course which the wind blows' (Darwin, 1876).

Genetic polymorphism in landrace populations is essential to permit adaptation to evolving selection pressures such as climate changes or changes in pathogen populations (Bellon *et al.*, 1997). Although the condition is

Box 10.1. Key areas of research needed for on-farm conservation

Lenné *et al.* (1997) highlighted the following key areas of research:

- Identification of benchmark sites: ecological conditions; cropping system characteristics; varieties used; communal tradition of varietal maintenance and experimentation; and sociocultural factors.
- Varietal and genetic characterization: for establishing baseline datasets to monitor changes in diversity.
- Varietal demography: movement of varieties into and out of farms and localities.
- Changes in genetic diversity: dependent on selection pressure and the breeding system of the crop.

In the context of climate change, Mercer and Perales (2010) posed many research questions that need to be tackled to understand how the genetic structure of landraces may respond to climate change:

- Is available genetic variation appropriate for evolutionary response to climate change, especially for selfing or clonal crops?
- At what rate will evolution proceed, given heritability of traits and strength of selection?
- Might there be constraints on evolution to multiple environmental changes, given the genetic correlations among traits?
- Is there capacity for evolution of plasticity?
- Might populations be plastic in response to climate change, especially for selfing or clonal crops?
- Will different types within a species, or landraces from different regions, respond differently?
- Will adaptive or novel variation be available to populations for evolution, based on patterns of gene flow and mutation rates?
- Would gene flow from improved varieties improve or reduce the evolutionary potential or plastic response of landrace populations?

necessary, it is not sufficient alone if the variability present does not permit the appropriate response to the selection pressure, e.g. increased adaptation to heat stress or resistance to the pathogen. This means that cultivated landraces are not necessarily the ones best adapted to local conditions where they are grown.

There seems to be no sound evidence that farmers increase the specific local adaptation of varieties, except – perhaps – in marginal, stress-prone environments (Wood and Lenné, 1997). A more realistic view is that farmers have created and managed the environments where crop varieties could evolve under a range of changing and often contrasting selective pressures. Given the lack of research, lack of evidence and uncertainties about the extent of local adaptation, the concept should not be used to justify on-farm conservation.

The farmers' role in on-farm conservation is fundamental since much of the cost of on-farm conservation strategies will be borne by them (Smale and Bellon, 1999; Smale, 2006). *De facto* conservation of diverse landraces continues in many parts of the world and for many crops (Brush, 2004) as farmers lack access to improved and more productive varieties, especially in marginal conditions (see Chapter 6, this volume). But these landrace populations will not necessarily be the most valuable to conserve for future crop improvement and food security. Again, this important issue has not been studied. The feasibility of on-farm conservation of potentially useful crop populations depends on whether farmers are able to tangibly benefit from it (Bellon, 2004). Furthermore, landraces that are highly valued by farmers will probably be conserved but these will not necessarily be those of most value for future food security.

It is difficult to cast aside the reality that it will be the farmer who foots the bill, whether he/she knows it or not (Frankel *et al.*, 1995). Smale *et al.* (2004) noted that it makes no economic sense to trade productivity for conservation or to thwart the opportunities that farmers may want to grow and benefit from modern varieties. Therefore, sustainable on-farm conservation of landraces without ongoing financial support is questionably feasible, especially in changing economic and social conditions. We should not, therefore, overestimate our ability to promote on-farm conservation.

Just as was noted above for *in situ* conservation of crop wild relatives with respect to diseases, conserving diversity on-farm will entail some sort of cost (Brown, 1999), even a loss of landraces. This is rarely highlighted in on-farm conservation projects. Evolution of disease resistance requires the presence of strong pathogen selection pressure. The crop population will probably suffer, resulting in yield losses for farmers. Furthermore, if the objective is the evolution of new, useful characters in the population, selection will no doubt eliminate less-desirable individuals from the population. This will result in diversity being lost. Similarly, high levels of gene flow between landraces and/or varieties could result in significant losses in genetic diversity on-farm, as shown for avocado in Costa Rica (Birnbaum *et al.*, 2003). In addition, Mercer and Perales (2010) noted that if climate change exposes the landrace populations to strong bouts of selection, this is likely to lead to extreme narrowing of genetic diversity in the populations, just as breeders select superior lines from breeding populations.

These concerns are likely to be similar for any abiotic or biotic selection pressure on any landrace population under on-farm conservation. This reinforces the need for detailed and ongoing monitoring of the genetic structure of the target landrace populations to facilitate capture of valuable alleles and traits for complementary *ex situ* conservation so that they are not lost; but for how long? On-farm conservation projects may need a time scale of 50–100 years to be of reasonable value (Frankel *et al.*, 1995).

Our analysis strongly suggests that the current approach to on-farm conservation of landraces is unlikely to yield valuable traits either for the farmer or for wider food security. If farmers want to maintain landraces, there are proven options available to improve their productivity while maintaining their diversity, for example, through client-oriented breeding (Witcombe *et al.*, 2005; also see Chapter 6, this volume) or through participatory plant breeding where some of the evaluation is done off-farm for efficiency and safety, especially when dealing with diseases. The opportunities for interaction and complementarity between formal breeding work on-station and farmer-managed crop populations on-farm also need much more attention (Lenné *et al.*, 1997). Once local germplasm with characters of value has been identified by formal evaluation, it can be multiplied and fed back into the cropping systems. Just as for *in situ* conservation of wild relatives, there is an urgent need for a critical assessment of on-farm conservation and its value both to the farmers implementing it and to local, national and international food security. Attempts should be made to enhance the farmers' abilities to recognize, promote and utilize genetic diversity for future evolution. For resources so important, it is surprising how little we know about the recent evolution of crop landraces on-farm (Lenné *et al.*, 1997; Mercer and Perales, 2010).

On-farm and *ex situ* Conservation of Agrobiodiversity: Complementarities

Farmers are proven experts at evaluating (based on their criteria) and managing variation; their bottleneck is in obtaining sufficient diversity to evaluate. In contrast, the formal *ex situ* system has in store enormous resources of plant diversity, but faces a bottleneck to adequately evaluate samples for a wide range of conditions. We need to combine the varietal management ability of farmers with the resources of samples in genebanks. There is opportunity for *ex situ* stores to return germplasm to farmers when farming communities have lost varieties through war, drought or other catastrophe (e.g. the Seeds of Hope initiative after the Rwandan genocide (see Anon., 1994); the 'Arche Noah' vegetable seed network in Austria to integrate *ex situ* and on-farm approaches in the management of local diversity, see web link: www.arche-noah.at). As an absolute right, farm communities should have easy and continued access to germplasm collected from the community and now held *ex situ* (this is now threatened by the FAO Seed Treaty). In addition, genetic resources threatened on-farm should be collected and stored *ex situ*. Productive interaction will depend on a greatly enhanced documentation capability – an obvious role for formal genebanks. If the intention is to transfer local knowledge and germplasm to other areas, then the ability and willingness of farmers to act as trainers will be important.

References

Alexander, E.P and Alexander, M. (2008) *Museums in Motion: an Introduction to the History and Functions of Museums*. AltaMira Press, Lanham, Maryland.

Alexander, H.M. (1992) Fungal pathogens and the structure of plant populations and communities. In: Caroll, G.C. and Wicklow, D.T. (eds) *The Fungal Community: Its Organization and Role in the Ecosystem*. Marcel Dekker, New York, pp. 482–496.

Allen, D.J., Lenné, J.M. and Waller, J.M. (1999) Pathogen biodiversity: its nature, characterization and consequences. In: Wood, D. and Lenné, J.M. (eds) *Agrobiodiversity: Characterization, Utilization and Management*. CAB International, Wallingford, UK, pp. 123–153.

Altieri, M.A. and Merrick, L.C. (1987) In situ conservation of crop genetic resources through maintenance of traditional farming systems. *Economic Botany* 41, 86–96.

Andersen, R. (2008) *Governing Agrobiodiversity: Plant Genetics and Developing Countries*. Ashgate, Farnham, UK.

Anikster, Y., Feldman, M. and Horovitz, A. (1997) The Ammiad experiment. In: Maxted, N., Ford-Lloyd, B.V. and Hawkes, J.G. (eds) *Plant Genetic Conservation*. Chapman and Hall, London, pp. 237–253.

Anon. (1994) Agricultural Centers launch 'Seeds of Hope' to save Rwanda's food crops, avert famine. *CGIAR News* Vol. 1.

Aoki, K. (2007) Distributive and syncretic motives in intellectual property law (with special reference to coercion, agency, and development). *University of Davis Law Review* 40, 717–801.

Bellon, M.R. (2004) Conceptualizing interventions to support on-farm genetic resource conservation. *World Development* 32, 159–172.

Bellon, M.R. (2009) Do we need crop landraces for the future? Realizing the global option value of *in situ* conservation. In: Kontoleon, A., Pascual, U. and Smale, M. (eds) *Agrobiodiversity, Conservation and Economic Development*. Routledge, Taylor and Francis Group, London, pp. 51–61.

Bellon, M.R., Pham, J.-L. and Jackson, M.T. (1997) Genetic conservation: a role for rice farmers. In: Maxted, N., Ford-Lloyd, B.V. and Hawkes, J.G. (eds) *Plant Genetic Conservation*. Chapman and Hall, London, pp. 263–289.

Birnbaum, K., DeSalle, R., Peters, C.M. and Benfey, P.N. (2003) Integrating gene flow, crop biology, and farm management in on-farm conservation of avocado (*Persea americana*, Lauraceae). *American Journal of Botany* 90, 1619–1627.

Bradshaw, J. and Ramsay, G. (2005) Utilisation of the Commonwealth Potato Collection in potato breeding. *Euphytica* 146, 9–19.

Brockway, L.H. (1979) *Science and Colonial Expansion: the Role of the British Royal Botanic Gardens*. Academic Press, New York.

Brown, A.H.D. (1999) The genetic structure of crop landraces and the challenge to conserve them *in situ* on farms. In: Brush, S.B. (ed.) *Genes in the Field*. International Plant Genetic Resources Institute, International Development Research Center, and Lewis Publishers, Rome, Ottawa, Canada and Boca Raton, Florida, pp. 29–48.

Browning, J.A. (1974) Relevance of knowledge about natural ecosystems to development of pest management programs for agro-ecosystems. *Proceedings of the American Phytopathological Society* 1, 191–199.

Brush, S.B. (1995) In situ conservation of landraces in centers of crop diversity. *Crop Science* 35, 346–354.

Brush, S.B. (ed.) (1999) *Genes in the Field*. International Plant Genetic Resources Institute, International Development Research Center, and Lewis Publishers, Rome, Ottawa, Canada and Boca Raton, Florida.

Brush, S.B. (ed.) (2000) *Genes in the Field: the On-Farm Conservation of Crop Diversity*. International Plant Genetic Resources Institute, Rome.

Brush, S.B. (2004) *Farmers' Bounty: Locating Crop Diversity in the Contemporary World*. Yale University Press, New Haven, Connecticut.

Bunting, A.H. and Curtis, D.L. (1968) Local adaptation of sorghum varieties in northern Nigeria. In: *Agroclimatological Methods*, Proceedings of the Reading Symposium. UNESCO, Paris, pp. 101–106.

Burdon, J.J. (1987) *Diseases and Plant Population Biology*. Cambridge University Press, Cambridge.

Burdon, J.J. and Jarosz, A.M. (1989) Wild relatives as sources of disease resistance. In: Brown, A.D.H., Marshall, D.R., Frankel, O.H. and Williams, J.T. (eds) *The Use of Plant Genetic Resources*. Cambridge University Press, Cambridge, pp. 280–296.

Burdon J.J. and Leather, S.R. (eds) (1990) *Pest, Pathogens and Plant Communities*. Blackwell Scientific Publications, Oxford.

CBD (1992) Handbook of the Convention on Biological Diversity. CBD, Montreal, Canada.

Charles, D. (2001) Seeds of discontent. *Science* 294, 772–775.

Cleveland, D.A., Soleri, D. and Smith, S.E. (1994) Do folk crop varieties have a role in sustainable agriculture? *BioScience* 44, 740–751.

Committee on Managing Global Genetic Resources (1991) *Managing Global Genetic Resources: The US National Plant Germplasm System*. National Academy Press, Washington, DC.

Damania, A.B. (1994) In situ conservation of biodiversity of wild progenitors of cereal crops in the Near East. *Biodiversity Letters* 2, 56–60.

Darwin, C. (1876) *The Life And Letters Of Charles Darwin including an Autobiographical Chapter Edited By His Son Francis Darwin*, Vol. I, Chapter 1, viii.

de Boef, W.S., Berg, T. and Haverkort, B. (1995) *Farmers, Crops and Landraces. Farmers' Roles in the Development and Conservation of Crop Diversity*. CPRO-DLO Centre for Genetic Resources, Wageningen, the Netherlands.

Dickson, J.D. (1977) *Check List and Uses of Plants in the Wilson Popenoe Botanical Garden, Lancetilla, Honduras*. SIATSA, La Lima, Honduras.

Dinoor, A. and Eshed, N. (1997) Plant conservation *in situ* for disease resistance. In: Maxted, N., Ford-Lloyd, B.V. and Hawkes, J.G. (eds) *Plant Genetic Conservation*. Chapman and Hall, London, pp. 323–336.

Duvick, D.N. (1986) The nobility of seed research and its critics. *The World and I* 7, 275–281.

Ellis, J., Mago, R., Kota, R., Dodds, P., McFadden, H., Lawrence, G., Spielmeyer, W. and Lagudah, E. (2007) Wheat rust resistance research at CSIRO. *Australian Journal of Agricultural Research* 58, 507–511.

Engels, J.M.M. and Wood, D. (1999) Conservation of agrobiodiversity. In: Wood, D. and Lenné, J.M. (eds) *Agrobiodiversity: Characterization, Utilization and Management*. CAB International, Wallingford, UK, pp. 355–386.

ETC Group (2001) *The Law of the Seed!* ETC Group Translator Vol. 3.

ETC Group (no date) Biopiracy. Available at: www.etcgroup.org/en/issues/biopiracy (accessed 22 September 2010).

ETC (2004) From Global Enclosure to Self Enclosure: Ten Years After – A Critique of the CBD and the 'Bonn Guidelines' on Access and Benefit Sharing (ABS). *ETC Group Communiqué* 83, pp. 14.

Fairchild, D. (1938) Reminiscences of early plant introduction work in South Florida. *Proceedings of the Florida State Horticultural Society* 51, 11–33.

Falcon, W.P. and Fowler, C. (2002) Carving up the commons – emergence of a new international regime for germplasm development and transfer. *Food Policy* 27, 197–222.

Frank, S.A. (1993) Coevolutionary genetics of plants and pathogens. *Evolutionary Ecology* 7, 45–75.

Frankel, O.H., Brown, A.D.H. and Burdon, J.J. (1995) *The Conservation of Plant Biodiversity*. Cambridge University Press, Cambridge.

GCDT (2006) Background on the development of the 'Global Strategy for the Ex Situ Conservation of Potato'. Available at: www.croptrust.org/documents/cropstrategies/Potato.pdf (accessed 22 September 2010).

Gregorio, G.B., Senadhira, D., Mendoza, R.D., Manigbas, N.L., Roxas, J.P. and Guerta, C.Q. (2002) Progress in breeding for salinity tolerance and associated abiotic stresses in rice. *Field Crops Research* 76, 91–101.

Harlan, J.R. (1977) Sources of genetic defense. *Annals of the New York Academy of Science* 287, 345–355.

Harper, J.L. (1990) Pest, pathogens and plant communities: an introduction. In: Burdon, J.J. and Leather, S.R. (eds) *Pest, Pathogens and Plant Communities*. Blackwell Scientific Publications, Oxford, pp. 3–14.

Heywood, V.H. and Dulloo, M.E. (2006) In situ conservation of wild plant species – a critical global review of good practices. *IPGRI Technical Bulletin No. 11*, IPGRI, Rome.

Heywood, V., Casas, A., Ford-Lloyd, B., Kell, S. and Maxted, N. (2007) Conservation and sustainable use of crop wild relatives. *Agriculture, Ecosystems and Environment* 121, 245–255.

Hoyt, E. (1988) *Conserving the Wild Relatives of Crops*. IBPGR/IUCN/WWF, Rome.

Jarvis, D.I. and Hodgkin, T. (1998) *Strengthening the scientific basis of in situ conservation of agricultural biodiversity on-farm: options for data collecting and analysis*. IPGRI, Rome.

Lenné, J.M. (1998) The biodiversity and conservation of crops for disease resistance. *Seventh International Congress of Plant Pathology*, Edinburgh, abstract 4.1.1S.

Lenné, J.M. and Wood, D. (1991) Plant diseases and the use of wild germplasm. *Annual Review of Phytopathology* 29, 35–63.

Lenné, J.M., Weltzien, E. and Stenhouse, J. (1997) Institutional reflections: a role for ICRISAT in enhancing and maintaining genetic resources on-farm. In: Sperling, L. and Loevinsohn, M. (eds) *Using Diversity: Enhancing and Maintaining Genetic Resources On-farm*. IDRC, New Delhi, India, pp. 322–326.

Leppik, E.E. (1970) Gene centers of plants as sources of disease resistance. *Annual Review of Phytopathology* 8, 323–344.

Maxted, N. and Kell, S. (2009) Establishment of a global network for the *in situ* conservation of crop wild relatives: status and needs. Commission on Genetic Resources for Food and Agriculture. *Background Study Paper No. 39*. FAO, Rome.

Maxted, N., Ford-Lloyd, B.V. and Hawkes, J.G. (eds) (1997) *Plant Genetic Conservation*. Chapman and Hall, London.

Maxted, N., Guarino, L., Myer, L. and Chiwona E.A. (2002) Towards a methodology for on-farm conservation of plant genetic resources. *Genetic Resources and Crop Evolution* 49, 31-46.

Maxted, N., Iriondo, J.M., De Hond, L., Dulloo, M.E., Lefèvre, F., Asdal, A., Kell, S.P. and Guarino, L. (2008) Genetic reserve management. In: Iriondo, J.M., Dulloo, M.E. and Maxted, N. (eds) *Conserving Plant Genetic Diversity in Protected Areas: Population Management of Crop Wild Relatives*. CAB International, Wallingford, UK, pp. 65–87.

Mercer, K.L. and Perales, H.R. (2010) Evolutionary response of landraces to climate change in centers of crop diversity. *Evolutionary Applications* 3, 480–493.

Mercer, K., Martınez-Vasquez, A. and Perales, H.R. (2008) Asymmetrical local adaptation of maize landraces along an altitudinal gradient. *Evolutionary Applications* 1, 489–500.

Newton, A.C., Akar, T., Baresel, J.P., Bebeli, P.J., Bettencourt, E., Bladenopoulos, K.V., Czembor, J.H., Fasoula, D.A., Katsiotis, A., Koutis, K., Koutsika-Sotiriou, M., Kovacs, G., Larsson, H., Pinheiro de Carvalho, M.A.A., Rubiales, D., Russell, J., Dos Santos, T.M.M. and Vaz Patto, M.C. (2010) Cereal landraces for sustainable agriculture: A review. *Agronomy for Sustainable Development* 30, 237–269.

NordGen (2010) Seed Portal of the Svalbard Global Seed Vault. Available at: www.nordgen.org/sgsv (accessed 22 September 2010).

Park, T.-H., Foster, S., Brigneti, G. and Jones, J.D.G. (2008) Two distinct potato late blight resistance genes from *Solanum berthaultii* are located on chromosome 10. *Euphytica* 165, 269–278.

Paul, B. (1984) Third world battles for fruit of its seed stocks. *Wall Street Journal* 15 June.

Peleg, Z., Fahima, T. and Saranga, Y. (2007) A century of wheat research – from wild emmer discovery to genome analysis. *Israel Journal of Plant Sciences* 55, 289–296.

RAFI (1994) Bioprospecting/Biopiracy and Indigenous Peoples. Available at: www.etcgroup.org/en/node/482 (accessed 22 September 2010).

Roelfs, R.P. (1988) Genetic control of phenotypes in wheat stem rust. *Annual Review of Phytopathology* 26, 351–367.

Royal Norwegian Ministry of Agriculture and Food (2008) http://www.nordgen.org/sgsv/scope/sgsv/files/SGSV_Deposit_Agreement.pdf

Shiva, V. (1996) *Biopiracy: the Plunder of Nature and Knowledge*. South End Press, New York.

Smale, M. (2006) *Valuing Crop Biodiversity: On-Farm Genetic Resources and Economic Change*. CAB International, Wallingford, UK.

Smale, M. and Bellon, M.R. (1999) A conceptual framework for valuing on-farm genetic resources. In: Wood, D. and Lenné, J.M. (eds) *Agrobiodiversity: Characterization, Utilization and Management*. CAB International, Wallingford, UK, pp. 387–408.

Smale, M., Bellon, M.R., Jarvis, D. and Sthapit, B. (2004) Economic concepts for designing policies to conserve crop genetic resources on farms. *Genetic Resources and Crop Evolution* 51, 121–135.

Sthapit, B.R., Upadhyay, M.P., Shrestha, P.K. and Jarvis, D.I. (2005) *On-farm Conservation of Agricultural Biodiversity in Nepal*. IPGRI, Rome.

UNEP (1992) *Convention on Biological Diversity*. UNEP, Geneva, Switzerland.

Veteläinen, M., Negri, V. and Maxted, N. (2009) *European Landraces: On-farm Conservation, Management and Use*. Bioversity International, Rome.

Weigl, E. (2003) Acclimatization: The Schomburgk brothers in South Australia. HiN Vol. IV, 7. Available at: www.uni-potsdam.de/u/romanistik/humboldt/hin/hin7/inh_weigl_1.htm (accessed 20 September 2010).

Wills, C. (1996) Safety in diversity. *New Scientist* 2002, 38–42.

Witcombe, J.R., Steele, K.A., Hash, C.T., Mottram, A., Bourai, V.A., Singh, D.N., Prasad, S.C. and Virk, D.S. (2005) Farmers, plant breeding and molecular markers: improving livelihoods through better client-orientation. In: Harris, D., Richards, J.I., Silverside, P., Ward, A.F. and Witcombe, J.R. (eds) Pathways out of Poverty. *Aspects of Applied Biology* No. 75, pp. 67–70.

Wood, D. (1992) Talking point: a matter of good breeding. *New Scientist* 18 January.

Wood, D. and Lenné, J.M. (1997) The conservation of agrobiodiversity on-farm: questioning the emerging paradigm. *Biodiversity and Conservation* 6, 109–129.

11 Can the International Assessment of Agricultural Knowledge, Science and Technology for Development (IAASTD) Approach Ensure Future Food Security?

D. Wood and J.M. Lenné

There is nothing more dangerous than blind passion in science. This is a direct path to unjustified self-confidence, to loss of self-criticalness, to scientific fanaticism, to false science. Given support from someone in power, it can lead to suppression of true science and, since science is now a matter of state importance, to inflicting great injury on the country.

Nikolay Semyonov (1965) (in Reiter, 2009)

The International Assessment of Agricultural Knowledge, Science and Technology for Development (IAASTD)

With great promise, the IAASTD began a 5-year assessment in 2003 to develop a future roadmap for agricultural knowledge, science and technology (AKST) to ensure future global food security (McIntyre *et al.*, 2009). Interestingly, this was stimulated by discussions at the World Bank with the private sector and nongovernmental organizations (NGOs) *on the state of the scientific understanding of biotechnology and, more specifically, genetically modified (GM) crops* rather than on the need to reduce hunger and poverty. The process was initiated through 11 consultation meetings on five continents. Over the next 5 years, numerous consultations generated five regional reports and a global synthesis report at a cost of US$15 million. The process was cosponsored and supported by the United Nations Food and Agriculture Organization, the Global Environment Facility, the United Nations Development and Environment Programmes, the World Health Organization and the United Nations Educational, Scientific and Cultural Organization as well as many bilateral donors, especially the UK Department for International Development. Its overall purpose was 'to assess agricultural knowledge, science and technology in order to use it more effectively to reduce hunger and poverty, improve rural livelihoods, and facilitate equitable, environmentally, socially and economically sustainable development' (McIntyre *et al.*, 2009; see also www.agassessment.org).

This global synthesis report on AKST is now being promoted as an '*evidence-based guide for future policy and decision-making*' on poverty and livelihoods, food security, environmental sustainability, human health and nutrition, equity and investments (McIntyre *et al.*, 2009a). It focuses on eight AKST themes of critical interest to meeting development and sustainability goals:

bioenergy, biotechnology, climate change, human health, natural resource management, trade and markets, traditional and local knowledge and community-based innovation, and women in agriculture. The key message of the report is that small-scale farming and agroecological methods provide the way forward (Scoones, 2009). The IAASTD report specifically stresses that it makes no recommendations, only key findings and options for action. In addition, the options for action are not prioritized because different options are considered actionable by different stakeholders with different priorities and responsibilities. Unfortunately, this approach fails to acknowledge the inter-dependency of many options and, also, that successful actions on some options are dependent on outputs from the achievements of other options. Just as for the MDGs (see Chapter 2, this volume), progress in responding to particular options is likely to be constrained if their inter-dependency is not addressed.

Although strongly endorsed as a multi-thematic, multi-spatial, multi-temporal intergovernmental process, the IAASTD had a very complex governance structure and stakeholder involvement (McIntyre *et al.*, 2009a). The multi-stakeholder Bureau was comprised of 30 government representatives from all major regions; 22 representatives from the private sector, NGOs, consumer and producer groups; representatives from eight institutions; and two co-chairs. Problems could be expected from this management and editing panel. The IAASTD Director was an atmospheric physicist, one of the two co-chairs an entomologist with biocontrol experience, the other co-chair a petroleum geologist: a strange team for an agricultural assessment. With over 800 stakeholders (from grassroots to scientists to global corporations) from 110 countries as well as well over 500 experts, authors and review editors, the IAASTD was unprecedented in the scale and complexity of its inclusiveness and participation. However, it is far from clear whether the IAASTD process genuinely allowed alternative voices to be included, created a new mode of engagement in global debate, or allowed collective understandings of global views on how AKST can best ensure future global food security (Scoones, 2009).

In bringing together such a diverse stakeholder community, it was inevitable that the IAASTD process revealed many and varied views and exposed considerable tensions among stakeholders: many issues were hotly debated; there were fraught scenes; some stakeholders found the process intimidating; other concerned stakeholders withdrew from the assessment (Scoones, 2009). For example, CropLife International withdrew due to major concerns about the inadequate treatment of the role that modern science and technology has played in supporting agriculture as well as the superficial and negative assessment of biotechnology, crop protection chemistry and the role of the private sector (Nature Biotechnology, 2008; also see www.croplife.org). Considerable guidance, cajoling and facilitation were needed by the director and the co-chairs in order to achieve even a basic level of consensus for the global synthesis report (Scoones, 2009).

Radically different opinions about the process have been expressed by stakeholders. At one extreme, the assessment director, Robert Watson, highlighted the 'inclusion of hundreds of experts from all relevant stakeholder groups'; an 'intellectually consistent framework'; a 'global, multi-scale and long term approach' resulting in 'plausible scenarios' to 2050; the 'integration of local and institutional knowledge' and a multi-thematic approach, encompassing nutrition, livelihoods, human health and linking science and technology issues to policies and institutions (see www.agassessment.org). At the other extreme, many stakeholders noted that 'the end result is a bit of a fudge: what someone described as the lowest common denominator analysis, with bits of everything mixed up in an unsavoury cocktail' (Scoones, 2009).

It was only to be expected that the global synthesis report findings and proposed options were greeted by radically different responses. On one hand, the NGO community was exuberantly supportive of many of the options proposed, particularly the emphasis on agroecological methods, organic agriculture, greater use of traditional knowledge and 'farmer breeders' as well as the strong criticism of genetically modified crops (Greenpeace, 2008; PANNA, 2008; Scoones, 2009). On the other hand, many scientists and scientific bodies were highly critical of the erroneous equating of 'modern biotechnology' with GM crops and the questionable evidence to support the extremely negative view of the value of modern biotechnology (see www.croplife.org). Other scientists were critical of the proposed options for support to complex organic agriculture and agroecological approaches rather than simple, proven seed-based technologies to more effectively reduce hunger and poverty (Kirchmann *et al.*, 2008; Goulding and Trewavas, 2009). Of particular note, Dr Chris Leaver, Emeritus Professor of Plant Science, St John's College, Oxford University, highlighted the 'clashing views' of the World Development Report (WDR) (World Bank, 2008) and the IAASTD report. Whereas the WDR concluded that science and technology has an important role to play in agricultural development for the benefit of poor farmers, the IAASTD 'missed the opportunity' by promoting ecological agriculture (Dano, 2008). He said that the IAASTD is advocating a 'formula for world starvation'. Furthermore, some governments, including Australia, Canada and the USA, indicated significant reservations and specific and substantive concerns about some of the assertions made in the report. In particular, China and the USA dismissed the entire section on biotechnology as unbalanced and superficial (McIntyre *et al.*, 2009a; Scoones, 2009).

In its search for a roadmap for AKST to ensure future global food security in the face of unprecedented challenges, the IAASTD report rightly highlights the critical need for significant increases in investment in agriculture both domestically and internationally. Three decades of under-investment in agricultural research for development has been a major contributing factor to the decline in the rate of yield increase for the major staple food crops rice and wheat and the recent food crisis and food price rises (Pardey *et al.*, 2006; Pardey and Pingali, 2010). Globally, there is a strong consensus on the urgent need for increased investment in well-targeted agricultural research to increase food production and improve rural livelihoods under increasingly constrained conditions of less land and inputs, especially water, and climate change.

Yet, a major conclusion of the IAASTD report is that reliance on what it calls 'industrial agriculture' (McIntyre *et al.*, 2009b, p. 7) is risky and unsustainable, particularly in the face of worsening climate, energy and water crises (Herren and Ishii-Eiteman, 2010). It appears to dismiss a role for the key agricultural production system which has continued to effectively feed most of the world's population as it grew from less than 2 billion to more than 6 billion over the past century. The report also emphasizes that expensive, quick fixes – including GM crops – fail to address the complex challenges that farmers face, and often exacerbate already bad conditions. Again, it appears to dismiss the fact that millions of small-scale, poor farmers are already willingly and successfully growing GM crops because of the benefits to themselves and their families. Is it not logical to expect that future AKST options would be more likely to succeed if built on the foundations of proven successes in increasing food production?

Such proven successes in increasing local, national and global food production were founded on simple, seed-based technologies through improved, high-yielding, pest and disease resistant, abiotic stress-tolerant and resource-use efficient staple food crops, e.g. rice, wheat and maize, supported by appropriate inputs. Far from being 'resource-extractive' agriculture (as defined by McIntyre *et al.*, 2009a), millions of hectares of irrigated rice have been cultivated on the same land for centuries (IRRI, 2010). Such technologies massively increased food production and reduced poverty, especially in Asia. In addition, returns on investment were substantial (see Chapters 2 and 5, this volume). Not only were these approaches the basis of the Green Revolution during the 1960s to the 1980s, they have also been increasingly improved and refined by AKST organizations over the past 20 to 30 years and continue to be the major approach to meeting global food security. Yet, the so-called '*evidence-based*' IAASTD global report ignores these gains and strongly promotes future food production scenarios based on complex multifunctional agriculture, small-scale farming based on traditional knowledge and 'farmer breeders', agroecological approaches, organic farming, and agroforestry (McIntyre *et al.*, 2009a; Herren and Ishii-Eiteman, 2010).

It appears that in order to meet the demands and campaigning of certain stakeholder groups, the IAASTD report has given greatest emphasis to complex AKST options, which are highly knowledge- and labour-intensive, to increase food production. But in most cases, the capacity to address these options does not exist in developing countries. Moreover, the ability of the experienced international and national agricultural research and extension systems to provide the wide-scale capacity building that will be required is limited. More importantly, the ability of the suggested AKST options to achieve the massive increases in food production required to feed not only the existing 6.7 billion people but also the predicted 9 billion people by 2050 is largely unproven. These approaches are yet to be tested. Through the IAASTD, the science of agriculture seems to have taken a back seat to ideology (Wager, 2008; see web link: http://web.viu.ca/wager).

A paradigm, therefore, appears to have emerged from the IAASTD global synthesis report due to a series of highly challengeable assertions based on largely unfounded and blanket criticisms of many existing AKST approaches, assumptions of questionable technical merit, and much incorrect or flawed evidence. The key elements of this paradigm will now be critically examined in the next section.

The IAASTD Paradigm

Shortcomings of Green Revolution approaches

Although the IAASTD report acknowledges the important role that the Green Revolution played in substantially increasing food production and food security globally during the past 50 years, it tends to give undue emphasis to some of its perceived shortcomings in order to justify the need for a fundamental rethink of the role of AKST in achieving equitable development and sustainability (McIntyre *et al.*, 2009). For example, the report notes that: during the past 50 years, 75% of the crop genetic base of agricultural crops has been lost; the Green Revolution has had negative consequences on environmental sustainability; people benefited unevenly from these yield increases; and the poor in developing countries have generally benefited the least, among other questionable statements.

The IAASTD also claims, although no evidence is given, that there is increasing recognition within 'formal science and technology organizations' that the current AKST model requires revision. This implies that major staple crop research institutes such as IRRI for rice and CIMMYT for wheat and maize have restricted themselves to follow the Green Revolution approach of 50 years ago. In reality, science and technology organizations have continued to improve, revise and refine their approaches to cereal production driven by changes in target environments, changing socio-economic conditions and evolving local needs since their beginnings in the 1960s (IRRI, 2008a, b). For example, in the 1980s and 1990s, improved varieties accounted for 50% of yield growth, compared to 21% in the 1960s and 1970s (World Bank, 2008). In the first decade of this millennium, IRRI and CIMMYT are already developing new crop varieties with tolerances to heat and drought to adapt staple cereal production to the expected effects of climate change. There is therefore no basis to call for a 'fundamental rethink' or even a revision; rather there is an urgent need for increased investment in and application of proven, adaptable, continuously improving, existing approaches.

In highlighting the shortcomings of the Green Revolution, the IAASTD regresses to the same type of criticisms made by many social scientists in the 1970s and which has since become NGO dogma (Evans, 1998). Some even deny the massive increases in food production (Shiva, 1993). No attempt appears to have been made to review the vast literature responding to them or even to check whether such criticisms have any validity whatsoever. It is highly disingenuous of the IAASTD to continue to emphasize these criticisms when they have not only been shown to be unfounded by many comprehensive analyses but also addressed where relevant (see below).

The Green Revolution has to be seen as one of the great achievements of our time (DeGregorio, 2004; Jain, 2010), and there is overwhelming evidence of its proven success (Evans, 1998; Hazell, 2009). The approach was designed to increase food production and reduce food insecurity with a focus on the most important staple cereals. During 1960–2000, AKST through the Green Revolution substantially increased yields of rice, wheat and maize and, as a result, fed the global population doubling from 3 to 6 billion on much the same land area. Estimates of land saving due to agricultural intensification amount to more than 400 million ha (DeGregorio, 2004). The global population is better fed in terms of both basic caloric needs and also basic nutritional needs (DeGregorio, 2004). The technologies spread far beyond the favourable irrigated areas to rain-fed farming, benefiting even more small-scale farmers. Contrary to claims made by some NGOs, 75% of the crop genetic base *was not* lost. In fact in many areas of substantial uptake of Green Revolution improved, high-yielding varieties, genetic diversity in farmers' fields increased (Witcombe, 1999; see Chapter 6, this volume). Furthermore, wide-scale collection in centres of diversity ensured that most crop diversity was conserved for future use (see Chapter 10, this volume). The global poor benefited through cheaper food. In Asia, millions of small-scale farmers benefited as much as larger farmers and rural employment opportunities increased (Hanumantha Rao, 1994; Jain, 2010). History records no increase in food production that was remotely comparable in scale, speed, spread and duration (Lipton and Longhurst, 1989). Thus, in spite of the Green Revolution being damned by social scientists and many NGOs, millions of small-scale farmers in developing countries voted for it through their cereal fields (Evans, 1998). As Lipton and Longhurst (1989) concluded: 'If social scientists had designed a blueprint in the 1950s for pro-poor innovation, it would have been like modern varieties.'

Due to the legitimate focus on the world's most important food crops – to feed more people, small-scale farmers growing rice, wheat and maize benefited more than small-scale farmers growing other crops, at least initially. Furthermore, there were – and still are – considerable non-AKST barriers and bottlenecks that prevent millions of small-scale poor farmers from benefiting from yield-improving technologies (discussed in more detail below). However, millions of poor people in all developing countries benefited from cheaper food (Hazell, 2009). And, as AKST was applied to other crops, e.g. sorghum, millet, grain legumes, potatoes, cassava and sweet potato etc., more small-scale farmers benefited, including in Africa. At the same time, the Green Revolution was not designed to meet the needs of all global poor and hungry or stand proxy for social reform (Evans, 1998).

Concerns about possible negative environmental effects from excess use of fertilizers and pesticides and expansion of irrigation are common to all forms of agricultural intensification (Evans, 1998). Unwelcome side-effects of excess use of some chemical inputs, e.g. nitrate contamination of groundwater and insect pest resistance to pesticides as well as salinity problems from excess irrigation, have occurred where intensification has not been well managed. But research to ameliorate these problems began in the 1980s. Improvements included: slow release, better placement, informed timing of application for fertilizers; improved techniques for applying irrigation water; and integrated pest management to reduce pesticide use. In addition, plant breeders also targeted improvements in major food crops in their efficiency of water and fertilizer use and resistance to insect pests. Today's improved, high-yielding staple food crop varieties require much less fertilizer and water per unit of output than the early Green Revolution varieties (DeGregorio, 2004). For example, there has been a 36% increase in nitrogen use efficiency in maize in the USA in the past 21 years as a result of AKST. These improved crop varieties also have polygenic resistances to diseases and pests through the accumulation of diverse, multiple genes controlling different mechanisms of resistance within single varieties, thus reducing the need to use pesticides. Furthermore, cereal production is more stable, e.g. the coefficient of variation in rice production has been steadily decreasing for the past 40 years (Evenson and Gollin, 1997; Wood and Lenné, 1999). Hence, even the legitimate concerns about some of the shortcomings of the Green Revolution have been substantially addressed through AKST of the past 20–30 years and continue to be an important target for further improvements today. However, the key proponents of the IAASTD process continue to promote this ideology, ever more stridently, for example: 'the ill-fated "Green Revolution" ... trapped millions of farmers on a pesticide treadmill while devastating the functioning of the ecosystems on which we depend' (Herren and Ishii-Eiteman, 2010).

Criticism of GM crops

The IAASTD report devotes almost the entire section on 'Biotechnology' to a comprehensive criticism of GM crops, especially in developing countries (McIntyre *et al.*, 2009a, pp. 40–45). Considering that GM crops currently only make up about 7% of global crop cultivated area and do not include rice or wheat, the two most important food crops, the focus of this section of the report is excessively narrow and unbalanced as noted by the USA and China. By broadly defining 'biotechnology' to embrace the manipulation of living organisms, the term encompasses a large range of activities from 'farmer breeding' to conventional plant and animal breeding to modern biotechnologies such as genetic manipulation to produce GM crops. But by focusing mainly on GM crops, the IAASTD report missed an important opportunity to present clear, practical options for increasing food crop production on the other 93% of cultivated land to feed future populations. Furthermore, although modern biotechnologies include a wide range of technologies such as tissue culture, genomics and marker assisted selection, increasingly being used to facilitate and accelerate conventional crop breeding, modern biotechnology is erroneously equated with genetic modification to produce GM crops only (for example see Figure SR-BT1, p. 41, Biotechnology section of McIntyre *et al.*, 2009). This enables the report to present an extremely negative view of the value of modern biotechnology in general. It was to be expected that many scientists and scientific organizations reacted strongly and swiftly to this travesty, for example: the IAASTD process 'has developed an astigmatism so severe with regard to genetically modified organisms (GMOs) that it comes close to blindness' (Nature Biotechnology, 2008).

Some examples of the criticisms of modern biotechnology/GM crops raised in the IAASTD report include: doubts about the adequacy of safety testing and regulatory frameworks; IPR instruments preventing farmer-saving of seed; suitability of GM crops to meet most farmers' needs; potential human health risks from unsafe food; potential risks from transgene flow into traditional varieties and weeds; and potential risks from pollen contamination in certified organic systems, among others. Even where benefits have been demonstrated, e.g. decreased insecticide use being beneficial to farmers and the environment (James, 2010), the IAASTD remains negative, casting doubt on their sustainability in the long term and extension to most agroecosystems.

Such concerns are expected for most new AKST options. After all, the first GM crop Roundup Ready® soybean was only commercialized 15 years ago. It has been described as a 'Model T Ford' technology (see Chapter 7, this volume), strongly implying that the next generations of GM crops will be much improved. The IAASTD claims to be 'an evidence-based guide for policy and decision-making', however, many of the examples given of 'potential' risks are not supported by sound evidence and, furthermore, existing evidence supporting the lack of risks has been ignored (Wager, 2008). For example, in 2003, the International Council for Science (ICSU), made up of most National

Academies of Science and over 150 scientific organizations, published an extensive review of GM crops and food (ICSU, 2003). The ICSU report clearly stated 'there is no evidence of any ill effects from the consumption of foods containing genetically modified ingredients'. In addition, it stated that 'Pest tolerant crops can be grown with lower chemical pesticides, resulting in reduced chemical residues in food and less exposure to pesticides'. With respect to the environment, the ICSU report states: 'there is no evidence of any deleterious environmental effects having occurred from the trait/species combinations currently available'. This report appears to have been ignored by the authors and reviewers of the IAASTD (McIntyre *et al.*, 2009a). There is no place in the 'evidence-based' IAASTD report for 'perceived risks' based on no evidence (Wager, 2008).

Furthermore, concerns were also raised about pollen contamination of certified organic agriculture from neighbouring GM crops (McIntyre *et al.*, 2009a). This is pure rhetoric from the organic food industry (Wager, 2008). During a time of unprecedented growth of both GM and organic agriculture there has not been a single case of loss of certification of an organic farmer as a result of pollen flow from GM crops. In fact, the International Federation of Organic Agriculture Movements does not advocate any testing for GM content.

Of the 134 million ha of GM crops globally, 46% (61.5 million ha) are grown in developing countries (James, 2010). Although Brazil and Argentina account for over 42 million ha, India, China, Paraguay and South Africa are increasing rapidly. Each year sees an average increase of 9 million ha globally. Already, over 13 million farmers in developing countries are planting and benefiting from GM crops and this is growing every year. This adoption rate indicates that farmers want GM crops as they offer real benefits: higher and more reliable yields and lower production costs/higher profits due to savings on inputs such as pesticides (Wager, 2008; James, 2010).

In some instances the IAASTD convoluted arguments against biotechnological improvement of farming use reasoning that applies to *any* type of improvement; for example: 'It may not be enough to use biotechnology to increase the number or types of cattle, for instance, if this reduces local genetic diversity or ownership, the ability to secure the best adapted animals, or they further degrade ecosystem services' (McIntyre *et al.*, 2009a, p. 43). This is overtly top-down: placing a whole string of hurdles in the way of animal breeding and the adoption of improvements by farmers. Some of these hurdles are based on questionable reasoning. For example, high genetic diversity is presented as an absolute good for livestock. But farmers worldwide sensibly breed from a limited number of preferred sires – rams, bulls and the rest. This favours quality of offspring over quantity of diversity. Farmers will also know how to choose the best adapted animals – either retaining their old stock or accepting the new. This patronizing approach of the IAASTD is pervasive – using questionable and certainly top-down reasoning to stand between the farmer and improvements.

Virtually every mention of GM crops is grudging and hedged about with doubts unsupported by data (Nature Biotechnology, 2008). The Public Research and Regulation Initiative (PRRI), an international forum for public researchers involved in biotechnology, has cited nearly 20 examples of such equivocation in the IAASTD synthesis report (see web link: www.pubresreg.org). It concludes that the biotechnology chapter:

> is written from a perspective that is so fundamentally different from what we believe should have been the perspective of such an evaluation, that a submission of comments on the many technical omissions and errors would not be meaningful.

We leave you with what should be the key message from the final paragraph of the IAASTD Executive Summary of the Synthesis Report. We question whether this can be done with any measure of effectiveness and efficiency.

> A problem-oriented approach to biotechnology research and development (R&D) would focus investment on local priorities identified through participatory and transparent processes, and favor multifunctional solutions to local problems. These processes require new kinds of support

> for the public to critically engage in assessments of the technical, social, political, cultural, gender, legal, environmental and economic impacts of modern biotechnology. Biotechnologies should be used to maintain local expertise and germplasm so that the capacity for further research resides within the local community. Such R&D would put much needed emphasis onto participatory breeding projects and agroecology.

Agroecological Approaches

One of the surprising and unacceptable outcomes of the IAASTD process was not its general content, but on the way its content has been selectively cited, paraphrased, and even twisted, to support factional interests. Nowhere is this more apparent than in the discussion (and subsequent promotion) of 'agroecology', which is seriously over-promoted by the IAASTD and subsequently was a major feature of reports from a host of NGOs. For example, from the National Family Farm Coalition in the USA, a link to which appeared on the IAASTD's own website a full 9 months after the publication of the IAASTD:

> The recent landmark report of the International Assessment of Agricultural Knowledge, Science and Technology, backed by United Nations Agencies and the World Bank and comprising over 400 scientists, showed that commercial agricultural practices are endangering the planet while also failing to rectify the hunger of millions. To reverse this, the report said investments in ecological practices and science that encourages participatory knowledge creation and the integration of indigenous knowledge shows more promise than relying on transgenic crops and other chemical-intensive Green Revolution tactics.
>
> (Naylor, 2009)

The following is also linked on the IAASTD website: 'The key message of the report is that small-scale farmers and agro-ecological methods provide the way forward to avert the current food crisis and meet the needs of local communities' (Anon., 2008). Greenpeace International's press release was headed: 'Urgent changes needed in global farming practices to avoid environmental destruction ...World's leading scientists condemn industrial farming methods and see no role for GE as a solution to soaring food prices and hunger crisis fears'. 'Modern farming solutions champion biodiversity, are labour intensive and work with nature, not against it', says Benny Härlin from Greenpeace International, who was on the IAASTD's governing body. 'This report is a call for governments and international agencies to redirect and increase their funding towards a revolution in agriculture that is firmly agro-ecological' (Greenpeace, 2008).

Dr Hans Herren, founder and president of the Swiss BioVision Foundation (BioVision Foundation, 2008), says: 'The approach of the ecological development projects, which BioVision has been supporting and promoting for 10 years in Africa, follow exactly the recommendations set out by these international experts.' As Herren has since shown himself to be a strong opponent of the Green Revolution, one is entitled to ask if Herren's prior 'ecological' approach for 10 years with BioVision unduly influenced the outcome of the IAASTD, of which Herren was co-chair and editor.

In a similar way, the environmentalist/climate change background of the Director, Robert Watson, seems to have influenced the outcome of the IAASTD. For example, in his Testimony to the Financial Services Committee of the US House of Representatives (Do we have a food crisis: Are the recent prices increases a harbinger of the future?), Watson promoted 'strategies that combine productivity with protecting natural resources such as soils, water, forests, and biodiversity by supporting biologically diverse agro-ecological farming and grazing methods'. The attempt across the IAASTD process to link a supposed 'food crisis' (actually a temporary spike in commodity prices) to apparent failings of modern agriculture, then to bring many environmental NGOs to the writing of the IAASTD reports, and finally to claim a positive role for the ecosystem services of biodiversity to support farming, we find to be fundamentally anti-developmental. If developed countries want to follow this path – and the US rejection of the IAASTD report shows that at least one does not – so be it, but it is a recipe for agricultural stagnation in most developing countries.

The wholesale post-IAASTD parroting of support for agroecology underlines the dangers to agricultural development caused by the uncritical compilation of the IAASTD. For example, a recent press release from the United Nations Office at Geneva (www.unog.ch) was entitled: '"Agroecology outperforms large-scale industrial farming for global food security," says UN expert' (UN Special Rapporteur on the right to food, Olivier De Schutter). The sole source for this false claim is given as:

> The widest study ever conducted on agroecological approaches (Jules Pretty, Essex University, UK) covered 286 projects in 57 developing countries, representing a total surface of 37 million hectares: the average crop yield gain was 79 per cent.

This same Pretty *et al.* (2006) report is used by the IAASTD as a justification for agroecology: 'A recent study reports 286 projects with agroecological interventions that include 12.6 million producers on approximately 37 million ha, or the equivalent of 3% of the land in non-industrialized countries (Pretty *et al.*, 2006)' (McIntyre *et al.*, 2009c, p. 52). But the subject of Pretty *et al.* (2006) was 'resource-conserving agriculture' and not agroecology in the usage of the IAASTD report. The full list of such agricultural approaches listed by Pretty *et al.* (2006) were: integrated pest management; integrated nutrient management; conservation tillage; agroforestry; aquaculture; water harvesting; and livestock integration into farming systems. These are all features of present conventional agricultural research (indeed there are several CGIAR research centres devoted to these topics and three of the co-authors of the Pretty *et al.* report work for them). These approaches also include sound traditional practices such as water harvesting (one of us, D.W., once had the pleasure of working for a year in Yemen, where a multitude of ancient water- and silt-harvesting techniques of decided genius provides fertile soil and food in a very hostile environment).

Where then is the 'Third Way' of agroecology, vaunted by the IAASTD, to replace both traditional and conventional farming? Incidentally, the projects contributing to the Pretty *et al.* (2006) report included the use of pesticides, herbicides, inorganic fertilizer and, for the many projects on soybean in Latin America, a distinct probability of using GM varieties: these projects are not even organic, never mind 'agroecological', in the IAASTD context. A major design flaw of the Pretty *et al.* (2006) report is that it measured the yield difference between 'no project' versus 'project' intervention (see Phalan *et al.*, 2007 for this and other criticisms of Pretty *et al.*, 2006). A more valid method to demonstrate the value of resource-conserving agriculture would have been to compare 'current best practice' versus 'resource conservation'.

It is not even certain from the various IAASTD reports just what authors mean by 'agroecology' and how it differs from other approaches. For example, it may be bundled inextricably with other approaches; 'Sustainable agriculture approaches come under many names: agroecology, organic farming, low external input farming, ecological agriculture, biodynamic agriculture and permaculture' (McIntyre et al., 2009c, p. 48). The glossary of the each of the five Sub-global reports describes rather than defines agroecology. Agroecology is:

> The science of applying ecological concepts and principles to the design and management of sustainable agroecosystems. It includes the study of the ecological processes in farming systems and processes such as: nutrient cycling, carbon cycling/sequestration, water cycling, food chains within and between trophic groups (microbes to top predators), lifecycles, herbivore/predator/prey/host interactions, pollination etc. Agroecological functions are generally maximized when there is high species diversity/perennial forest-like habitats.

This is certainly inadequate as a definition, as all science-based conventional farming applies ecological principles and concepts to operations. *We suggest that there is no new science in 'agroecology' and seemingly no place for 'agroecology' in the continuous spectrum from traditional to conventional approaches.*

However, our concern – and it is a major one – is with the last sentence of the definition, with its promotion of high species diversity and perennial 'forest-like' habitats. There are three major errors with this sentence. It confuses '*ecological functions*' of high biodiversity with '*ecological services*'. Ecological *functions* may be maximized in natural species-diverse forests, but, importantly, are *not* under our control and may be exceedingly negative, as with interspecies competition, and with pest and disease outbreaks. In contrast, ecological *services* of high diversity *are* under our control for farming, and have been, since the dawn of farming, based on ecological knowledge. Such ecological services have been manipulated to benefit farming to increase the positive services and to decrease or control the negative aspects of farm ecology. The second error – and it is a disastrously bad one – is that any attempt to apply ecological concepts and principles to the design of farming to encourage high species diversity/forest-like habitats is ignoring just what the word 'agriculture' means. Agriculture for ten thousand years has been 'field-tilling'. Preparing seedbeds and controlling weeds is not just a form of *al fresco* exercise for farm families, they have an important purpose. And the purpose is dictated by the absolute ecological fact of plant competition: more weeds, less crop yield (we have discussed the ecology of succession, competition, and the evolution of farmers' practices in Chapter 3, this volume).

The third error is one of priority. The term 'agroecological' has long been used to compare farming under similar ecological conditions, as in 'agroecological zone' and 'agroecologies' – used across the IAASTD Sub-Global reports. For example, 'the world's agroecological zones' and that 'countries share similar agroecological characteristics' and that development needs to be adapted to the 'specific agroecological and socioeconomic conditions of the farm enterprises' (Central and West Asia and North Africa Sub-Global Report 2008). It is simply not possible to transfer the term to an entirely different meaning and then in the same report mix the two meanings.

Perhaps the most questionable discussion of 'agroecology' can be found in the IAASTD Latin America and the Caribbean Sub-Global Report. This divides all agriculture into: 'three main systems of production in the region: the indigenous/traditional, the conventional/productivist and the emerging agroecological system.' (However, the correct terminology of 'agroecological zones' continued to be used in the LAC report – a problem of multi-authored and poorly edited reports.) In this same report any useful concept of 'agroecology' gets bogged down with multiple social issues, for example, 'the rise of very strong rural social movements and indigenous movements that propose alternatives for autonomy, food sovereignty, agroecology and peasant networks.' It is worth noting that 'indigenous movements' did not propose 'agroecology': it was mainly promoted by academics based in the USA.

We believe that the continual promotion of what is a new and untried approach to agriculture in the LAC report and beyond is highly questionable and undermines the entire IAASTD process. For example, the statement that 'organic or agroecological foods are of significantly better quality than conventional ones' has been shown not to be true (FSA, 2009) and should never have passed the editors unquestioned. It is stated that transgenes are prohibited in 'organic or agroecological foods' (McIntyre et al., 2009b, p. 62). While this is true for organic food it cannot be true for 'agroecological foods' as they are not yet recognized or supported in international or national legal food standards.

We have criticized the unfounded assumption that agriculture should mimic complex natural ecosystems for many years (Wood, 1998; Lenné and Wood, 1999a, b; Wood and Lenné, 2001), yet the IAASTD LAC (McIntyre et al., 2009b, p. 171) continues to claim that the basic paradigm of agroecology:

> is that the more similar the agricultural, forestry and cattle-farming ecosystems are to the natural ecosystem the more sustainable are medium- and long-term production and other environmental services, such as the recycling of nutrients, carbon sequestration in soils, and water percolation, detoxification, regulation and storage.

This – based as it is on a belief in the necessity of biological diversity in natural ecosystems

– is an unproven and dangerous paradigm for the future global food production. It ignores sound ecological science on plant succession and on the status of climax vegetation. It ignores the genius of traditional farming in controlling tropical forest through shifting cultivation tropics-wide, and it shows no knowledge whatever of the vast scale and success of decidedly unnatural yet highly productive traditional terrace agriculture to capture water and silt. Despite providing an uncritical platform for 'agroecology' and despite the close involvement of NGOs known to be hostile to modern farming, the IAASTD has failed to establish the 'Third Way' of 'agreoecology' as an alternative to the functional and effective blend of traditional and conventional farming.

In contrast to over-promoting by the IAASTD report of the dubious pseudoscience of 'agroecology' there is an outstanding ignoring of gardens (two mentions only in a section on women in agriculture – McIntyre *et al.*, 2009a; p. 78). Given the vast importance of gardens of many kinds (and in many ecological settings) in providing for crop diversity, nutritional value, market opportunities, wildlife habitats, security from crop theft, woodlots, opportunities to work from home, child care and much more, the IAASTD neglect of gardens is a reflection of its operational blindness, a partiality to a dubious 'agroecology' and a distinctly inadequate approach to feeding people in developing countries. As the original objective of the IAASTD was food security, gardening should have been centre stage. Instead there was a formidable bias towards the untried claims of 'agroecology'. We again refer to the neglect of gardens in Chapter 13, this volume.

Reliance on organic agriculture to feed the world

The IAASTD report repeatedly highlights and promotes organic agriculture as a key approach to increasing yields and supporting future global food security (McIntyre *et al.*, 2009). Organic systems based on ecological approaches are promoted for their potential to enhance ecosystem functionality, environmental quality and social equity. But, it is far from clear how organic agriculture will reduce hunger and poverty. By concentrating on organic approaches to farming, the IAASTD misses important opportunities just as it did by concentrating on GM crops in the biotechnology section (see above). In 2006, global certified organic production encompassed 31 million ha and 600,000 farmers, that is, only 2% of the global crop cultivated area. The IAASTD strangely ignores other dominant, widespread and proven options for increasing production on the other 91% of the cultivated area (less the 7% used for GM crops) through the further improvement of existing technologies based on high-yielding, fertilizer-use efficient crops and non-organic input technologies. Certified organic agriculture is proposed as an attractive rural development pathway – a chance to expand the global market and extend economic opportunities to small-scale farmers – although no evidence is given to support the feasibility of this approach. Furthermore, the very strict regulations covering organic food exported into Europe from East Africa, for example, make it very difficult for even the major export companies to meet the quality standards (Mr Tiku Shah, Director, Sunripe Ltd., Kenya, 2005, personal communication).

Furthermore, the IAASTD fails to acknowledge that for most of the past 10,000 years, farming has been through organic methods. This type of low yield agriculture resulted in the many famines that the human population constantly suffered until the 1960s and the widespread adoption of Green Revolution technologies (Large, 1940; Carefoot and Sprott, 1969). Hence, to suggest that widespread promotion of organic agriculture is a sound option to reduce hunger and poverty 'defies logic and demonstrates how the so-called "science- and evidence-based" assessment of the IAASTD has been completely over-ridden by ideological-based green-washing' (Wager, 2008).

The strong support given by the IAASTD report for greatly increasing organic farming globally appears to be based on an analysis by Badgley *et al.* (2007), as two of the authors on this paper were also contributors to the

IAASTD report (McIntyre *et al.*, 2009). In this paper, Badgley, who is a palaeontologist and not an agricultural scientist, claimed that organic farming, if used globally, would provide sufficient food for a growing world population. This claim is based on a survey of selected literature, some of which was unpublished, on comparisons of organic and conventional yields and assessments of nitrogen fixation by legumes. This information was analysed and used to calculate potential food production. This paper has since been severely criticized by scientists for: (i) omitting many published papers that show organic yields to be substantially lower than conventional yields; (ii) misreporting yield results and multiple use of the same data from different sources; and (iii) for wrongly equating mineralization levels with nitrogen in seed yield (Avery, 2007; Goulding and Trewavas, 2009). As a result, the conclusions are fatally flawed due to misinterpretation of data and erroneous calculations (Avery, 2007; Kirchman *et al.*, 2008; Connor, 2009).

On average, many studies have shown that organic agriculture produces about 60–70% of the yield produced under conventional agriculture (Mader *et al.*, 2002; Stockdale *et al.*, 2002; Kirchmann *et al.*, 2008; Goulding and Trewavas, 2009). Nobel Laureate the late Dr Norman Borlaug noted that organic agriculture can only feed 4 billion people and he did not believe that 2 billion people would volunteer to starve to death (Wager, 2008). If organic agriculture was to be increased on a global scale as promoted by the IAASTD, it would be necessary to cultivate most of the remaining wilderness areas and to substantially increase numbers of cattle and sheep for manure. For example, if Europe tried to feed itself organically, it would need an additional 28 million ha of cropland, equal to all remaining forest cover in Britain, Denmark, France and Germany. In order to produce enough manure to farm organically, the USA would need to increase its animal population fivefold (Paarlberg, 2010). This would severely threaten global biodiversity and have profoundly negative impacts on the environment through increases in methane production (Kirchmann *et al.*, 2008; Wager, 2008; Goulding and Trewavas, 2009). Furthermore, animal manure is bulky and very labour-intensive to transport and apply as well as being a valuable source of fuel in many developing countries – Bangladesh and India – which is unlikely to be sacrificed for fertilizer (Meisner, 2007). The world increasingly needs highly productive agriculture that can save land and biodiversity – not further increased land use for food production.

Badgley *et al.* (2007) also claim that legume nitrogen fixation would be sufficient to replace the current use of nitrogen fertilizer based on a selected literature on legume cover crops. However, reliable values of annual nitrogen fixation rates vary across countries, cropping systems and climates over an order of magnitude or more (Smil, 2001). In addition, biologically fixed nitrogen is not necessarily released in synchrony with crop demand, harvesting grain legumes can remove more nitrogen than is fixed, and legumes, e.g. soybean, can also be nitrogen plunderers (Kirchmann *et al.*, 2008). Furthermore, in practice, all existing cropland cannot be provided with nitrogen through an additional legume cover crop without significant disruption to crop area and food production (Connor, 2009). For example, in many developing countries with cropping intensity well over two crops per year, replacing one crop with a legume cover crop would effectively halve food production (Meisner, 2007). Finally, Badgley *et al.* (2007) confuse the soil nitrogen available to the plant with the amounts eventually taken up by plants, often 50–80% less than the data used in their calculations (Goulding and Trewavas, 2009). The insistence that the mineralization of soil organic matter and crop residues is the only way to provide nutrient to crops misses the best opportunity of using organic matter strategically and efficiently *with* fertilizers to synchronize applications to when the crop needs it most *and* to improve soil structure and water holding capacity for good root development. Soluble chemical fertilizers provide a readily available form of the *same ions* that plants would take up from mineralized organic matter (Goulding and Trewavas, 2009); there is absolutely no difference in the biochemical make-up of plants grown in pure organics compared to chemical fertilizers (Meisner, 2007). And, there is no scientific evidence that organic food is more nutritious than non-organic food (Williams, 2002; FSA, 2009; Rosen, 2010).

Comparative studies of organic versus conventional agriculture must be based on objective science and not on ideological bias, political correctness or environmental opinions (Kirchmann *et al.*, 2008). Most importantly, a critical analysis of the nature and use of organic versus conventional agriculture does not support the proposition that large-scale organic agriculture would be sufficient to feed the world (Connor, 2009). Those who use the conclusions of the Badgley *et al.* (2007) to promote organic agriculture – as the IAASTD (McIntyre *et al.*, 2009a) appears to have done – will be misled and limited resources for research and development will be wasted. There is considerable potential to increase production on the 91% of conventional cultivated area through the further improvement of science-based technologies of high yielding, fertilizer use-efficient crops and improved fertilizer application and management that seems less likely through organic farming with its arbitrary, often ideological regulations (Goulding and Trewavas, 2009). Furthermore, food produced under conventional methods is far cheaper than food produced under organic farming and this is likely to remain so for many years due to the lower yields and higher risks associated with organic production. Any attempt to convert world agriculture to organic would increase food prices significantly, placing millions of poor people at risk.

Increased Reliance on Small-scale Farmers for Future Food Production

The IAASTD report and the summary report for decision makers place considerable emphasis on the need to increasingly rely on small-scale farmers for future food production (McIntyre *et al.*, 2009). For example: 'Significant pro-poor progress requires creating opportunities for innovation and entrepreneurship which explicitly target resource poor farmers and rural labourers.' There is no doubt that future global food production strategies must place emphasis on those farming systems in all countries with the greatest potential to increase food production. As the majority of farmers in developing countries, especially Asia and Africa, are small-scale farmers, they have an important role to play, particularly for national food security. In Africa, for example, there are 33 million small-scale farmers – 80% of all farms – with an average size of 1.6 ha (Wiggins, 2009). In spite of claims by many NGOs that small farm agriculture globally has been bypassed, neglected and disenfranchised by modern AKST, millions of small-scale, poor farmers have benefited. There is overwhelming evidence that Green Revolution technologies substantially benefited small-scale rice and wheat farmers and rural labourers in Asia (Hanumantha Rao, 1994; Evans, 1998; Hazell, 2009; Li *et al.*, 2009; Jain, 2010; discussed above). There is also strong evidence that Green Revolution technologies have benefited small-scale maize and cassava farmers in East and Southern Africa and West Africa, respectively (Nweke, 2009; Spielman and Pandya-Lorch, 2009). These interventions have also shown that agriculture can be a key driver of growth and development for many of the world's poorest countries. However, it is also acknowledged that small-scale farmers have benefited unevenly from modern AKST. The reasons for this are many and varied, and often beyond the direct reach of AKST.

First, as the potential to increase yields of food crops other than rice, wheat and maize has generally proven more difficult, small-scale farmers growing such crops have generally benefited less. However, this varies from crop to crop and region to region. For example, small-scale farmers growing sorghum and millet in India have benefited considerably more than farmers growing the same crops in Africa because of the widespread use of hybrids, growth of private sector seed companies and better development of markets in India (Pray and Nagarajan, 2009; Spielman and Pandya-Lorch, 2009). There are no major reasons why African farmers could not also benefit under the same conditions. Second, and far more importantly, millions of small-scale farmers have not benefited from modern AKST due to the failure of technology promotion systems, inefficiencies in commodity and value chains and the lack of national enabling policies (Hazell *et al.*, 2007; Wiggins, 2009; Lenné and Ward, 2010 for vegetables in East Africa). These non-AKST barriers and bottlenecks that prevented millions of small-scale poor farmers from benefiting from yield-improving technologies in the past still persist in most developing countries today, continuing to severely reduce the potential impact of modern AKST. Finally, the severe erosion of funding for AKST in the past 30 years and the continued lack of both international and national investment continue to shackle agricultural growth based on small-scale farming systems in most developing countries (Pardey *et al.*, 2006; Pardey and Pingali, 2010).

Hence, there remains tremendous potential for many more small-scale farmers to benefit from existing AKST providing the non-AKST challenges facing small-scale farmers are addressed. Hazell *et al.* (2007) recommended three key elements necessary for promoting growth and equity for small-scale farmers in developing countries: (i) an enabling environment must be created which should include: a stable macro-economy; state-funded infra-structure support for rural roads, rural education and health care and agricultural research and extension; and good governance for agricultural rural development; (ii) farmers need to be encouraged to follow demand and market systems need to be improved and made more transparent; and (iii) institutional innovation is needed in providing inputs and services, for example, improved coordination in the delivery of input, financial, technical and output marketing services to enable small-scale farm intensification. Such actions are mainly within the jurisdiction of national governments rather than international bodies, although support from the latter is likely to be necessary. Unless key national policy makers adopt a more assertive agenda towards small-scale agriculture, there is a growing risk that rural poverty will increase dramatically (Ashley and Maxwell, 2001; Hazell *et al.*, 2007).

There are many good examples of successes in addressing some of the above constraints, both with individual farmer groups and at the national scale. A detailed assessment of the opportunities for farmers, traders, processors and consumers to improve the efficiency of domestic vegetable marketing systems in East Africa by Lenné and Ward (2010) is supported by a number of innovative and successful initiatives that have benefited many small-scale farmers. All of these initiatives have potential to be scaled-up both nationally and regionally to benefit many more small-scale farmers. In addition, successful promotion of improved AKST through small seed- and fertilizer-packs has been achieved through support networks of NGOs and CBOs in Malawi (Blackie and Ward, 2005) and is ongoing in East Africa (Farm Input Promotions Africa – see web link: www.fipsafrica.org). Furthermore, Operation Flood, a dairy development project that integrated over 6 million small-scale, marginal and landless dairy farmers in India, brought significant technological advances into the rural milk sector, commercialized small-scale dairy production, and transformed the policy environment in support of dairy industry growth (Cunningham, 2009). Harris *et al.* (2005) and Spielman and Pandya-Lorch (2009) provide more examples of successful, small-scale farmer food production initiatives.

In a detailed analysis of small farm agriculture in Africa, Wiggins (2009) showed that there is considerable differentiation amongst small farms. He warned that those who advocate the potential of small farm development need to recognize that most of the increased production, and hence the benefits – financially and otherwise – will accrue to only a minority of small-scale farmers – probably the upper 25% of small-scale farmers with better resources and land. The reality is that although improved small-scale farm development will produce more food, provide more opportunities for rural labour, foster rural development through investment in local goods and services and, possibly, reduce food prices, it is not likely to be sufficient on its own to deliver food security nationally and globally. There is little doubt that future global food security strategies must place emphasis on those farming systems in all countries with the greatest potential to increase food production – whether small, medium or large.

Increased Emphasis on Knowledge- and Labour-intensive AKST

The IAASTD report places considerable emphasis on the recognition of the complex 'multi-functional' nature of agricultural systems as the key platform for future implementation of AKST to reduce hunger and poverty and improve nutrition and livelihoods (McIntyre *et al.*, 2009). Such a multifunctional approach to AKST is expected to achieve social, environmental and economic sustainability in contrast to the perceived failures of previous mono-functional, single commodity approaches such as monocultures of major staple cereals. The report stresses that increased attention needs to be directed towards new and successful existing approaches to maintain and restore soil fertility and to maintain sustainable production through practices such as low-input, resource-conserving technologies based on integrated management systems, agroecological approaches, conservation agriculture, organic agriculture and permaculture. Such systems include mixed cropping, polycultures and agroforestry but not – it appears – staple cereal monocultures. The experts consider that these technologies are 'socially appropriate' for small-scale agriculture (McIntyre *et al.*, 2009). The scientific ability of these approaches to feed billions has already been substantially questioned and criticized (see above). Here, we question the wisdom of the 'multi-functional' view of agriculture and the practical and equitable issues of imposing complex, knowledge- and labour-intensive technologies on small-scale farmers in developing countries.

Surely the principal objective of agricultural practice is to cultivate the land to produce food, especially in developing countries where the need is greatest? Even in developed countries, where agri-environmental schemes are well-established, mono-functional, single commodity systems such as monocultures of staple cereals are well-accepted as the most practical and efficient way of producing most of our food. Such systems are integrated with set-aside, conservation strips and field boundaries and other agri-environmental options. Mono-functional, single commodity approaches to agriculture have successfully fed billions and are likely to continue to do so in the foreseeable future, as has been highlighted many times already throughout this volume. In the light of this, it seems rather unwise to give such strong endorsement to unproven, questionable 'multi-functional' approaches.

But far more contentious is the expectation that poor, small-scale farmers will voluntarily adopt complex, knowledge- and labour-intensive, and often risky AKST in preference to simple seed- or plant-based technologies with manageable labour requirements because it is 'socially appropriate'. This apparent social appropriateness of multi-functional approaches seems to be based on a distorted image of farmers, common in developed countries. This mythic image depicts them as romantically but demeaningly backward, tradition-loving, innocent and helpless creatures carrying on their occupation for love of the land and the soil, as practitioners of a 'way of life' rather than a toilsome income-earning occupation (Omvedt, 1998). The reality is that small-scale farms are commercial, profit-seeking units that seek to maximize their production as efficiently as possible (Lipton, 2005). Under-utilized labour is then available to generate further income through rural employment or value-adding activities. It is therefore unlikely that small-scale farmers will voluntarily adopt knowledge- and labour-intensive AKST unless the benefits – as food and income – are substantially greater than their existing practices.

Deficiencies in the 'Global Assessment' Approach

In recent years, global assessments have become the popular method to address issues of major international significance (Scoones,

2009). The IAASTD follows on from the FAO Global Plan of Action, the International Panel on Climate Change and the Millennium Ecosystem Assessment to name a few. Recently, the Global Conference on Agricultural Research for Development, GCARD 2010, brought together farmers, civil society, scientists, development agencies and policy makers in Montpellier to 'pull together solutions and ways forward to achieve agricultural development goals' (www.egfar.org/egfar/website/gcard/2010-conference). Such assessments attempt to combine 'expert assessment' with processes of 'stakeholder consultation' in a globally inclusive and participatory manner in response to critiques of past top-down, northern-dominated expert assessments (Scoones, 2009).

Global assessments are 'brave' attempts at engaging diverse groups of stakeholders on a key topic with major global ramifications (Scoones, 2009). It is argued that it is a more democratic and accountable system of governance and policy making. However, going beyond the well-rehearsed rhetoric of participation, inclusion and citizen engagement, significant shortcomings emerged in both the process and the outcomes driven by the underlying politics. The IAASTD process was a highly political setting, dominated by powerful groups with particular perspectives and interests set on undermining established views. Through campaigning and selective drafting, it was just another case of one stakeholder group's views being over-represented in the synthesis report. For example, the authoring and reviewing processes were captured by NGOs including the Pesticides Action Network North America (PANNA, 2008) and Greenpeace (Greenpeace, 2008). The end result was an inevitably partial, political and value-laden exercise (Scoones, 2009). It has been said that deliberative forms of democratic practice such as the global assessment approach are profoundly mistaken (Mouffe, 2005). Their aim to establish a world 'beyond left and right', 'beyond hegemony' and 'beyond antagonism' reveals a complete lack of understanding of what is at stake in democratic politics (and, for the IAASTD, what farmers actually do). The unfortunate result is to exacerbate the antagonistic potential existing in society, as clearly demonstrated by the questionable results of the IAASTD process (Scoones, 2009; discussed above).

Above all, the IAASTD was not, as claimed, a 'scientific review'. While the NGO lauding of the IAASTD attempts to capture the high ground by talking of a consensus of 400 scientists, the IAASTD was no such thing: it involved 400+ 'experts' in various roles, from many backgrounds and interests. Many were *not* scientists. The IAASTD certainly was too negative over the value of conventional agriculture, the immense success of the Green Revolution and the potential of transgenic crops. And, as a supposedly scientific review, the IAASTD should not have attempted to foist on the world a distinctly second-hand and, we think, second-rate 'agroecology' of questionable value, nor should the IAASTD ask for or respect some decidedly anti-development and anti-science views expressed by the many vocal NGOs involved in the process. On a personal point, we cannot find a single citation in the entire corpus of the IAASTD of our first 'Agrobiodiversity' book (Wood and Lenné, 1999), which was a comprehensive review written around the theme of agricultural knowledge for development (it appears only in the references, not the text, to the SSA report – apparently retained in error).

Due to its inability to deal effectively and realistically with the really tough issues and choices confronting future agriculture based on the extensive scientific evidence available, the IAASTD report is clearly not an appropriate roadmap for AKST to ensure future global food security. However, the IAASTD is substantially 'good in parts', with some valuable analysis masked by the flawed editing. The IAASTD can and should be withdrawn and re-edited by a panel of agricultural scientists to provide a more realistic and practical outcome for future AKST for development, building on the sound legacy of past and current successes in feeding increasing billions of people.

References

Anon. (2008) Civil Society Statement from Johannesburg, South Africa: a new era of agriculture begins today. Available at: www.agassessment.org/docs/Civil_Society_Statement_on_IAASTD-28Apr08.pdf (accessed 15 May 2010).

Ashley, C. and Maxwell, S. (2001) Rethinking rural development. *Development Policy Review* 19, 395–425.

Avery, A. (2007) 'Organic abundance' report: fatally flawed. *Renewable Agriculture and Food Systems* 22, 321–323. Available at: www.thetruthaboutorganicfoods.org/2007/09/14/%E2%80%9Corganic-abundance%E2%80%9D-report-fatally-flawed (accessed 15 May 2010).

Badgley, C., Moghtader, J., Quintero, E., Zakern, E., Chapell., J., Aviles-Vazquez, K., Samulon, A. and Perfecto, I. (2007) Organic agriculture and the global food supply. *Renewable Agricultural Food Systems* 22, 86–108.

BioVision Foundation (2008) Media Release: Launch of World Agriculture Report. Available at: www.agassessment.org/docs/BioVisionFoundation_IAASTD.pdf (accessed 15 May 2010).

Blackie, M.J. and Ward, A. (2005) Breaking out of poverty: lessons from harmonizing research and policy in Malawi. In: Harris, D., Richards, J.I., Silverside, P., Ward, A.F. and Witcombe, J.R. (eds) Pathways Out of Poverty. *Aspects of Applied Biology* 75, 115–126.

Carefoot, G.L. and Sprott, E.R. (1969) *Famine on the Wind: plant diseases and human history.* Angus and Robertson, London.

Connor, D.J. (2009) Organic agriculture cannot feed the world. *Field Crops Research* 106, 187–190.

Cunningham, K. (2009) Rural and urban linkages: Operation flood's role in India's dairy development. IFPRI Discussion Paper 00924, 2020 Vision Initiative.

Dano, N. (2008) The truth behind the IAASTD report: misquote or misinformed. Sustainable Development Issues Network Outreach Issues 8 May 2008, p. 5. Available at: www.sdin-ngo.net/publications/oi/080508-05.html (accessed 15 May 2008).

DeGregorio, T.R. (2004) Green Myth vs. the Green Revolution. Butterflies and Wheels 5 February. Available at: www.butterfliesandwheels.org/2004/green-myth-vs-the-green-revolution (accessed 4 May 2010).

Evans, L.T. (1998) *Feeding the Ten Billion.* Cambridge University Press, Cambridge.

Evenson, R.E. and Gollin, D. (eds) (1997) *Crop Variety Improvement and its Effect on Productivity: The Impact of International Agricultural Research.* CAB International, Wallingford, UK.

FSA (2009) *Comparison of composition (nutrient and other substances) of organically and conventionally produced foodstuffs: a systematic review of available literature.* Report for the Food Standards Agency. Nutrition and Public Health Intervention Research Unit, London School of Hygiene & Tropical Medicine, London. Available at: www.food.gov.uk/multimedia/pdfs/organicreviewappendices.pdf (accessed 7 July 2010).

Goulding, K.W.T. and Trewavas, A.J. (2009) Can organic agriculture feed the world? AgBioView Special Paper. Available at: www.agbioworld.org/newsletter_wm/index.php?caseid=archive&newsid=2894 (accessed 25 April 2010).

Greenpeace (2008) Press release: Urgent changes needed in global farming practices to avoid environmental destruction. Available at: www.agassessment-watch.org/docs/greenpeace_15_april.pdf (accessed 15 June 2010).

Hanumantha Rao, C.H. (1994) *Agricultural Growth, Rural Poverty, and Environmental Degradation in India.* Oxford University Press, Delhi and New York.

Harris, D., Richards, J.I., Silverside, P., Ward, A.F. and Witcombe, J.R. (2005) Pathways out of poverty. *Aspects of Applied Biology* 75.

Hazell, P.B.R. (2009) Transforming agriculture: the Green Revolution in Asia. *IFPRI Discussion Paper* 00911, 2020 Vision Initiative.

Hazell, P.B.R., Poulton, C., Wiggins, S. and Dorward, A. (2007) The future of small farms for poverty reduction and growth. *International Food Policy Research Institute 2020 Vision Discussion Paper* 42.

Herren, H. and Ishii-Eiteman, M. (2010) Genetically modified crops are not the answer (see web link: http://thehill.com/opinion/op-ed/93907-genetically-modified-crops-are-not-the-answer/).

ICSU (2003) New Genetics, Food and Agriculture: Scientific Discoveries – Societal Dilemmas. ICSU web site, see web link: http://www.icsu.org/2_resourcecentre/INIT_GMOrep_1.php4.

IRRI (2008a) *Background Paper: The rice crisis: What needs to be done?* International Rice Research Institute (IRRI), Los Baños, the Philippines. Available at: www.irri.org (accessed 15 June 2010).

IRRI (2008b) *Responding to the rice crisis: How IRRI can work with its partners.* International Rice Research Institute (IRRI), Los Baños, Philippines. Available at: www.irri.org (accessed 15 June 2010).

IRRI (2010) GRiSP International Rice Research Institute (IRRI), Los Baños, Philippines. Available at: www.irri.org (accessed 10 November 2010).
Jain, H.K. (2010) *The Green Revolution: History, Impact and Future.* Studium Press LLC, Houston, Texas.
James, C. (2010) Global status of commercialized biotech/GM crops in 2009: ISAAA Brief 41. Available at: www.isaaa.org/resources/publications/briefs/41/executivesummary.
Kirchmann, H., Bergström, L., Kätterer, T., Andrén, O. and Andersson, R. (2008) Can organic crop production feed the world? In: Kirchmann, H. and Bergström, L. (eds) *Organic Crop Production – Ambitions and Limitations.* Springer Science and Business Media, B.V., pp. 39–72.
Large, E.C. (1940) *The Advance of the Fungi.* Johnathan Cape, London.
Lenné, J.M. and Ward, A.F. (2010) Improving the efficiency of domestic vegetable marketing systems in East Africa: constraints and opportunities. *Outlook on Agriculture* 39, 31–40.
Lenné, J. and Wood, D. (1999a) Vegetational diversity in agroecosystems: a mixed blessing for successful pest management? In: Terry, P.J. (ed.) *International Crop Protection: Achievements and Ambitions.* BCPC Symposium Proceedings No. 73, British Crop Protection Council, Farnham, UK, pp. 75–98.
Lenné, J.M. and Wood, D. (1999b) Optimizing biodiversity for productive agriculture. In: Wood, D. and Lenné, J.M. (eds) *Agrobiodiversity: Characterization, Utilization and Management.* CAB International, Wallingford, UK, pp. 447–470.
Li, Jiming, Xin, Yeyun, Yuan, Longping (2009) Pushing the yield frontier: hybrid rice in China. *IFPRI Discussion Paper* 918, 2020 Vision Initiative.
Lipton, M. (2005) The family farm in a globalizing world. *International Food Policy Research Institute 2020 Vision Policy Brief* 74.
Lipton, M. and Longhurst, R. (1989) *New Seeds and Poor People.* Unwin Hyman, London.
Mader, P., Fliessbach, A., Dubois, D., Fried, P. and Niggli, U. (2002) Soil fertility and biodiversity in organic farming. *Science* 296, 1694–1697.
McIntyre, B.D., Herren, H.R., Wakhungu, J. and Watson, R.T. (eds) (2009a) *Agriculture at the Crossroads. The global report of the International Assessment of Agricultural Knowledge, Science and Technology.* Island Press, Washington, DC.
McIntyre, B.D., Herren, H.R., Wakhungu, J. and Watson, R.T. (eds) (2009b) *Agriculture at the Crossroads.* Vol. III *Latin America and the Caribbean (LAC). Report of the International Assessment of Agricultural Knowledge. Science and Technology,* Island Press, Washington, DC.
McIntyre, B.D., Herren, H.R., Wakhungu, J. and Watson, R.T. (eds) (2009c) *Agriculture at the Crossroads.* Vol. II *East and South Asia and the Pacific (ESAP). Report of the International Assessment of Agricultural Knowledge. Science and Technology,* Island Press, Washington, DC.
Meisner, C. (2007) Why organic food can't feed the world. *Cosmos Online* 24 September. Available at: www.cosmosmagazine.com/features/online/1601/why-organic-food-cant-feed-world (accessed 4 May 2010).
Mouffe, C. (2005) *On the Political.* Routledge, London.
Nature Biotechnology (2008) Off the rails. *Nature Biotechnology* 26, 247.
Naylor, G. (2009) Agriculture does not need 'business as usual'. *Chicago Tribune* 22 January.
Nweke, F. (2009) Resisting viruses and bugs: cassava in sub-Saharan Africa. *IFPRI Discussion Paper* 00912, 2020 Vision Initiative.
Omvedt, G. (1998) Terminating choice. *The Hindu* 14 December, p. 12.
Paarlberg, R. (2010) Attention Whole Foods Shoppers: Stop obsessing about arugula. Your 'sustainable' mantra – organic, local and slow – is no recipe for saving the world's hungry millions. *Foreign Policy* 4 May. Available at: www.foreignpolicy.com/articles/2010/04/26/attention_whole_food_shoppers (accessed 5 June 2010).
PANNA (2008) UN assessment of agriculture, poverty, hunger and the environment. Pesticide Action Network North America. Available at: www.panna.org/jt/agAssessment (Accessed 25 April 2010).
Pardey, P.G. and Pingali, P.L. (2010) *Reassessing International Agricultural Research for Food and Agriculture.* Global Conference on Agricultural Research for Development 2010, Background Paper. Available at: www.gcard2010.net.
Pardey, P.G., Alston, J.M. and Piggott, R.R. (eds) (2006) *Agricultural R&D in the Developing World: Too Little, Too Late?* International Food Policy Research Institute, Washington, DC.
Phalan, B., Rodrigues, A.S.L., Balmford, A., Green, R.E. and Ewers, R.M. (2007) Comment on 'Resource-Conserving Agriculture Increases Yields in Developing Countries'. *Environmental Science and Technology* 41, 1054–1055.

Pray, C.E. and Nagarajan, L. (2009) Pearl millet and sorghum improvement in India. *IFPRI Discussion Paper* 00919, 2020 Vision Initiative.

Pretty, J., Noble, A., Bossio, D., Dixon, J., Hine, R.E., Penning de Vries, P. and Morison, J.I.L. (2006) Resource conserving agriculture increases yields in developing countries. *Environmental Science and Technology* 40, 1114–1119.

Reiter, P. (2009) Against the prevailing wind. *The Economist* 16 December.

Rosen, J.D. (2010) A review of nutrition claims made by proponents of organic food. *Comprehensive Reviews in Food Science and Food Safety* 9, 270–277.

Scoones, I. (2009) The politics of global assessments: the case of the International Assessment of Agricultural Knowledge, Science and Technology for Development (IAASTD). *The Journal of Peasant Studies* 36, 547–571.

Shiva, V. (1993) *Monocultures of the Mind: Perspectives on Biodiversity and Biotechnology.* Zed Books and Third World Network, London and Penang, Malaysia.

Smil, V. (2001) *Feeding the World: A Challenge for the Twenty-first Century.* MIT Press, Cambridge, Massachusetts.

Spielman, D.J. and Pandya-Lorch, R. (2009) *Millions Fed: Proven Successes in Agricultural Development.* International Food Policy Research Institute, Washington, DC.

Stockdale, E.A., Shepherd, M.A., Fortune, S. and Cuttle, S.P. (2002) Soil fertility in organic farming systems – fundamentally different? *Soil Use & Management* 18 (Suppl.), 301–308.

Wager, R. (2008) Why the IAASTD failed. Available at: http://web.viu.ca/wager (accessed 10 April 2010).

Wiggins, S. (2009) Can the smallholder model deliver poverty reduction and food security for a rapidly growing population in Africa? *Paper prepared for the FAO Expert Meeting 'How to Feed the World in 2050'*, FAO, Rome, 24–26 June, 2009.

Williams, C.M. (2002) Nutritional quality of organic food: shades of grey or shades of green? *Proceedings of the Nutritional Society* 61, 19–24.

Witcombe, J.R. (1999) Does plant breeding lead to a loss of genetic diversity? In: Wood, D. and Lenné, J.M. (eds) *Agrobiodiversity: Characterization, Utilization and Management.* CAB International, Wallingford, UK, pp. 245–272.

Wood, D. (1998) Ecological principles in agricultural policy: but which principles? *Food Policy* 23, 371–381.

Wood, D. and Lenné, J.M. (1999) Agrobiodiversity and natural biodiversity: some parallels. In: Wood, D. and Lenné, J.M. (eds) *Agrobiodiversity: Characterization, Utilization and Management.* CAB International, Wallingford, UK, pp. 425–445.

Wood, D. and Lenné, J.M. (2001) Nature's Fields: a neglected model for increasing food production. *Outlook on Agriculture* 30, 165–174.

World Bank (2008) *Meeting Growing Demand for Agriculture through Innovations in Science and Technology.* World Development Report 2008, World Bank, Washington, DC.

12 Agrobiodiversity Management for Climate Change

R. Ortiz

Introduction

The world faces an increasing demand for its finite resources. There will be 1.7 billion more people to feed by 2030, but with a declining ratio of arable land between 40% and 55% and about 1.8 billion people living under water scarcity (CropLife International, 2009). Furthermore, a recent scenario analysis suggests that on average about 3000 kcal per capita daily will need to be available worldwide in 2050 to feed the growing human population (Hubert *et al.*, 2010). This goal may be seen as attainable but the world in the mid-21st century will be facing water shortages, flooding and global warming as a result of climate change (Baetghen, 2009). Increasingly, more wealthy and healthy people will demand greater dietary diversity in a global bio-based economy. Global economic growth and sustainable intensification of crop–livestock agroecosystems remain therefore as major challenges for feeding this growing human population. In this regard, today's farming worldwide needs high yielding crops that can grow more efficiently, such as those requiring less inputs or adapting to water and heat stresses or new epidemics of emerging pests at a time of global climate change.

In this chapter, innovations on agrobiodiversity management that reduce vulnerability to climate change (e.g. mitigation through management and adaptation through the genetic improvement of resilient and climate-proof crops) are considered in detail. Such innovations will greatly assist in addressing these challenges and will ensure enough food, feed, fibre and biofuel supply in the next decades. Furthermore, learning from today's agrobiodiversity management that buffers crops and cropping systems against annual extreme weather variations could help to improve their adaptation to future climate. Nelson *et al.* (2009) argued recently that crops and livestock that perform reasonably well in a range of production environments are needed rather than those doing extremely well in a narrow set of climates. And, as indicated by Challinor *et al.* (2007), crop cultivars should adapt to both means and extremes of temperature stresses under climate change.

Climate Change Impacts on Agrobiodiversity and Food Security

Global yield losses due to global warming have amounted to 40 million t or US$5 billion yearly for wheat, maize and barley since 1981 (Lobell and Field, 2007). Furthermore, crop modelling shows that climate change will continue to reduce agricultural production, thus reducing food availability and thereby affecting food security and farm incomes (Schmidhuber and Tubiello, 2007; Lobell *et al.*, 2008; Battisti and Naylor, 2009). The Intergovernmental Panel on Climate Change in its 4th Assessment Report confirms that indeed changing climate will bring a high intensity and frequency of storms, drought and flooding, weather extremes, altered hydrological cycles and precipitation, which, without doubt, will affect agricultural production. These impacts will depend on region, growing season, weather patterns and crops. For example, severe crop losses are expected for cotton, maize and soybean in the USA by the end of this century due to warmer temperatures (Schlenker and Roberts, 2009). Grain harvests in China and South Asia may also drop by 37% and 30%, respectively, by 2050 due to weather extremes, whereas extreme drought (i.e. doubling severity and frequency) in north-east China could result in 12% crop losses (or 13.8 million t) by 2030 (Bloomberg News, 2009). Although models provide an important tool for understanding and assessing future climate impacts, results from modelling should be taken with caution because their spatial scales could fail to capture topographical or microclimatic buffering, and they do not often consider the wide acclimation capacity of animals and plants (Willis and Bhagwat, 2009). Hence, as stated by Tubiello *et al.* (2007), understanding the key dynamics characterizing interactions between elevated CO_2 and changes in climate variables (e.g. extremes, soil and water quality, pests, pathogens) and ecosystem vulnerability remains as priority research for quantifying better the impacts of climate change on crops and pastures.

Changes in climate could also rapidly shift plant distributions because some species will expand in newly favourable areas and others will decline in increasingly adverse locations (Kelly and Goulden, 2008). For example, models suggest that at least 50% of the plant species in Europe could be vulnerable or threatened by 2080 (Thuiller *et al.*, 2005). In this regard, Lane and Jarvis (2007) using the Ecocrop model (http://ecocrop.fao.org) projected the impact of climate change by 2055 on suitable areas for most important staples and cash crops, including those of the multilateral system of the International Treaty on Plant Genetic Resources for Food and Agriculture. The largest gain in suitable areas is likely to be in Europe (3.7%) whereas sub-Saharan Africa and the Caribbean may suffer 2.6% and 2.2% declines of land area, respectively. Although their modelling suggests some crop gains in suitable areas (e.g. 31% for pearl millet, 18% for sunflower, 15% for chickpea and 14% for soybean), these 'new crop lands' are in regions where they are not important local food staples, e.g. 10% increase for pearl millet in Europe and the Caribbean rather than in sub-Saharan Africa and India.

As this chapter will show, agrobiodiversity remains the main raw material for agroecosystems to cope with climate change because it can provide traits for plant breeders and farmers to select resilient, climate-ready crop germplasm and release new cultivars. However, modelling research suggests that some crop wild relatives may become extinct by 2055 (Jarvis *et al.*, 2008), e.g. 8% of *Vigna*, 12% of tuber-bearing *Solanum* and 61% of *Arachis* species. Collecting samples of endangered species to be preserved in genebanks will be the first step, but also protecting the habitats where they thrive should be a must to ensure the *in situ* evolutionary processes of wild species contributing to agrobiodiversity. Furthermore, as noted by recent research of maize, pearl millet and sorghum genetic resources in sub-Saharan Africa (Burke *et al.*, 2009), available genetic resources for these crops in genebanks may not be the most useful for adapting them to climate change in this continent. Hence, analogue crop areas for many future climates should be promising locations to focus future collecting and conserving of crop genetic resources.

Inter-governmental Panel on Climate Change (IPCC) and Agrobiodiversity Management

Although the world can cope with climate change by maintaining and using agrobiodiversity, IPCC has not given enough attention to the value of biodiversity for food and agriculture, which will increase with global warming, drought and other stresses. The chapter on agriculture of the 4th IPPC Assessment (Metz *et al.*, 2007) does not mention agrobiodiversity (or refer properly to agricultural biodiversity) and how it can contribute to climate change adaptation. There are, in this and other chapters, a few references to biodiversity at large and mostly related to mitigation or losses brought by climate change, particularly in forests or the soil biota. However, agrobiodiversity maintenance through use plays an important role for climate change adaptation. In the past, crop and livestock diversity has traditionally been an important part of farmer risk management. An increase of agrobiodiversity use is further expected and necessary as a result of climate change.

Agrobiodiversity at the gene, species and agroecosystem levels increases resilience to the changing climate. Promoting agrobiodiversity remains therefore crucial for local adaptation and resilience of agroecosystems (FAO Interdepartmental Working Group on Climate Change and the Stockholm Environment Institute, 2007). Adapting agriculture to climate change will indeed rely on matching crop cultivars to future climates and plant breeding for coping both with climate variability and extremes, but also on promoting farmer resilience and adaptability. Hence, agrobiodiversity is not a victim of climate change but provides the raw resource for adapting to this global challenge.

The United Nations Environment Programme considers that breeding stress-resistant crop cultivars, along with provision of crop and livestock insurance, social safety nets, new irrigation schemes and local management form the core of short-term responses for adapting to climate change (UNEP, 2008). Likewise, local agrobiodiversity is an important coping mechanism, especially for most vulnerable people. However, the locally available agrobiodiversity in some areas may not be able to adapt quickly to the changing climates. Hence, new crop cultivars, livestock breeds or other species better suited to these new environments will be needed to cope with climate change.

Coping with Climate Change through Knowledge-base Agricultural Research Answers

Howden *et al.* (2007) advocate a multidisciplinary approach to address climate change. This integrated rather than disciplinary approach also considers strengthening the interface with decision makers. Recently, the Food and Agriculture Organization of the United Nations provided a summary of potential changes in agroecosystems that have been proposed to increase agricultural production, as well as to decrease output variability due to climate variability and extreme climate events (FAO, 2009). The suggested options advocate an adaptation approach to climate change focusing on an increase in agroecosystem resilience that reduces the impacts brought by extreme climate events on food supply. In this regard, any adaptation strategy should aim to minimize the agroecosystem's vulnerability to climate change. Adapting agriculture to climate change will depend on the affordability of the adaptive measurements, technology access and biophysical characteristics (land and water availability, soil, topography) and useful agrobiodiversity for crop and livestock breeding.

Cropland management for climate change

Sustainable land management involves changes that increase natural capital and reduce negative environmental impacts, and offers a means for mitigating climate change through carbon sequestration in soils and biomass, as well as reducing emissions from degradation and inappropriate farming practices (Various, 2008). New cultivars, conservation agriculture practices (e.g. minimum tillage) and increased input efficiency are among those adaptation options for cropland management (Reynolds and Ortiz, 2010). Conservation agriculture can increase soil organic carbon, thereby improving soil fertility, and also helps to sequester carbon in agricultural soils. Crop breeding (including modern biotechnology such as genomics and transgenics) provides genetically enhanced seed embedded technology (GESET) that adapts crops to both abiotic and biotic stresses, whereas conservation agriculture practices assist in both adapting and mitigating climate change, especially in intensive agroecosystems. Likewise, geographical information systems (GIS) and modelling utilize available crop and cropping systems datasets to define most suitable germplasm and agrobiodiversity management practices to deploy appropriate GESET to target environments. For example, productivity of crops, livestock and pastures in Africa is predictably associated with the El Niño Southern Oscillation and the North Atlantic Oscillation (Stige *et al.*, 2006). Further analysis suggests that maize yield in Africa may be severely reduced if the global climate changes towards more El Niño-like conditions. Increased use of irrigation and changes in land use – including the planting of alternative crops (e.g. cassava or sorghum) – may assist ensuring enough African food production under such a scenario. Similarly, Kumar (2008) indicates that the Indo-Gangetic Plains may remain as a food basket if newly bred cultivars are grown in location-wise, judiciously selected planting schedules in this intensive agroecosystem that may be affected both by heat and water stresses.

Adapting livestock to changing climate

There will be significant impacts on livestock and livestock-based systems as the climate changes (Thornton *et al.*, 2007). Options are therefore needed to adapt livestock to climate change. For example, local breeds – which appear to be better than exotic germplasm for coping with climate change – and community-based, participatory breeding could assist in adapting livestock to global warming and drought (CGIAR, 2009). In this approach, the entire community herd is regarded as a single breeding pool for genetically enhancing target traits such as milk yield or growth rate. Likewise, shrub and other species adapted to drought- and heat-prone environments will help in replanting grazing lands whereas fodder banks with legumes such as *Stylosanthes* may ensure feed availability during scarcity periods due to drought. Feeding better quality diets (e.g. forage legumes with low tannin content) to ruminants will also reduce methane emissions, of which livestock and paddy rice fields are among the largest contributors in agriculture. Similarly, recuperating degraded pasture lands with silvopastoral systems that combine highly productive African forage grasses, such as *Brachiaria* species (Fisher *et al.*, 1994), and trees can capture significant amounts of carbon from the atmosphere and retain it due to their deep root systems. Furthermore, increasing the carbon sequestered by grasslands may assist pastoralists adapting to climate change (Neely *et al.*, 2009). Adding carbon will improve the capacity for water retention of the soil, thereby enhancing its ability to withstand drought.

Genetic enhancement for a changing climate

The use of GIS tools and passport data facilitates identification of accessions for stress-prone environments whereas the available characterization (including DNA fingerprinting) and evaluation data assist in selecting promising accessions for further screening against specific stresses. The Focused Identification of Germplasm Strategy (FIGS) can assist in this endeavour (Mackay *et al.*, 2004). FIGS uses geo-coordinates of collecting sites (passport descriptors), environmental data (including agroecological data) and GIS technology to select 'best-bet' genebank accessions that could have evolved under selection pressures for the trait(s) of interest. For example, El Bouhssini *et al.* (2009) found new sources of resistance at the vegetative stage to overwintered Sunn pest adults after sampling wheat genebank accessions using FIGS. Similarly, Bhullar *et al.* (2009) used FIGS to select a set of 1320 bread wheat landraces (from a database of 16,089 accessions) for large-scale allele mining, which led to identifying resistance alleles of the powdery mildew resistance gene *Pm3*, doubling the known functional allelic diversity at this locus.

Information on multi-site testing locations and crop performance (or any other assessment) therein provides means for modelling 'stress-impacts' on crops in target populations of environments, as well as for finding areas where climate change scenarios already occur or can be mimicked to set screening sites to select promising genotypes from large segregating populations. Furthermore, reliable phenotypic data and appropriate statistical techniques (variance associated with genotype by environment interactions) allows partitioning to discrete environmental variables in time. This can then be interpreted in terms of the unique response of a genotype, at a given phenological stage, to year-to-year variation in weather patterns (Crossa *et al.*, 2004). Such approaches can help better to define the environments affected by heat or water stresses, and identify sensitive growth stages to both. For example, new crop cultivars in the Indo-Gangetic Plains and west-central regions of South Asia should adapt to high temperature and water stress throughout their life cycle (Kumar, 2008). Early flowering (photo- and temperature-insensitivity, but development-related onset of flowering) and early maturity and high yield, as well as tolerance to both heat and drought, are among the traits to be bred to keep this region as a major food grain-producer under climate change.

A knowledge-intensive approach for climate-proof GESET should be guided by crop physiology that defines the ideotypes to be pursued. Precise phenotyping remains very important for success and it may rely on instrumentation tools that span remote sensing to trait recording in the experimental fields or greenhouses. Breeding populations derived from this approach can be used to understand stress tolerance and to develop potential mapping populations for further molecular analysis and association genetics. Genetic insights on flowering time, inflorescence architecture and adaptation to abiotic stresses are becoming available in model plant systems and comparative genomics may assist improving such traits in GESET that feed the world. Association mapping helps further to identify alleles associated with an adaptation range to various stresses, and new alleles of previously identified genes that can be screened in genebank accessions and characterized to determine their relative value. High-throughput DNA marker systems can be also used for both monitoring the available genetic diversity in gene pools and for whole-genome-aided selection in crops, forestry and livestock.

Transgenic crops and climate change

Knowledge about stress-adaptive mechanisms ensuing from research with elite genetic resources and their offspring could be incorporated into conceptual models for adaptation to stress-prone environments, and used to identify candidate genes for crop genetic engineering. Ainsworth *et al.* (2008) suggested some crop biotechnology targets for a future high-CO_2 and high-O_3 environment, e.g. manipulating ribulose-1,5-bisphosphate carboxylase/oxygenase or Rubisco (the key carboxylating enzyme and frequently the rate-limiting factor for photosynthesis, Zhu *et al.*, 2004), or increasing the apoplastic antioxidant capacity of crops. Ortiz (2008) gives a recent overview on genetic engineering for improving traits such as heat tolerance, water productivity and better use of nutrients that may enhance crop adaptation to the changing climate. Both review articles agree on the progress for breeding crop germplasm with greater abiotic stress resistance. However, engineering complex traits for adapting to climate change is likely to be much more challenging than the first generation of biotech crops such as herbicide tolerance or host-plant resistance to pests, which manipulated single transgenes.

Transgenic or genetically modified (GM) crops with enhanced environmental stress tolerance are also likely to require substantial advances in biosafety assessment and regulatory approval that are very different to the first generation of commercial transgenic crops. This second generation of transgenic crops for abiotic stress-prone environments poses new safety questions because they may lead to increased competitiveness if the transgenes are introgressed into wild populations (Ortiz *et al.*, 2007a). Biosafety frameworks will need to be based on a comprehensive molecular characterization of the expression of these new transgenes and their environmental impact assessment, addressing both the increased direct and indirect weediness; i.e. potential escape of the tolerant transgenic crop or the escape of the gene into wild relatives, respectively.

Adaptation and Mitigation to Climate Change – an International Endeavour

Mitigation and adaptation should be regarded as complementary strategies to manage risks and opportunities caused by climate change in agroecosystems. Mitigation, from the perspective of agriculture, should aim at reducing and stabilizing atmospheric greenhouse gas concentrations by increasing carbon sequestration and diminishing CO_2 and N_2O emissions. High yielding cultivars, crop rotations and less fallow periods, leaving crop residues in the field and organic manures may increase carbon sequestration in farmers' fields. The use of renewable energy in farming or conservation tillage will reduce CO_2 emissions whereas N use efficiency in crops and cropping systems will be the most important approach for diminishing N_2O emissions. Adaptation management, which should depend on local conditions, should consider site-specific crops, breeding climate-proof cultivars for stress-prone environments due to heat, water scarcity, floods and salinity, integrated management for existing and emerging pathogens and pests, adjusting operational farm management, and developing insurance systems for farmers exposed to extreme weather events. The Consultative Group on International Agricultural Research (CGIAR) and partners are using this knowledge and the technology generated to shift current practices in developing country agroecosystems that will help farmers and other land users therein to adapt to climate change (CGIAR, 2009). The CGIAR and the Earth System Science Partnership launched recently the Challenge Program on Climate Change, Agriculture and Food Security with the aim of ensuring sustainable production of sufficient food, fodder and fibre for a growing global population under a changing climate. This Challenge Program builds on and complements research already done by the CGIAR Centers.

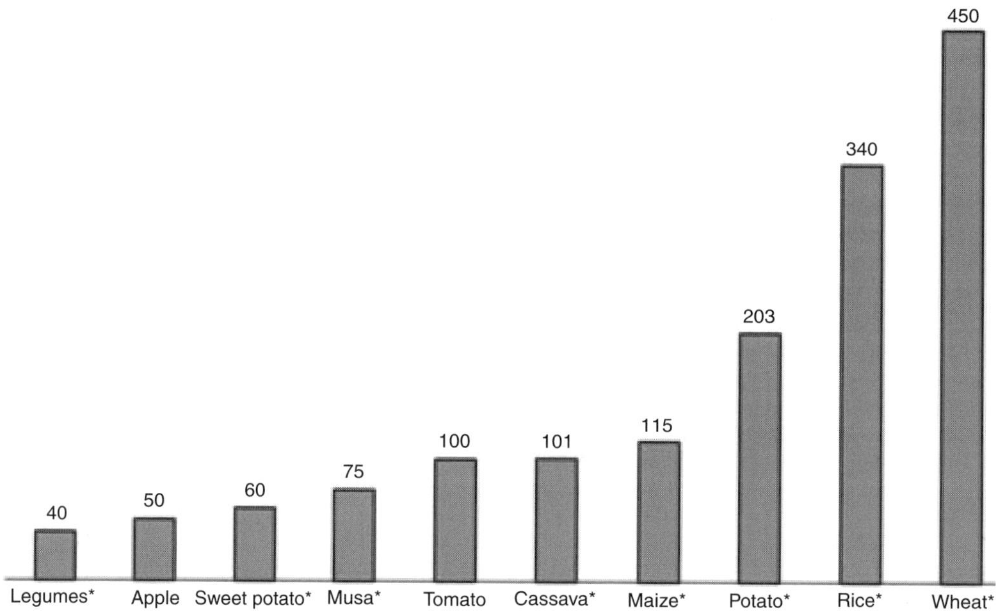

Fig. 12.1. Main crops that feed the world's human population (in million t; * crops for which a CGIAR Center has a breeding programme).

CGIAR GESET and other crop research advances for adapting to climate change

CGIAR Centers have been engaged in genetic enhancement through breeding new cultivars since the 1960s. Staple crop cultivars were bred for improved performance against major abiotic and biotic stresses to increase food production (Fig. 12.1). In their appraisal of the Green Revolution, Evenson and Golin (2003) clearly showed that crop yields in developing countries would have been at least 20% lower without the CGIAR (Fig. 12.2). Furthermore, their model indicates that equilibrium prices for all crops combined would have been at least 19% higher in 2000

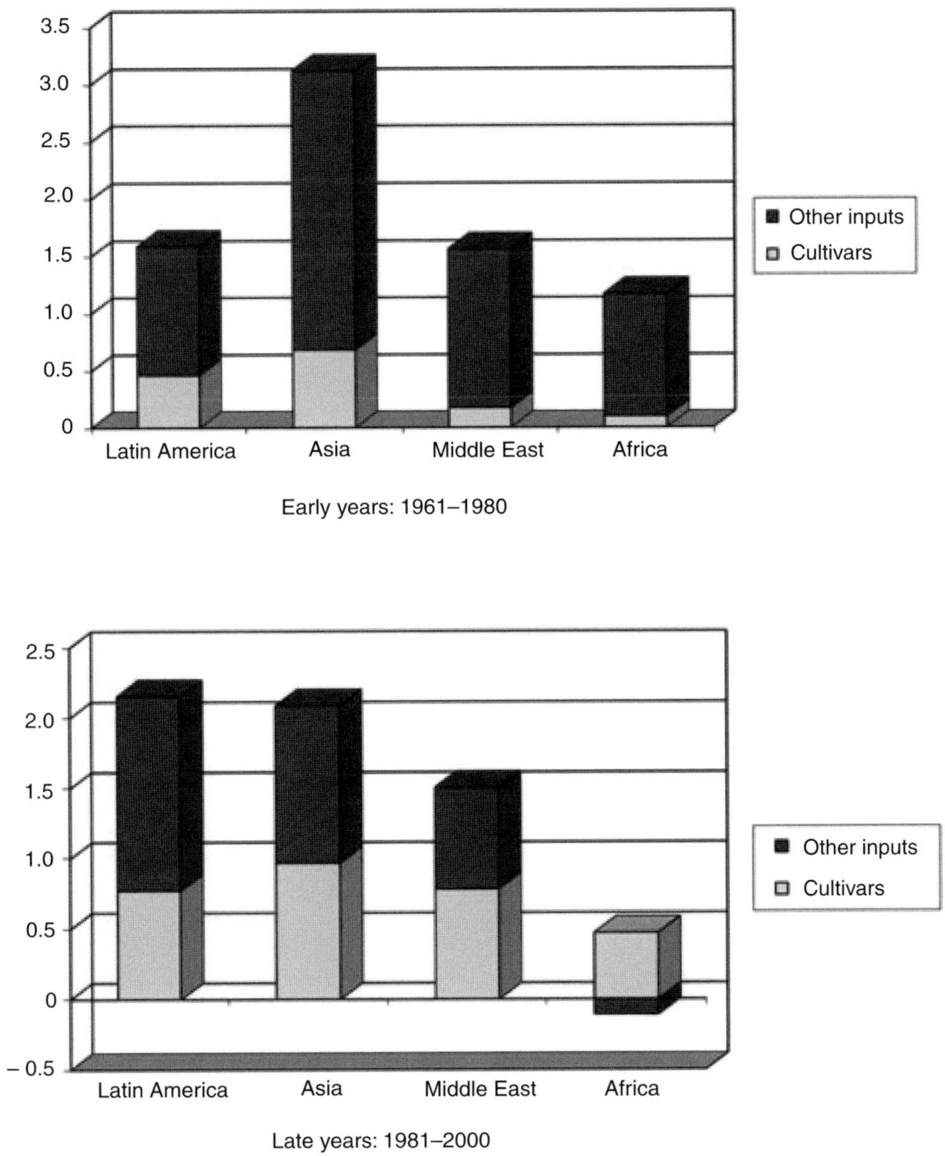

Fig. 12.2. Growth rates of yield due to the Green Revolution (After Evenson and Gollin, 2003).

without CGIAR research. There would have been a drop of 5% in calorie consumption and 2% more malnourished children in the developing world without CGIAR-bred germplasm in main staples.

Crop-related biodiversity is the founding asset of the CGIAR and remains the basic raw material for their breeding programmes and partners. The CGIAR Centers continue therefore to genetically enhance crop germplasm that is likely to allow developing country farmers to meet the main challenges of climate change such as drought, flood, heat and more damaging endemic pests. Their GESET (shared as populations, lines and clones with public and private sector breeders worldwide) and advances in companion production technology will provide more and affordable food, improve poor farmers' income through the sale of crop surpluses and combat malnutrition with new micronutrient-dense cultivars. The examples given below from breeding wheat, rice and maize – the main cereals in the human diet – and other crops by the CGIAR and partners illustrate recent advances in developing GESET that addresses climate change.

Wheat

Global warming could strongly affect the wheat crop in the developing world, particularly in the food basket of South Asia (Ortiz *et al.*, 2008b). Due to potential climate shifts in the Indo-Gangetic Plains, as much as 51% of its favourable high potential lands might be reclassified as a heat-stressed, irrigated, short-season production mega-environment. Such a shift will bring lower wheat yields, unless South Asian farmers adopt appropriate cultivars and crop management practices to adapt their wheat farming to climate change.

Research at the Centro Internacional de Mejoramiento de Maíz y Trigo (CIMMYT, Mexico) was able to disaggregate grain yield under water stress into distinct components and to apply these findings to wheat genetic enhancement. Root architecture and several physiological traits, resistance to soil-borne pests, tolerance to heat and salinity and zinc-deficient and boron toxic soils are among the target traits having major effects on water productivity in dryland wheat areas (Reynolds *et al.*, 2007). Important traits for drought-prone environments are available in the wild relatives of wheat. Re-synthesizing hexaploid wheat with wild ancestors has therefore been used at CIMMYT for tapping this useful variation and incorporating it into wheat-bred germplasm (Dreccer *et al.*, 2007). Lines deriving from re-synthesizing wheat yielded 8–30% higher than the best local check in multi-site trials (Ogbonnaya *et al.*, 2007). CIMMYT also pursues transgenic approaches for incorporating stress-inducible regulatory genes that encode proteins such as transcription factors (e.g. *DREB1A*) into the wheat cultigen pool (Ortiz *et al.*, 2007a and references therein). Contained field trials for evaluating transgenic *DREB*-wheat lines are underway in a drought testing site in Mexico.

Reynolds *et al.* (1994) found wheat cultivars capable of maintaining high 1000-kernel weight under heat stress, which also appear to possess tolerance to hot environments. Canopy temperature depression, membrane thermostability, leaf chlorophyll content during grain filling, leaf conductance and photosynthesis are physiological traits that are associated with wheat yield in heat-prone environments (Reynolds *et al.*, 1998). Canopy temperature depression was used to select for yield under a hot, dry, irrigated wheat environment in Mexico (Amani *et al.*, 1996), whereas leaf chlorophyll content was correlated with 1000-kernel weight while screening Mexican wheat landraces (Hede *et al.*, 1999).

The fungi *Cochliobolus sativus* causing spot blotch and *Pyrenophora tritici-repentis* inducing tan spot are pathogens responsible for leaf blight in humid and hot areas, particularly in the Indo-Gangetic Plain. Their increasing severity with growth stage depends on crop resilience to heat stress. Improvement of spot blotch resistance in these areas requires a crop physiology adapted to stressed environments and host plant resistance to leaf blight. This has been achieved by crossing resistance sources or wild relatives to high-yielding cultivars (Duveiller, 2004). Similarly, new threats such as wheat head blast (due to the fungus *Magnaporthe grisea*) that may induce grain yield losses over 50% in warm (25–28°C), humid environments of South America's Southern Cone – a major grain basket in the world – is likely to be a target for wheat breeding under climate change. In Brazil, Prestes *et al.* (2007) observed head infection ranging from 10% to 86% among cultivars and breeding lines. A few breeding materials and cultivars displayed lower head infection than BR18, a moderately resistant cultivar in the field. These may be potential sources for breeding host plant resistance to head blast.

Some of the above traits are helping to define a new wheat ideotype that can provide the basis for genetically enhanced wheat by design in heat- and drought-prone environments. Ortiz *et al.* (2008a) suggest that climate-proof GESET technology should ensue from regional partnerships that more specifically address the needs of warmer and drier areas because improved adaptation of local wheat cultivars could result from selective breeding using resistant and agronomically superior genotypes. In this regard, Ortiz *et al.* (2007b) used selection percentage from nursery sets in Bangladesh to show the benefits of decentralizing wheat breeding with materials carrying the desired traits.

Rice

Analysis of multi-season datasets from irrigated long-term field trials at the International Rice Research Institute (IRRI), Los Baños, the Philippines revealed that rice grain yields declined by 10% for each 1°C increase in growing-season minimum temperature in the dry season, whereas the effect of maximum temperature on crop yield was insignificant (Peng *et al.*, 2004). This finding confirms that rice yields will decline with global warming due to higher night temperature. Wassmann *et al.* (2009b) indicated that South and East Asia are prone to heat stress because they are already approaching critical levels during susceptible development stages of the rice plant. They also stated that drought stress will be further aggravated in rice due to climate change, particularly in Thailand and the eastern Ganges of South Asia. As Asia accounts for 90% of global rice land, effects on global rice supply could be very serious.

Direct selection for yield under drought seems to be feasible (Kumar *et al.*, 2008). Parental sources to breed in drought-prone environments are available for upland rice (Atlin *et al.*, 2006), but only a few are known for the rainfed lowland system (Wassmann *et al.*, 2009a). Cultivars exhibiting heat tolerance during reproductive development, high harvest index, small leaves and low leaf area per unit ground are proposed for rice production under high temperatures.

The length of basal dehiscence (Matsui *et al.*, 2005) and highly efficient transpirational cooling – a heat avoidance mechanism (Wassmann *et al.*, 2009a) – could be used as phenological traits for breeding high-temperature tolerance in rice. Shifting the time of peak flowering to cooler periods will also help to overcome high temperatures in rice because it could protect rice fertility from future adverse effects of climate change. Heat-resistant cultivars will provide yield reliability where many crops are grown at or near their thermal optimum, when any increase in temperature causes photosynthesis to slow and eventually cease. Jena and Mackill (2008) suggest the feasibility for transferring major putative quantitative trait loci (QTL) for high temperature tolerance into locally adapted or other genotypes using either conventional or molecular breeding approaches.

IRRI is also incorporating the *submergence 1* (*sub1*) gene into popular cultivars, which can be immediately used by farmers, to allow the rice crop to survive prolonged periods of submergence due to increased rainfall and flooding, likely under climate change in many parts of Asia. The full rice genome sequence (Matsumoto *et al.*, 2005) and intensive QTL mapping for several traits (Ismail *et al.*, 2007) are facilitating and accelerating the genetic gains in rice breeding. For example, Xu *et al.* (2006) fine mapped and sequenced *sub1* in an FR13A-derived tolerant line. DNA marker-aided backcrossing can speed up the breeding of submergence tolerance in popular rice mega-cultivars, e.g. IR64 (Septiningish *et al.*, 2008), which are preferred both by farmers and consumers due to their quality traits.

Rice is also at risk in Africa due to climate change. In the early 1990s, the Africa Rice Centre (known previously by its acronym as WARDA) started cross-breeding, with the aid of embryo rescue, for producing hybrids between Asian rice *Oriza sativa* and African rice *Oriza glaberrima*, which is also a source for submergence tolerance (Futakuchi *et al.*, 2001). Progeny with robust fertility were bred as a result of several cycles of backcrossing with the *O. sativa* parent (Jones *et al.*, 1997b). Anther culture was used to double the chromosome number for producing true-breeding lines for further testing (Jones *et al.*, 1997a). By the mid-1990s, the new rice for Africa (NERICA) was being widely tested. NERICA cultivars benefit from the high yields of Asian rice combined with the adaptation of African rice to harsh growing environments (Johnson *et al.*, 1998; Dingkuhn *et al.*, 1999). Such environments are likely to become more widespread under climate change. A participatory varietal selection (PVS) approach was adopted to identify the NERICA cultivars best suited to the growers after consulting with rice stakeholders, i.e. scientists from national programmes, extension workers, farmers and nongovernmental organizations. PVS was complemented by a community-based seed system (CBSS), which was built on farmers' own seed-saving practices with complementary training in seed technologies. This approach made NERICA quality seeds available to farmers in just 4 years, as opposed to the 7 years normally required for formal seed system release. NERICA cultivars are already being grown on 200,000 ha in rainfed uplands across 30 African countries (CGIAR, 2009). African rice farmers are keen on early-maturing NERICA cultivars which mature 30–50 days earlier than available local cultivars. They also out-yield other cultivars with little or no fertilizer, permit more intensive cropping and may escape intermittent droughts occurring at critical stages in crop development. Very recently, Nuitjen *et al.* (2009) noticed hybridization between African and Asian rice in West African farmers' fields, thereby resulting in novel genotypes that may be of further use by plant breeders for adapting rice to climatic uncertainty.

Maize

CGIAR research – coupling climate simulation models with data from various sources to simulate the growth, development and yield of maize – shows that an overall 10% reduction for smallholder rainfed maize production in Africa and Latin America should be expected by 2055 due to climate change (Jones and Thornton, 2003), i.e. a loss in maize grain worth approximately US$2 billion yearly. Conventional and molecular breeding are being used for developing new maize cultivars for drought-prone environments (Bänziger and Araus, 2007). Selection for component traits such as kernel set, rapid silk exertion and reduced barrenness in multi-environment trials has led to significant progress in grain yield under drought stress (Campos *et al.*, 2004). CIMMYT, using a client-oriented breeding approach, targeted at improving maize for the drought-prone mid-altitudes of sub-Saharan Africa (Bänziger *et al.*, 2006), bred in excess of 50 new cultivars, which are now grown on at least 1 million ha in drought-prone environments of Southern Africa (Spielman and Pandya-Lorch, 2009). Some of these new maize cultivars also show tolerance to infertile soils, host plant resistance to the parasitic weed *Striga* and to other endemic pathogens and pests affecting the crop in sub-Saharan Africa. Building on this and other maize breeding successes in sub-Saharan Africa (e.g. in West and Central Africa by the International Institute of Tropical Agriculture – IITA, Nigeria – Ortiz and Hartmann, 2003), a new initiative named 'Drought Tolerant Maize for Africa' (or DTMA) was launched a few years ago in partnership with local public and private breeders in Africa and researchers from advanced research institutes in the northern hemisphere (http://dtma.cimmyt.org). DTMA aims to generate maize GESET that may yield at least 20% more than today's cultivars grown by African smallholders.

Marker-assisted selection (MAS) has been used for grain yield and quality, tolerance to abiotic stresses and host plant resistance to major pathogens and pests affecting maize (Xu and Crouch, 2008 and references therein). It is claimed that commercial maize breeding programmes have achieved twice the rate of genetic gain through MAS vis-à-vis phenotypic selection (Crosbie *et al.*, 2006). DNA marker-aided analysis also provides further genetic insights on maize performance under drought (Ribaut and Ragot, 2007). Marker-aided backcrossing can be used for introgressing a few target QTL into an elite maize line but this breeding approach does not seem to be very effective when many QTL of small effect are involved. Moreover, QTL may be germplasm-specific and MAS costs for many QTL of small effect may be higher than those from conventional crossbreeding of maize. Identifying QTL of major effect and independent of genetic background as well as devising more efficient DNA marker-aided breeding approaches than pedigree selection remain a challenge for maize molecular breeding for drought-prone environments (Ortiz *et al.*, 2007a). Recently, CIMMYT started using selective genotyping (from the two tails of the phenotypic distribution of a population) together with pooled DNA analysis as a highly cost effective alternative to analysis of the entire population of individual genotypes for genetic mapping (Xu *et al.*, 2008). Likewise, single nucleotide polymorphism (SNP) markers are becoming publicly available and will assist on genome-wide association mapping in maize (Ortiz *et al.*, 2010 and references therein). Advances in sequencing both the 2.3 giga-base genome of popular US inbred line B73 (Barsh *et al.*, 2009; Schnable *et al.*, 2009) and the smaller genome of the landrace Palomero from Mexico (Vielle-Calzada *et al.*, 2009), the haplotype map (Gore *et al.*, 2009), genome-wide transcript analysis on gene expression patterns (Swanson-Wagner *et al.*, 2009), comprehensive association genetics research using methods such as nested association mapping (Buckler *et al.*, 2009; McMullen *et al.*, 2009), seed-DNA-based genotyping systems (Gao *et al.*, 2008) and precise phenotyping will accelerate the discovery of functional alleles and allelic variation that are associated with traits of interest for enhancing adaptation of maize to climate change.

CIMMYT has advocated a new user-led philanthropy–private–public partnership paradigm for the development and deployment of transgenic solutions for maize in the drought-prone environments of the developing world (Ortiz *et al.*, 2007a). Table 12.1 lists most recent advances on genetically engineering maize for drought-prone environments. Access to proprietary technology can lead to stable grain yields in complex drought-prone areas and could allow resource-poor African maize farmers to harvest a reasonable crop in most years. An example of this partnership is the project 'Water Efficient Maize for Africa' (WEMA, 2010). The African Agriculture Technology Foundation (AATF) – the organization leading WEMA – works with CIMMYT, the private agricultural company Monsanto and the agricultural research systems in eastern and southern Africa in this effort. AATF contributes its leadership, unique experience in public–private partnership management, technology stewardship and project management expertise. CIMMYT provides high-yielding maize cultivars that are adapted to African conditions and expertise in conventional breeding and testing for drought tolerance. Monsanto provides proprietary germplasm, advanced breeding tools and expertise, and drought-tolerance transgenes developed in collaboration with BASF. The cultivars bred through this project will be distributed to African seed companies through AATF without royalty and made available to smallholder farmers as part of their seed business. The national agricultural research systems, farmers' groups and seed companies participating in this project will contribute their expertise in field testing, seed multiplication and distribution. This project also involves local institutions, both public and private, and in the process expands their capacity and experience in crop breeding, biotechnology and biosafety. The Bill & Melinda Gates Foundation and the Howard G. Buffet Foundation are the funding partners of WEMA.

Other main staples

The Centro Internacional de Agricultura Tropical (CIAT) in Colombia has been assessing the impact of climate change on cassava. This research suggests that the global areas suitable for cassava will increase by

Table 12.1. Advances in transgenic maize technology for drought-prone environments.

Transgene	Reference
Escherichia coli's glutamate dehydrogenase (*gdhA*) gene	
Germination and grain biomass production were increased in *gdhA* transgenic maize in the field during seasons with significant water scarcity. Water deficit tolerance under controlled conditions was also increased	Lightfoot *et al.* (2007)
Cold shock proteins (CSPs) from bacteria	
CspA from *E. coli*, and CspB from *Bacillus subtilis*, promote stress adaptation in multiple plant species. The expression of CSP proteins in maize is not associated with negative pleiotropic effects; i.e. stress tolerance without crop yield penalty under limiting water	Castiglioni *et al.* (2008)
Phosphatidylinositol-specific phospholipase C (PI-PLC)	
Phospholipase C1 gene (*ZmPLC1*) cloned from maize encoded a PI-PLC and up-regulated the expression in maize roots under dehydration	Zhai *et al.* (2005)
Enhanced expression of *ZmPLC1* improves drought tolerance in transgenic maize, which showed higher relative water content, better osmotic adjustment, increased photosynthesis rates, lower percentage of ion leakage, less lipid membrane peroxidation and higher grain yield than the control under water scarcity	Wang *et al.* (2008)
Orthologous maize transcription factor (ZmNF-YB2)	
Transgenic maize plants with increased ZmNFYB2 expression show tolerance to drought as measured by chlorophyll content, stomatal conductance, leaf temperature, reduced wilting and maintenance of photosynthesis under limiting water, all of which will contribute to grain yield, when this transgenic maize grows in drought-prone environments	Nelson *et al.* (2007)

5.1% on average by 2050 but many areas of Latin America will suffer negative impacts. For example, about 1.6 million ha growing cassava in South America may be affected and ~30% of cassava fields will need to grow cassava cultivars with tolerance to water stresses, both drought and flood (A. Jarvis, CIAT, Colombia, 2009, personal communication). The cassava cultigen pool could therefore be broadened by interspecific hybridization with wild *Manihot* relatives that possess desired genes to enhance adaptation to water stressful environments (Nassar and Ortiz, 2009). Similarly, IITA and partners are investigating the adaptation mechanisms of cassava in African drought-prone environments. Very recently, an international consortium led by the University of Arizona and including CIAT, announced the first draft of the cassava genome. The annotated draft genome sequence (416 Mb of the ~760 Mb estimated size of cassava) is available (www.phytozome.net/cassava). This genome sequencing opens a new chapter in the genetic enhancement of cassava. Geneticists and breeders will be able to access large DNA marker databases that can be used for identifying genes of many important traits and further enhancing this crop, which remains the daily primary food source for more than 750 million people in the tropics.

The vulnerability of both potato and sweet potato to climate change has been analysed by the Centro Internacional de la Papa (CIP) in Peru. The models used suggest that potato yield may reduce between 20% and 30% in the tropics and subtropics (CGIAR, 2009). Such losses could be mitigated with adaptation options such as stress-tolerant cultivars and improved crop management, e.g. short-season cultivars that avoid unfavourable hot or dry periods and adapt to new rainfall patterns as climate changes. CIP research also shows the impacts of a warmer, wetter world on the late blight pathogen *Phytophthora infestans* (Forbes and Simon, 2007). They advocate gene broadening by using new sources of host-plant resistance from wild species for achieving durable resistance to this pathogen. For example, a major QTL on chromosome 11 of the species *Solanum paucissectum* will be a newly 'mined' species for potato breeding (Villamon *et al.*, 2005). Germplasm enhancement methods involving ploidy manipulations with 2*n* gametes and haploids are used to transfer genes for important traits from wild tuber-bearing *Solanum* species and diploid landraces to the tetraploid potato cultigen pool (Ortiz *et al.*, 2009). Diploid potato genetic resources (including a landrace) are also facilitating the genome sequencing enterprise for this crop. The Potato Genome Sequencing Consortium, led by the Plant Breeding Department of Wageningen University & Research, the Netherlands, with CIP as one of its members, was initiated in 2006. This Consortium released the first draft of the potato genome to the public domain in 2009 (www.potatogenome.net). Visser *et al.* (2009) point out that annotated data will facilitate characterization of accessions held in potato and wild species germplasm based on allelic variance, and assist potato breeders to fully exploit their genetic potential in the tetraploid cultigen pool. None the less, reliable tuber–seed systems – following a user-needs approach – will remain a key factor for shortening the time period for making pathogen-tested propagules of newly bred cultivars available to potato farmers adapting to global warming, water stresses and new pest threats.

Blomme and Ortiz (2000) indicated that there was a great variability of root development between and within *Musa* (banana and plantain) groups. The triploid cooking banana cultivar Fougamou and tetraploid hybrid FHIA 3 had best early root development in a degraded humid forest location. Cooking bananas (ABB) such as Bluggoe or Fougamou are also promising triploid cultivars for transient dry conditions in West Africa (Ekanayake *et al.*, 1994). These cooking bananas possessed a high potential for restricting water use due to their ability to close stomata during the afternoon. Further IITA research showed that most of the drought-tolerant ABB cooking bananas had a higher rate of conductance and transpiration in the afternoon (Ekanayake *et al.*, 1998). The B genome (deriving from the wild diploid species *Musa balbisiana*) seems to be a valuable source of alleles for adapting cooking bananas to mild drought environments. Banana and plantain breeding efforts offer a means to initiate a new phase in the evolution of triploid *Musa* (Vuylsteke, 2001). The increased use of molecular markers will also accelerate the process of recurrent selection of improved *Musa* germplasm and facilitate the development of new hybrids. In 2009, Genoscope (France) therefore initiated the sequencing of the *Musa* genome under the framework Global Musa Genomics Consortium, in which Bioversity International in Italy participates actively. They are sequencing the diploid accession Pahang HD of *Musa acuminate*, one of the ancestors of triploid dessert and cooking bananas. The results, which will become available in public databases, will facilitate genetics research in banana and plantain, which will provide useful information for further evolutionary breeding of new *Musa* cultivars (Ortiz, 1997), which should have a broad genetic base, high yield, appropriate fruit quality, host plant resistance to pests and pathogens, and better adaptation to the changing climate.

CGIAR Centers are also breeding hardy climate-proof germplasm of legume crops such as beans, broadbeans, chickpeas, cowpeas, groundnuts (or peanuts), lentils, pigeonpeas and soybeans, as well as dryland cereals such as barley, pearl millet and sorghum. Further research on drought and the sorghum genome (Paterson *et al.*, 2009) may assist in breeding more hardy and water-efficient maize, rice and wheat due to gene synteny among cereals.

Modifying photosynthesis

Photosynthesis – one of the top ten evolutionary milestones for helping make and keep the Earth lush (Leslie, 2009) – is the process by which plants, some bacteria and some protistans use sunlight to produce sugar that cellular respiration converts into ATP, which is the 'fuel' used by all living things. Although photosynthesis plays an important role as a provider of energy as well as assimilates for growth and reproduction, the influence of abiotic stresses (e.g. salinity or water deficits) and biotic stresses (e.g. insect herbivory) on photosynthesis remains unclear (Lawlor, 2009). The CGIAR, together with research partners worldwide, are therefore looking at ways to boost shrinking crop yields by reconfiguring the plant's photosynthetic engine so it can convert solar power and atmospheric carbon more efficiently into harvests of the main staples that feed the world.

Rubisco acts as the primary CO_2-fixing enzyme of C_3 photosynthesis in about 90% of terrestrial plants, including major staple crops such as barley, rice, soybean and wheat. C_3 photosynthesis' ability to use O_2 as a substrate instead of CO_2 results, however, in photorespiration, which is an energy-wasting process. C_4 plants – such as maize, pearl millet, sorghum and sugarcane – achieve higher photosynthetic capacities and better water- and nitrogen-use efficiencies than C_3 species (Black, 1973). Photorespiration is largely suppressed in the C_4 cycle, accumulating CO_2 at the site of Rubisco and inhibiting its oxygenase activity. However, C_4 photosynthesis is as sensitive to water stress as its C_3 counterpart or even more so (Ghannoum, 2009). Reduced photosynthetic efficiency may occur in C_3 plants because of a rapid rise in O_2 competition with CO_2 in the reaction catalyzed by Rubisco, when temperatures increase above 20°C. Moreover, low availability of atmospheric CO_2 to Rubisco under limiting water supply and an increase in competition from O_2 will significantly reduce photosynthetic efficiency.

Photosynthetic efficiency should be improved to increase input efficiency and position the most important staple crops to respond to climate change. Hubbart *et al.* (2007) showed that leaf photosynthesis in rice may be systematically affected by breeding, whereas Fischer *et al.* (1998) demonstrated that historic progress in the yield potential of advanced bread wheat cultivars has been associated with increased stomatal conductance and light-saturated photosynthetic rate.

Recently initiated research by IRRI and partners worldwide aims to develop C_4 rice and thus increase by 50% the crop's grain yield. Proponents of this approach argue that this magnitude of yield increase will only be achieved by altering rice photosynthesis to the C_4 pathway (Hibberd *et al.*, 2008). The polyphyletic evolution of the C_4 pathway (Kellogg, 1999) suggests that the transition from C_3 to C_4 is relatively simple. Moreover, Kranz anatomy – the specialized leaf anatomy of C_4 plants – is not essential for terrestrial C_4 plant photosynthesis, as found by Voznesenskaya *et al.* (2001) in *Borszowia aralocaspica*. This Chenopodiaceae plant, which lacks Kranz anatomy, accomplishes C_4 photosynthesis through spatial compartmentation of photosynthetic enzymes and by separation of two chloroplast types and other organelles in distinct positions within the chlorenchyma cell cytoplasm. Preliminary observations suggest that variation available within *Oryza* genetic resources may be a source of traits to breed an efficient C_4 pathway in rice. However, there are still various challenges to overcome in leaf morphology and metabolism to successfully accomplish this long-term research task.

It will be important to develop a stepwise plant breeding strategy that integrates conventional and advanced genetic enhancement approaches and examines afresh ways to increase the crop's efficiency at converting sunlight to energy and grain (Fig. 12.3). Table 12.2 lists target traits and early-generation methods for improving photosynthetic efficiency and crop yields in C_3 crops. Genetic engineering also has the potential for improving photosynthesis as shown by transgenic tobacco plants with an increased sedoheptulose-1,7-bisphosphatase activity that leads to higher photosynthetic rates (and growth) at an early development stage (Lefebvre *et al.*, 2005). This result was attributed to an increase in Rubisco regenerative capacity, which shows the potential of genetic manipulation of Rubisco in chloroplasts through transgenics. The overexpression of C_4 cycle enzymes in transgenic C_3 plants (Häusler *et al.*, 2002) will be another transgenic approach to improve photosynthesis, especially after some promising research results in potato, rice and tobacco. Furthermore, photosynthesis in water-rich areas may be enhanced by bioengineering stomata that stay open for a longer period.

Mitigation through crop germplasm enhancement and transgenics

Although agriculture is a significant source of greenhouse gas (GHG) emissions, agricultural research also offers means for climate change mitigation. In this regard, crop productivity gains avoid conversion of native landscapes for food, feed, fibre or fuel production. Such a land conversion practice often involves tree or plant burning that generates CO_2 and other GHG. Borlaug (2007) indicated that the contribution of the Green Revolution's high-yield agriculture to environmental conservation can be measured by comparing today's crop yields vis-à-vis those from harvests during the 1950s and calculating the land save due to the improved technology. For example, the world cereal production in 1950 – largely produced using what will be regarded today as organic means – was 650 million t whereas it was 1.9 billion t in 2000

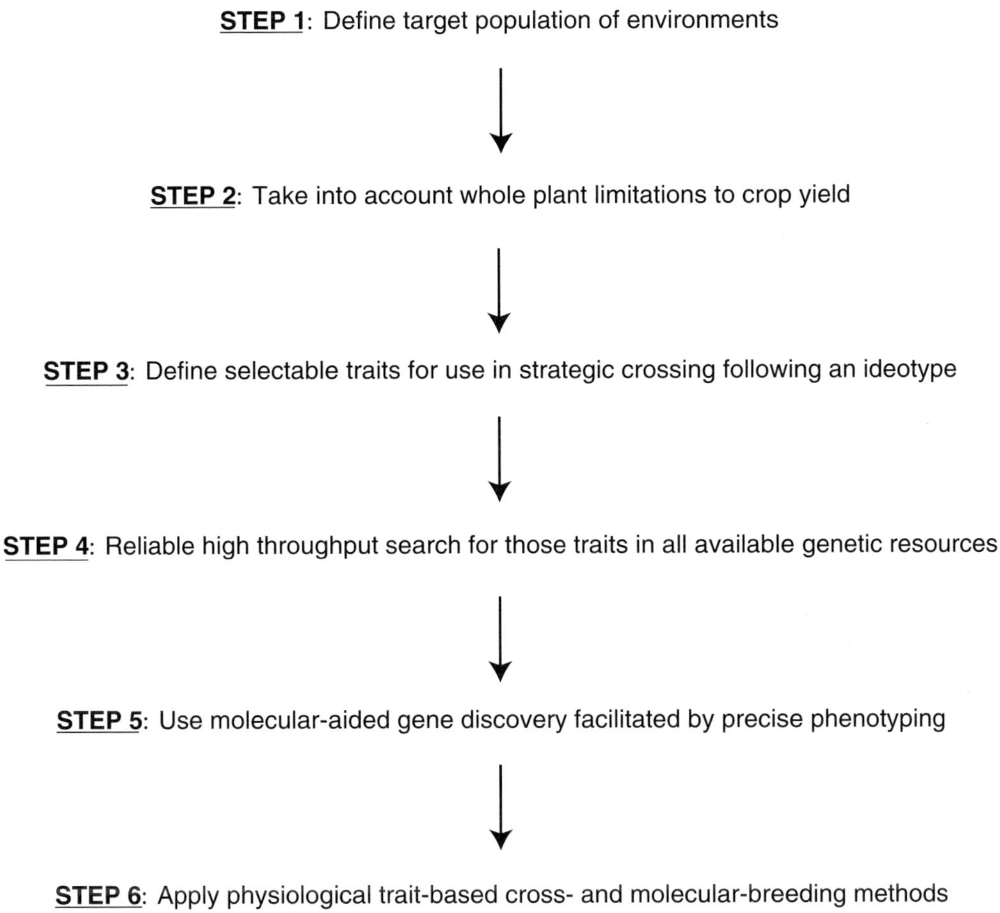

Fig. 12.3. Step-wise genetic enhancement approach to improve plant photosynthesis and crop yield.

Table 12.2. Some target traits and early selection methods for improving photosynthetic efficiency in C_3 crops.

Item	References
Target traits:	
Canopy	Murchie *et al.* (2008)
Leaf angle	Reynolds *et al.* (2000)
Radiation use efficiency	Horton (2000); Long *et al.* (2006)
Reducing the respiration rate	Sharma-Natu and Ghildiyal (2005)
Rubisco optimization	Parry *et al.* (2007)
Sink demand	Reynolds *et al.* (2009a)
Spike photosynthesis	Tambussi *et al.* (2007)
Early-generation selection methods for rapid proxy estimations of photosynthesis and crop yields:	
Canopy temperature	Reynolds *et al.* (2009b)
Leaf porometry	
Spectral reflectance	
Visual assessments	

but with only a 10% increase on planted area and the remaining due to gains in crop yields. About 1.1 billion ha of additional land of the same quality, by cutting large forest areas and ploughing various grasslands, would have been needed in 2000 to produce the same grain harvests using the 1950s technology. Of course, today's increased grain harvests also reflect how profitable it was for cereal growers to raise crop yields on their farms. Very recently Burney *et al.* (2010) estimated the net effect on GHG emissions of historical agricultural intensification between 1961 and 2005. They found that emissions from fertilizer production and application rose but the net effect of higher yields due to the Green Revolution avoided emissions of up to 161 gigatons of carbon (GtC) (590 $GtCO_2e$) since 1961. Their research shows that crop productivity gains should be prominent in the strategy to reduce GHG emissions. As shown by these results, modern intensive agriculture seems to be better for the environment than the 'old-fashioned' way of doing things, often advocated as 'natural' by some anti-science 'greenies'.

Herbicide-tolerant crops help to reduce ploughing in fields, thereby saving fuel costs and protecting the soil structure by reducing its erosion. Similarly, breeding host plant resistance to pathogens and pests leads to fewer pesticide sprays, which also means a reduction in fuel use and lower CO_2 emissions. Transgenic crops are therefore making important contributions to food production and sustainable farming, as well as mitigating climate change. Brookes and Barfoot (2009) estimated that transgenic crops significantly reduced the release of GHG emissions from agricultural practices: for example, in 2007 alone their impact amounted to removing 14.2 billion kg of CO_2 from the atmosphere (i.e. equal to removing nearly 6.3 million cars from the road for 1 year). Likewise, between 1996 and 2007, transgenic crops reduced pesticide spraying by 359 million kg – equivalent to 125% of the annual volume of pesticide active ingredient applied to arable crops in the European Union. And, since 1996, the soil carbon sequestered, facilitated by transgenic herbicide-tolerant crops coupled with other conservation agriculture practices, has been equivalent to 83.2 billion t of CO_2, which would otherwise have been released into the global atmosphere.

Leaf albedo bio-geoengineering

Very recently, Ridgwell *et al.* (2009) proposed a 'bio-geoengineering' approach to mitigate surface warming by selecting crop cultivars having specific leaf glossiness or canopy morphological traits that maximize solar reflectivity. They estimated that their approach could lead to a potential summertime cooling of more than 1°C throughout much of central North America and mid-latitude Eurasia, equivalent to seasonally offsetting approximately 20% of regional warming due to doubling of atmospheric CO_2.

Agrobiodiversity and Climate Change

This chapter gives an overview on the implications of climate change for agrobiodiversity, and how understanding and managing it can provide important means for coping with the changing climate. Genetically enhancing climate-resilient crop cultivars and breeds remains the core of the short term options for adapting to climate change, especially for those agroecosystems in regions that are or will be suffering temperature extremes, water scarcity or floods. Unfortunately, important international bodies such as the IPCC, policy makers and governments have largely failed to realize the critical role that agrobiodiversity plays on climate change adaptation and mitigation; i.e. understanding agrobiodiversity as insurance for both agriculture and the agri-food business in coming decades.

Adaptation to climate change requires traditional and modern breeding methods, focusing on tolerance to abiotic stresses or host plant resistance to emerging pathogen and pest epidemics. Tagging a price for conserving through use of agrobiodiversity (a form of 'agrobiodiversity credits') should also encourage better management options for enhancing functional diversity in agroecosystems. Likewise, agrobiodiversity will benefit from community-based management approaches that allow adaptive capacity to climate change, especially when dealing with new suitability and distribution patterns of crops and livestock as a result of global warming and water stresses.

Agrobiodiversity, which remains as the main resource base for food, seems to be undervalued in the global agenda. This lack of interest in agrobiodiversity may become an additional barrier for its effective use in contributing to adaptation and mitigation of farming systems to the changing climate. Better public awareness about the role of agrobiodiversity for dealing with climate change and achieving food security needs to be raised. In this way, agrobiodiversity conservation through use can be included as an important building block of adaptation to climate change.

References

Ainsworth E., Rogers, A. and Leakey, A.D.B. (2008) Targets for crop biotechnology in a future high-CO_2 and high-O_3 world. *Plant Physiology* 147, 13–19.

Amani, J., Fischer, R.A. and Reynolds, M.P. (1996) Canopy temperature depression association with yield of irrigated spring wheat cultivars in a hot climate. *Journal of Agronomy and Crop Science* 176, 119–129.

Atlin, G.N., Lafitte, H.R., Tao, D., Laza, A., Amante, A. and Courtois, B. (2006) Developing rice cultivars for high-fertility upland systems in the Asian tropics. *Field Crops Research* 97, 43–52.

Baetghen, W.E. (2009) La adaptación al cambio climático en el sector agropecuario. *Revista ARROZ* 58, 24–32.

Bänziger, M. and Araus, J.L. (2007) Recent advances in breeding maize for drought and salinity stress tolerance. In: Jenks, M.A., Hasegawa. P.M. and Jain, S.M. (eds) *Molecular Breeding towards Salinity and Drought Tolerance*. Springer, Dordretch, the Netherlands, pp. 587–601.

Bänziger, M., Setimela, P.S., Hodson, D. and Vivek, B. (2006) Breeding for improved drought tolerance in maize adapted to southern Africa. *Agricultural Water Management* 80, 212–224.

Barsh, G.S., Copenhaver, G.P., Ecker, J.R. and Malik, H.S. (eds) (2009) Maize genome collections. *PLOS Genetics*. Available at: http://collections.plos.org/plosgenetics/maize.php (accessed 9 November 2010).

Battisti, D.S. and Naylor, R.L. (2009) Future food insecurity with unprecedented seasonal heat. *Science* 323, 240–244.

Bhullar, N.K., Street, K., Mackay, M., Yahiaoui, N. and Keller, B. (2009) Unlocking wheat genetic resources for the molecular identification of previously undescribed functional alleles at the *Pm3* resistance locus. *Proceedings of the National Academy of Sciences* 106, 9519–9524.

Black, C.C. Jr (1973) Photosynthetic carbon fixation in relation to net CO_2 uptake. *Annual Review of Plant Physiology* 24, 253–286.

Blomme, G. and Ortiz, R. (2000) Preliminary assessment of root systems in *Musa. Acta Horticulturæ* 540, 259–266.

Bloomberg News (2009) China grain output may drop 37% on climate change. Available at: www.bloomberg.com/apps/news?pid=20601086&sid=alYpYphnXuEQ (accessed 9 November 2010).

Borlaug, N.E. (2007) Sixty-two years of fighting hunger: personal recollections. *Euphytica* 157, 287–297.

Brookes, G. and Barfoot, P. (2009) *GM Crops: Global Socio-economic and environmental impacts 1996-2007.* PG Economics Limited, Dorchester, Dorset, UK, 128 pp. Available at: www.pgeconomics.co.uk/pdf/2009globalimpactstudy.pdf (accessed 9 November 2010).

Buckler, E.S., Holland, J.B., Bradbury, P.J., Acharya, C.B., Brown, P.J., Browne, C., Ersoz, E., Flint-Garcia, S., Garcia, A., Glaubitz, J.C., Goodman, M.M., Harjes, S., Guill, K., Kroon, D.E., Larsson, S., Lepak, N.K., Li, H., Mitchell, S.E., Pressoir, G., Peiffer, J.A., Oropeza Rosas, M., Rocheford, T.R., Cinta Romay, M., Romero, S., Salvo, S., Sanchez Villeda, H., da Silva, H.S., Sun, Q., Tian, F., Upadyayula, N., Ware, D., Yates, H., Yu, J., Zhang, Z., Kresovich, S. and McMullen, M.D. (2009) The genetic architecture of maize flowering time. *Science* 325, 714–718.

Burke, M.B., Lobell, D.B. and Guarino, L. (2009) Shifts in African crop climates by 2050, and the implications for crop improvement and genetic resources conservation. *Global Environmental Change* 19, 317–325.

Burney, J.A., Davis, S.J. and Lobell, D.B. (2010) Greenhouse gas mitigation by agricultural intensification. *Proceedings of the National Academy of Sciences* doi:10.1073/pnas.0914216107.

Campos, H., Cooper, M., Habben, J.E., Edmeades, G.O. and Schussler, J.R. (2004) Improving drought tolerance in maize: a view from industry. *Field Crops Research* 90, 19–34.

Castiglioni, P., Warner, D., Bensen, R.J., Anstrom, D.C., Harrison, J., Stoecker, M., Abad, M., Kumar, G., Salvador, S., D'Ordine, R., Navarro, S., Back, S., Fernandes, M., Targolli, J., Dasgupta, S., Bonin, C., Luethy, M.H. and Heard, J.E. (2008) Bacterial RNA chaperones confer abiotic stress tolerance in plants and improved grain yield in maize under water-limited conditions. *Plant Physiology* 147, 446–455.

CGIAR (2009) *Climate, Agriculture and Food Security: A Strategy for Change.* CGIAR Challenge Program on Climate Change, Agriculture and Food Security, Copenhagen, Denmark, 48 pp.

Challinor, A.J., Wheeler, T.R., Craufurd, P.Q., Ferro, C.A.T and Stephenson, D.B. (2007) Adaptation of crops to climate change through genotypic responses to mean and extreme temperatures. *Agriculture, Ecosystems and Environment* 119, 190–204.

CropLife International (2009) *Facts and Figures – the status of global agriculture (2008-2009).* CropLife International, Brussels, Belgium, 12 pp.

Crosbie, T.M., Eathington, S.R., Johnson, G.R., Edwards, M., Reiter, R., Stark, S., Mohanty, R.G., Oyervides, M., Buehler, R.E., Walker, A.K., Dobert, R., Delannay, X., Pershing, J.C., Hall, M.A. and Lamkey, K.R. (2006) Plant breeding: past, present, and future. In: Lamkey, K.R. and Lee, M. (eds) *Plant Breeding: The Arnel R. Hallauer International Symposium.* Blackwell Publishing, Ames, Iowa, pp. 3–50.

Crossa, J., Yang, R.-C. and Cornelius, P.L. (2004) Studying crossover genotype x environment interaction using linear-bilinear models and mixed models. *Journal of the Agricultural, Biological and Environmental Statistics* 9, 362–380.

Dingkuhn, M., Johnson, D.E., Sow, A. and Audebert, A.Y. (1999) Relationships between upland rice canopy characteristics and weed competitiveness. *Field Crops Research* 61, 79–95.

Dreccer, M.F., Borgognone, M.G., Ogbonnaya, F.C., Trethowan, R.M. and Winter, B. (2007) CIMMYT-selected derived synthetic bread wheats for rainfed environments: Yield evaluation in Mexico and Australia. *Field Crops Research* 100, 218–228.

Duveiller, E. (2004) Controlling foliar blights of wheat in the rice-wheat systems of Asia. *Plant Disease* 88, 552–556.

Ekanayake, I., Ortiz, R. and Vuylsteke, D. (1994) Influence of leaf age, leaf surface and time of day on leaf conductance of various *Musa* genotypes. *Annals of Botany* 73, 173–178.

Ekanayake I.J., Ortiz, R. and Vuylsteke, D. (1998) Leaf stomatal conductance and stomatal morphology of *Musa* germplasm. *Euphytica* 99, 221–229.

El Bouhssini, M., Street, K., Joubi, A., Ibrahim, Z. and Rihawi, F. (2009) Sources of wheat resistance to Sunn pest, *Eurygaster integriceps* Puton, in Syria. *Genetic Resources and Crops Evolution* 56, 1065–1069.

Evenson, R.E. and Gollin, D. (2003) Assessing the Impact of the Green Revolution, 1960 to 2000. *Science* 300, 758–762.

FAO Interdepartmental Working Group on Climate Change and the Stockholm Environment Institute (2007) *Climate Change and Food Security – A Framework Document.* Food and Agriculture Organization of the United Nations, Rome, 80 pp.

FAO (2009) *Food Security and Agricultural Mitigation in Developing Countries: Options for Capturing Synergies.* Food and Agriculture Organization of the United Nations, Rome, 80 pp.

Fischer R.A., Rees, D., Sayre, K.D., Lu, Z.-M., Condon, A.G. and Larque-Saavedra, A. (1998) Wheat yield progress associated with higher stomatal conductance and photosynthetic rate, and cooler canopies. *Crop Science* 38, 1467–1475.

Fisher, M.J., Rao, I.M., Ayarza, C.E., Lascano, C.E., Sanz, J.I., Thomas, R.J. and Vera, R.R. (1994) Carbon storage by introduced deep-rooted grasses in the South American savannas. *Nature* 371, 236–238.

Forbes, G.A. and Simon, R. (2007) Implications for a warmer, wetter world on the late blight pathogen: how CIP efforts can reduce risk for low-input potato farmers. *Journal of Semi-Arid Tropics Agricultural Research* 4. Available at: http://ejournal.icrisat.org/SpecialProject/sp4.pdf.

Futakuchi, K., Jones, M.P. and Ishii, R. (2001) Physiological and morphological mechanisms of submergence resistance in African rice (*Oryza glaberrima* Steud.). *Japanese Journal of Tropical Agriculture* 45, 8–14.

Gao, S., Martinez, C., Skinner, D.J., Krivanek, A.F., Crouch, J.H. and Xu, Y. (2008) Development of a seed DNA-based genotyping system for marker-assisted selection in maize. *Molecular Breeding* 22, 507–515.

Ghannoum, O. (2009) C_4 photosynthesis and water stress. *Annals of Botany* 103, 635–644.

Gore, M.A., Chia, J.-M., Elshire, R.J., Sun, Q., Ersoz, E.S., Hurwitz, B.L., Peiffer, J.A., McMullen, M.D., Grills, G.S., Ross-Ibarra, J., Ware, D.H. and Buckler, E.S. (2009) A first-generation haplotype map of maize. *Science* 326, 1115–1117.

Häusler, R.E., Hirsch, H.-J., Kreuzaler, F. and Peterhänsel, C. (2002) Overexpression of C_4-cycle enzymes in transgenic C_3 plants: a biotechnological approach to improve C_3 photosynthesis. *Journal of Experimental Botany* 53, 591–607.

Hede, A., Skovmand, B., Reynolds, M.P., Crossa, J., Vilhelmsen, A.L. and Stølen, O. (1999) Evaluating genetic diversity for heat tolerance in Mexican wheat landraces. *Genetic Resources and Crop Evolution* 46, 37–45.

Hibberd, J.M., Sheehy, J.E. and Langdale, J.A. (2008) Using C_4 rice photosynthesis to increase the yield of rice – rationale and feasibility. *Current Opinion in Plant Biology* 11, 228–231.

Horton, P. (2000) Prospects for crop improvement through the genetic manipulation of photosynthesis: morphological and biochemical aspects of light capture. *Journal of Experimental Botany* 51, 475–485.

Howden, S.M., Soussana, J.S., Tubiello, F.N., Chhetri, N., Dunlop, M. and Meinke, H. (2007) Adapting agriculture to climate change. *Proceedings of the National Academy of Sciences* 104, 19691–19696.

Hubbart, S., Peng, S., Horton, P., Chen, Y. and Murchie, E.H. (2007) Trends in leaf photosynthesis in historical rice varieties developed in the Philippines since 1966. *Journal of Experimental Botany* 58(12), 3429–3428.

Hubert, B., Rosegrant, M., van Boekel, M.A.J.S. and Ortiz, R. (2010) The future of food: scenarios for 2050. *Crop Science* 50, S1–S18.

Ismail, A.M, Heuer, S., Thomson, M.J. and Wissuwa, M. (2007) Genetic and genomic approaches to develop rice germplasm for problem soils. *Plant Molecular Biology* 65, 547–570.

Jarvis, A., Lane, A. and Hijmans, R. (2008) The effect of climate change on crop wild relatives. *Agriculture, Ecosystems & Environment* 126, 13–23.

Jena, K.K. and Mackill, D.J. (2008) Molecular markers and their use in marker-assisted selection in rice. *Crop Science* 48, 1266–1276.

Johnson, D.E., Dingkuhn, M., Jones, M.P. and Mahamane, M.C. (1998) The influence of rice plant type on the effect of weed competition on *Oryza sativa* and *Oryza glaberrima. Weed Research* 38, 207–216.

Jones, M.P., Dingkuhn, M., Aluko, G.K. and Semon, M. (1997a) Interspecific *Oryza sativa* L. x *O. glaberrima* Steud. progenies in upland rice improvement. *Euphytica* 92, 237–246.

Jones, M.P., Mande, S. and Aluko, K. (1997b) Diversity and potential of *Oryza glaberrima* Steud. in upland rice breeding. *Breeding Science* 47, 395–398.

Jones, P.G. and Thornton, P.K. (2003) The potential impact of climate change on maize production in Africa and Latin America in 2055. *Global Environmental Change* 13, 51–59.

Kellogg, E.A. (1999) Phylogenetic aspects of the evolution of C_4 photosynthesis. In: Sage, R.F. and Monson, R.K. (eds) *C_4 Plant Biology.* Academic Press, San Diego, pp. 411–444.

Kelly, A.E. and Goulden, M.L. (2008) Rapid shifts in plant distribution with recent climate change. *Proceedings of the National Academy of Sciences* 105, 11823–11826.

Kumar, A., Bernier, J., Verulkar, S., Lafitte, H.R. and Atlin, G.N. (2008) Breeding for drought tolerance: direct selection for yield, response to selection and use of drought tolerant donors in upland and lowland-adapted populations. *Field Crop Research* 107, 221–231.
Kumar, S. (2008) Climate change and crop breeding objectives in the twenty first century. *Current Science* 90, 1053–1054.
Lane, A. and Jarvis, A. (2007) Changes in climate will modify the geography of crop suitability: agricultural biodiversity can help with adaptation. *Journal of Semi-Arid Tropics Agricultural Research* 4. Available at: www.icrisat.org/journal/SpecialProject/SP2.pdf (accessed 9 November 2010).
Lawlor, D.W. (2009) Musings about the effect of environment on photosynthesis. *Annals of Botany* 103, 543–549.
Lefebvre, S., Lawson, T., Fryer, M., Zakhleniuk, O.V., Lloyd, J.C. and Raines, C.A. (2005) Increased sedoheptulose-1,7-bisphosphatase activity in transgenic tobacco plants stimulates photosynthesis and growth from an early stage in development. *Plant Physiology* 138, 451–460.
Leslie, M. (2009) On the origins of photosynthesis. *Science* 323, 1286–1287.
Lightfoot, D.A., Mungur, R., Ameziane, R., Nolte, S., Long, L., Bernhard, K., Colter, A., Jones, K., Iqbal, M.J., Varsa, E. and Young, B. (2007) Improved drought tolerance of transgenic *Zea mays* plants that express the glutamate dehydrogenase gene (*gdhA*) of *E. coli*. *Euphytica* 156, 103–116.
Lobell D.B. and Field, C.B. (2007) Global scale climate–crop yield relationships and the impacts of recent warming. *Environmental Research Letters* 2, (doi):10.1088/1748-9326/2/1/014002.
Lobell D.B., Burke, M.B., Tebaldi, C., Mastrandrea, M.D., Falcon, W.P. and Naylor, R.L. (2008) Prioritizing climate change adaptation needs for food security in 2030. *Science* 319, 607–610.
Long, S.P., Zhu, X.-G., Naidu, S.L. and Ort, D.R. (2006) Can improvement in photosynthesis increase crop yields? *Plant, Cell and Environment* 29, 315–330.
Mackay, M.C., Street, K.A., Mitrofanova, O.P., Konopka, J. and Berger, J. (2004) Focused identification of germplasm strategy–FIGS. In: Black, C.K., Panozzo, J.F. and Rebetzke, G.J. (eds) *Cereals 2004, Proceedings of the 54th Australian Cereal Chemistry Conference and the 11th Wheat Breeders' Assembly*, 21–24 September 2004, Canberra, Australian Capital Territory. Royal Australian Chemical Institute, Melbourne, Australia, pp. 138–141.
Matsui, T., Kobayasi, K., Kagata, H. and Horie, T. (2005) Correlation between viability of pollination and length of basal dehiscence of the theca in rice under a hot-and-humid condition. *Plant Production Science* 8, 109–114.
Matsumoto, T. and [257] co-authors (2005) The map-based sequence of the rice genome. *Nature* 436, 793–800.
McMullen, M.D., Kresovich, S., Sanchez Villeda, H., Bradbury, P., Li, H., Sun, Q., Flint-Garcia, S., Thornsberry, J., Acharya, C., Bottoms, C., Brown, P., Browne, C., Eller, M., Guill, K., Harjes, C., Kroon, D., Lepak, N., Mitchell, S.E., Peterson, B., Pressoir, G., Romero, S., Oropeza Rosas, M., Salvo, S., Yates, H., Hanson, M., Jones, E., Smith, S., Glaubitz, J.C., Goodman, M., Ware, D., Holland, J.B. and Buckler, E.S. (2009) Genetic properties of the maize nested association mapping population. *Science* 325, 737–740.
Metz, B., Davidson, O., Bosch, P., Dave, R. and Mayer, L. (2007) *Climate Change 2007 Mitigation*. Cambridge University Press, Cambridge.
Murchie, E.H., Pinto, M. and Horton, P. (2008) Agriculture and the new challenges of photosynthesis research. *New Phytologist* 181, 532–552.
Nassar, N.M.A. and Ortiz, R. (2009) Cassava genetic resources: manipulation for crop improvement. *Plant Breeding Reviews* 31, 247–275.
Neely, C., Bunning, S. and Wilkes, A. (eds) (2009) *Review of Evidence on Dryland Pastoral Systems and Climate Change – Implications and Opportunities for Mitigation and Adaptation*. Food and Agriculture Organization of the United Nations, Rome.
Nelson D.E., Repetti, P.P., Adams, T.R., Creelman, R.A., Wu, J., Warner, D.C., Anstrom, D.C., Bensen, R.J., Castiglioni, P.P., Donnarummo, M.G., Hinchey, B.S., Kumimoto, R.W., Maszle, D.R., Canales, R.D., Krolikowski, K.A., Dotson, S.B., Gutterson, N., Ratcliffe, O.J. and Heard, J.E. (2007) Plant nuclear factor Y (NF-Y) B subunits confer drought tolerance and lead to improved corn yields on water-limited acres. *Proceedings of the National Academy of Sciences* 104, 16400–16455.
Nelson, G.C., Rosegrant, M.W., Koo, J., Robertson, R., Sulser, T., Zhu, T., Ringler, C., Msangi, S., Palazzo, A., Batka, M., Magalhaes, M., Valmonte-Santos, R., Ewing, M. and Lee, D. (2009) *Climate Change – Impact on Agriculture and Costs of Adaptation*. International Food Policy Research Institute, Washington, DC, 20 pp.
Nuitjen, E., van Treuren, R., Struik, P.C., Mokuwa, A., Okry, F., Teeken, B. and Richards, P. (2009) Evidence for

the emergence of new rice types of interspecific hybrid origin in West African farmers' fields. *PLoS ONE* 4(10), e7335. DOI:10.1371/journal.pone.0007335.
Ogbonnaya, F.C., Ye, G., Trethowan, R., Dreccer, F., Lush, D., Shepperd, J. and van Ginkel, M. (2007) Yield of synthetic backcross-derived lines in rainfed environments of Australia. *Euphytica* 157, 321–336.
Ortiz, R. (1997) Secondary polyploids, heterosis and evolutionary crop breeding for further improvement of the plantain and banana genome. *Theoretical and Applied Genetics* 94, 1113–1120.
Ortiz, R. (2008) Crop genetic engineering under global climate change. *Annals of Arid Zone* 47, 1–12.
Ortiz, R. and Hartmann, P. (2003) Beyond crop technology: the challenge for African rural development. In: *Vol. 2. Reference Material of the Sub-Saharan Africa Challenge Program 'Building Livelihoods through Integrated Agricultural Research for Development – Securing the Future for Africa's Children'*. Forum for Agricultural Research in Africa, Accra, Ghana, pp. 39–72.
Ortiz R., Iwanaga, M., Reynolds, M.P., Wu, H. and Crouch, J.H. (2007a) Overview on crop genetic engineering for drought-prone environments. *Journal of Semi-Arid Tropics Agricultural Research* 4. Available at: http://ejournal.icrisat.org/SpecialProject/sp3.pdf (accessed 9 November 2010).
Ortiz, R., Trethowan, R., Ortiz Ferrara, G., Iwanaga, M., Dodds, J.H., Crouch, J.H., Crossa, J. and Braun, H.J. (2007b) High yield potential, shuttle breeding and a new international wheat improvement strategy. *Euphytica* 157, 365–384.
Ortiz, R., Braun, H.J., Crossa, J., Crouch, J.H., Davenport, G., Dixon, J., Dreisigacker, S., Duveiller, E., He, Z., Huerta, J., Joshi, A.K., Kishii, M., Kosina, P., Manes, Y., Mezzalama, M., Morgounov, A., Murakami, J., Nicol, J., Ortiz-Ferrara, G., Ortiz-Monasterio, J.I., Payne, T.S., Peña, R.J., Reynolds, M.P., Sayre, K.D., Sharma, R.C., Singh, R.P., Wang, J., Warburton, M., Wu, H. and Iwanaga, M. (2008a) Wheat genetic resources enhancement by the International Maize and Wheat Improvement Center (CIMMYT). *Genetic Resources and Crop Evolution* 55, 1095–1140.
Ortiz, R., Sayre, K.D., Govaerts, B., Gupta, R., Subbarao, G.V., Ban, T., Hodson, D., Dixon, J.M., Ortiz-Monasterio, J.I. and Reynolds, M. (2008b) Climate change: can wheat beat the heat? *Agriculture, Ecosystems & Environment* 126, 45–58.
Ortiz, R., Simon, P., Jansky, S. and Stelly, D. (2009) Ploidy manipulation of the gametophyte, endosperm, and sporophyte in nature and for crop improvement – A tribute to Prof. Stanley J. Peloquin (1921-2008). *Annals of Botany* 104, 795–807.
Ortiz, R., Taba, S., Chávez Tovar, V.H., Mezzalama, M., Xu, Y., Yan, J. and Crouch, J.H. (2010) Conserving and enhancing maize genetic resources as global public goods – A perspective from CIMMYT. *Crop Science* 50, 13–28.
Parry, M.A.J., Madgwick, P.J., Carvalho, J.F.C. and Andralolc, P.J. (2007) Prospects for increasing photosynthesis by overcoming the limitations of Rubisco. *Journal of Agricultural Science* (Cambridge) 145, 31–43.
Paterson, A. and [44] co-authors (2009) The *Sorghum bicolor* genome and the diversification of grasses. *Nature* 457, 551–556.
Peng, S., Huang, J., Sheehy, J.E., Laza, R.C., Visperas, R.M., Zhong, X., Centeno, G.S., Khush, G.S. and Cassman, K.G. (2004) Rice yields decline with higher night temperature from global warming. *Proceedings of the National Academy of Sciences* 101, 9971–9975.
Prestes, A.M., Arendt, P.F., Fernandes, J.M.C. and Scheeren, P.L. (2007) Resistance to *Magnaporthe grisea* among Brazilian wheat genotypes. In: Buck, H.T., Nisi, J.E. and Salomón, N. (eds) *Wheat Production in Stressed Environments*. Springer, Dordrecht, the Netherlands, pp. 119–123.
Reynolds, M.P. and Ortiz, R. (2010) Adapting crops to climate change: a summary. In: Reynolds, M.P. (ed.) *Climate Change and Crop Production*. CAB International, Wallingford, UK. In press.
Reynolds, M.P., Balota, M., Delgado, M.I.B., Amani, J. and Fischer, R.A. (1994) Physiological and morphological traits associated with spring wheat yield under hot, irrigated conditions. *Australian Journal of Plant Physiology* 21, 717–730.
Reynolds, M.P., Singh, R.P., Ibrahim, A., Ageeb, O.A., Larqué-Saavedra, A. and Quick, J.S. (1998) Evaluating physiological traits to compliment empirical selection for wheat in warm environments. *Euphytica* 100, 85–94
Reynolds, M.P., van Ginkel, M. and Ribaut, J.-M. (2000) Avenues for genetic modification of radiation use efficiency in wheat. *Journal of Experimental Botany* 51, 459–473.
Reynolds, M.P., Dreccer, F. and Trethowan, R. (2007) Drought adaptive traits derived from wheat wild relatives and landraces. *Journal of Experimental Botany* 58, 177–186.
Reynolds, M.P., Foulkes, J.M., Slafer, G.A., Berry, P., Parry, M.A.J., Snape, J. and Angus, W.J. (2009a) Raising yield potential in wheat. *Journal of Experimental Botany* 60, 1899–1918.

Reynolds, M.P., Manes, Y., Inzaloo, A. and Langridge, P. (2009b) Phenotyping approaches for physiological breeding and gene discovery in wheat. *Annals of Applied Biology* 155, 309–320.

Ribaut, J.-M. and Ragot, M. (2007) Marker-assisted selection to improve drought adaptation in maize: the backcross approach, perspectives, limitations, and alternatives. *Journal of Experimental Botany* 58, 351–360.

Ridgwell, A., Singarayer, J.S., Hetherington, A.M. and Valdes, P.J. (2009) Tackling regional climate change by leaf albedo bio-geoengineering. *Current Biology* 19, 116–150.

Schlenker, W. and Roberts, M.J. (2009) Nonlinear temperature effects indicate severe damages to US crop yields under climate change. *Proceedings of the National Academy of Sciences* 106, 15594–15598.

Schmidhuber, J. and Tubiello, F.N. (2007) Global food security under climate change. *Proceedings of the National Academy of Sciences* 104, 19703–19708.

Schnable, P.S. and [156] co-authors (2009) The B73 maize genome: complexity, diversity, and dynamics. *Science* 326, 1112–1115.

Septiningish, E.M., Pamplona, A.M., Sanchez, D.L., Neeraja, C.N., Vergara, G.V., Heuer, S., Ismail, A.M. and Mackill, D.J. (2008) Development of submergence tolerant rice cultivars: The *Sub1* locus and beyond. *Annals of Botany* 103, 151–160.

Sharma-Natu, P. and Ghildiyal, M.C. (2005) Potential targets for improving photosynthesis and crop yield. *Current Science* 88, 1918–1928.

Spielman, D. and Pandya-Lorch, R. (2009) *Millions Fed: Proven Success in Agricultural Development.* International Food Policy Research Institute, Washington, DC.

Stige, L.C., Stave, J., Chan, K.-S., Ciannelli, L., Pettorelli, N., Glantz, M., Herren, H.R. and Stenseth, N.C. (2006) The effect of climate variation on agro-pastoral production in Africa. *Proceedings of the National Academy of Sciences* 103, 3049–3053.

Swanson-Wagner, R.A., DeCook, R., Jia, Y., Bancroft, T., Ji, T., Zhao, X., Nettleton, D. and Schnable, P.S. (2009) Paternal dominance of trans-eQTL influences gene expression patterns in maize hybrids. *Science* 326, 1118–1119.

Tambussi, E.A., Bort, J., Guiamet, J.J., Nogués, S. and Araus, J.L. (2007) The photosynthetic role of ears in C_3 cereals: metabolism, water use efficiency and contribution to grain yield. *Critical Reviews in Plant Sciences* 26, 1–16.

Thornton, P., Herrero, M., Freeman, A., Mwai, O., Rege, E., Jones, P. and McDermott, J. (2007) Vulnerability, climate change and livestock – research opportunities and challenges for poverty alleviation. *Journal of Semi-Arid Tropics Agricultural Research* 4 (http://ejournal.icrisat.org/SpecialProject/sp7.pdf).

Thuiller, W., Lavorel, S., Araújo, M.B., Sykes, M.T. and Prentice, I.C. (2005) Climate change threats to plant diversity in Europe. *Proceedings of the National Academy of Sciences* 102, 8245–8250.

Tubiello, F.N., Soussana, J.-F. and Howden, S.M. (2007) Crop and pasture response to climate change. *Proceedings of the National Academy of Sciences* 104, 19686–19690.

UNEP (2008) *Agriculture, Agro-biodiversity and Climate Change.* United Nations Environment Programme, Nairobi, Kenya, 2 pp.

Various (2008) How can sustainable land management contribute to mitigating climate change? *Natural Resources Forum* 32, 251–255.

Vielle-Calzada, J.-P., Martínez de la Vega, O., Hernández-Guzmán, G., Ibarra-Laclette, E., Alvarez-Mejía, C., Vega-Arreguín, J.C., Jiménez-Moraila, B., Fernández-Cortés, A., Corona-Armenta, G., Herrera-Estrella, L. and Herrera-Estrella, A. (2009) The Palomero genome suggests metal effects on domestication. *Science* 326, 1078.

Villamon, F.G., Spooner, D.M., Orrillo, M., Mihovilovich, E., Pérez, W. and Bonierbale, M. (2005) Late blight resistance linkages in a novel cross of the wild potato species *Solanum paucissectum* (series *Piurana*). *Theoretical and Applied Genetics* 111, 1201–1214.

Visser, R.G.F., Bachem, C.W.B., de Boer, J.M., Bryan, G.J., Chakrabati, S.K., Feingold, S., Gromadka, R., van Ham, R.C.H.J., Huang, S., Jacobs, J.M.E., Kuznetsov, B., de Melo, P.E., Milbourne, D., Orjeda, G., Sagredo, B. and Tang, X. (2009) Sequencing the potato genome: Outline and first results to come from the elucidation of the sequence of the world's third most important food crop. *American Journal of Potato Research* (doi): 10.1007/s12230-009-9097-8 (http://www.springerlink.com/content/h360657j23u33345/fulltext.html).

Voznesenskaya, E.V., Franceschi, V.R., Kllrats, O., Freitag, H. and Edwards, G.E. (2001) Kranz anatomy is not essential for terrestrial C_4 photosynthesis. *Nature* 414, 543–554.

Vuylsteke, D.R. (2001) Strategies for the utilization of genetic variation in plantain improvement. Ph.D thesis, Katholieke Universiteit Leuven, Belgium, 213 pp.

Wang, C.-R., Yang, A.-F., Yue, G.-D., Gao, Q., Yin, H.-Y. and Zhang, J.-R. (2008) Enhanced expression of phospholipase C 1 (*ZmPLC1*) improves drought tolerance in transgenic maize. *Planta* 227, 1127–1140.

Wassmann, R., Jagadish, S.V.K., Heuer, S., Ismail, A., Redona, E., Serraj, R., Singh, R.K., Howell, G., Pathak, H. and Sumfleth, K. (2009a) Climate change affecting rice production: the physiological and agronomic basis for possible adaptation strategies. *Advances in Agronomy* 101, 59–122.

Wassmann, R., Jagadish, S.V.K., Sumfleth, K., Pathak, H., Howell, G., Ismail, A., Serraj, R., Redona, E., Singh, R.K. and Heuer, S. (2009b) Regional vulnerability of climate change impacts on Asian rice production and scope for adaptation. *Advances in Agronomy* 102, 91–133.

WEMA (2010) Water Efficient Maize for Africa. Available at: http://www.aatf-africa.org/wema (accessed 9 November 2010).

Willis, K.J. and Bhagwat, S.A. (2009) Biodiversity and climate change. *Science* 326, 806–807.

Xu, K., Xu, X., Fukao, P., Canlas, R., Maghirang-Rodriguez, R., Heuer, S., Ismail, A.M., Bailey-Serres, J., Ronald, P.C. and Mackill, D.J. (2006) *Sub1 A* is an ethyleneresponse-factor-like gene that confers submergence tolerance. *Nature* 442, 705–708.

Xu, Y. and Crouch, J.H. (2008) Genomics of tropical maize, a staple food and feed across the world. In: Moore, P.H. and Ming, R. (eds) *Genomics of Tropical Crop Plants*. Springer Verlag, New York, pp. 333–370.

Xu, Y., Wang, J. and Crouch, J.H. (2008) Selective genotyping and pooled DNA analysis: an innovative use of an old concept. In: *Recognizing Past Achievement, Meeting Future Needs*, Proceedings of the 5th International Crop Science Congress, 13–18 April 2008, Jeju, Korea. Published on CDROM. Available at: www.cropscience2008.com (accessed 9 November 2010).

Zhai, S.M., Sui, Z.H., Yang, A.F. and Zhang, J.R. (2005) Characterization of a novel phosphoinositide-specific phospholipase C from *Zea mays* and its expression in *Escherichia coli*. *Biotechnology Letters* 27, 799–804.

Zhu X.G., Portis, A.R. and Long, S.P. (2004) Would transformation of C_3 crop plants with foreign Rubisco increase productivity? A computational analysis extrapolating from kinetic properties to canopy photosynthesis. *Plant, Cell & Environment* 27, 155–165.

13 Agricultural Revolutions and Their Enemies: Lessons for Policy Makers

J.M. Lenné and D. Wood

... government commitment to enabling policies and to sustained investments in agricultural research will provide the crucial building blocks for future successes in ... agriculture.

Haggblade and Hazell (2010)

Science policy has, above all else, been science budget policy.

Sarewitz (2010)

... he still believes that environmental activists and their allies in international agencies are a threat to progress on global food security.

Bailey (2009, writing on Norman Borlaug)

Introduction

This book has clearly shown that science-based agrobiodiversity management to increase global food production has led to enormous gains in agricultural productivity, food security and human well-being (Evenson and Gollin, 2003; Spielman and Pandya-Lorch, 2009; Raudsepp-Hearn *et al.*, 2010). From 1961 to 2007, gross world food production increased from 1.84 to 4.38 billion t (238%) from a land area increase of only 11% (4.51 to 4.93 billion ha) (Royal Society, 2009). Great progress has also been made in improving the nutritional quality of food. Importantly, these efforts have done more than just feed millions. The interventions of the past half century have also demonstrated that agriculture can be a powerful tool to reduce poverty as well as a key driver of growth and development for many of the world's poor countries (Byerlee *et al.*, 2009; Spielman and Pandya-Lorch, 2009; Haggblade and Hazell, 2010). In addition, successes in increasing productivity in one country have readily and dramatically spilled over to other countries with additional impacts on increased food security and incomes and reduced poverty (Alston, 2002; Pardey *et al.*, 2006; Pardey and Pingali, 2010).

During the period from the 1960s to the 1980s policy makers and donors strongly supported and invested in the science that enabled these advances and ensuing successes (see Chapter 2, this volume). But during the past 30 years, investment has drastically declined along with political will (Pardey *et al.*, 2006; Pardey and Pingali, 2010). Globally, Official Development Assistance (ODA) spent on agriculture fell dramatically from US$6.2 billion in 1980 to US$2.3 billion in 2002. Furthermore, commitments made since 2000, to increase investments in agricultural research both in the developing and developed world, have been lukewarm to woefully inadequate (Fan, 2010). Such lack of commitment and complacency by both policy makers and donors is partly based on the belief that the world is producing enough food to feed existing and growing populations (for an early claim that there is enough food for all, see Lappe and Collins, 1977 – this is rather like saying there is enough money for all: there certainly is, but will it ever be distributed equally?).

This book has also strongly argued that future growth in agricultural productivity and global food security can continue to be achieved through renewal and concerted application of science-based agrobiodiversity management, *provided there is enhanced policy support and significant increases in government and international donor investment*. But to do so sustainably in the face of climate change, equitably in the face of social and regional inequalities, and successfully in the current uncertain policy and investment environment is a major challenge (Evans, 1998). Substantial increases in commitment and support for research – sooner rather than later – are essential to enable the food system to cope with both known and unknown challenges in the coming decades to achieve food security.

More importantly, there has been a trend in recent years for modern intensive agriculture to be attacked by international and national groups, especially NGOs (as has been referred to many times in this book, see Chapters 5, 7, 8 and 11). It is likely that this anti-monoculture, anti-GM crop and anti-intensive agriculture lobby is having a growing influence on policy makers and investors, especially in Europe and the USA. More worryingly, there is now an 'alternative agriculture' lobby based on organic agriculture, agroecological approaches and small-farm agriculture that is threatening to divert already scarce funds for sound science to unproven, often extreme, approaches to future food production (see Chapter 11, this volume). However, the appeal for ecological approaches to agriculture and the need to mimic diverse natural vegetation now appears to be fundamentally flawed. From what can be read in the foregoing chapters about agrobiodiversity and how it functions to produce food, it is apparent that many of these supposedly 'ecological' attacks on modern agriculture are scientifically invalid and certainly could damage future food security.

However, it is useful to flesh-out our claim by characterizing and then classifying these damaging attacks to see if a pattern emerges that could be of value to policy makers. Agriculture has developed through an incremental series of 'agricultural revolutions' where the advantages of an earlier stage are incorporated in later stages (see Evans, 2003). These include: the 'Neolithic revolution' at the time of the origin of agriculture; the 'Agricultural Revolution' in farming in Europe (dating from 1750 to around 1850 in England at a time of rapid population growth); the 'Green Revolution' in South and East Asia in wheat and rice dating from the decades following the foundation of the International Agricultural Research Centres (IARCs) in the early 1960s; and, possibly, the ongoing 'Biotechnology Revolution'.

But agricultural revolutions, as with political revolutions, have never been simple affairs: conflict abounds. While proponents of revolutions try to maintain the impetus, opponents of various kinds try to derail the process; for example, 'counter-revolutionaries' desire to return to the old order; 'anti-revolutionaries' promote alternatives to the revolution with themselves in charge. We feel that there are demonstrable parallels between political and agricultural revolutions, with some factions (the 'contras') wanting to return to the old ways and some factions (the 'antis') promoting new alternatives. And as with political revolutions, there may be foreign interests trying to reverse or change the path of agricultural revolution in their own political or economic interests.

The 2008 food crisis and ongoing economic crisis has started to shock policy makers and investors in agriculture back to a semblance of reality. But there is still a long way to go in convincing policy makers, governments and international donors of the critical need to massively increase support for and investment in sound science for future food production, both nationally and internationally. Also, as we outline above and detail further below, there are very many attempts to derail productive agriculture for many and varied (and often technically spurious) reasons. The main objective of this chapter is to attempt to make policy makers and investors in agriculture more clearly aware of: (i) the value of supporting proven and promising i.e. 'good' approaches to increasing food production; and (ii) the serious pitfalls in supporting the unproven, flawed and failed i.e. 'bad and ugly' approaches.

Proven Approaches: the 'Good'

Knowledge of the ecology of crop domestication is key to food security

Crops, domestic animals and all the agrobiodiversity that became associated with them, originated with early farmers more than 10,000 years ago in several separate regions of the world (see Chapter 3, this volume). We believe that a greater knowledge of the origin of domesticates would be a valuable resource for the future management and sustainability of agriculture. To contribute to this resource we have sidestepped academic controversies over the origin of agriculture and instead concentrated on the ecology of agricultural origins – with a focus, as elsewhere in this book, on crops. An interesting and rather surprising pattern emerged, in particular for staple cereals. It is apparent that domestication was not a random selection of wild species growing in many different habitats and taking place over an indeterminate time scale. Rather, it was very highly selective as to: the type of species domesticated, especially their mode of seed production and ecology; their regions of domestication; and, significantly, given our concern over global warming, at least in some cases, at a time of dramatic and rapid climatic cooling and then warming over a time span of less than a thousand years.

An ecological view of plant domestication provides substantial lessons for present and future agriculture:

- Annual crops such as our cereals are ecologically valid. They invest a high proportion of their biomass in seed and also avoid pests, diseases and adverse climate by having a short growing season.
- Monocultures of large-seeded plants are ecologically valid under certain conditions, including annual disturbance (now achieved by tillage) and consequent lack of tree cover.
- Traditional forms of agriculture by fire (shifting cultivation and savannah burning) and silt (captured on flooded terraces) are a rational way to avoid competition from trees. Tree-free fields are close mimics of the ecology of some wild relatives of crops which, under similar conditions, can thrive and even form persistent climax vegetation.
- There is growing evidence that at least some of our important crops were domesticated at a time of global warming at the end of the Younger Dryas cold spell. Domestication was a highly successful human technical response to a period of unprecedented and very rapid climate change.
- Knowledge of the ecology of past domestications could help domesticate new food crops.

Many of the 'magic bullet' claims for the future of food production ignore these ecological foundations of domestication, foundations which have supported the 'Green Revolution' of large yield increases of annual cereals grown in monocultures. Following our earlier paper on this subject, Grime (2002) noted that: 'Wood & Lenné (2001) have argued persuasively that the origins of arable farming and perhaps also its future are to be found as adaptations of naturally-occurring, productive ecosystems dominated by few species.'

Monoculture agriculture has massively increased food security

The ecology of plant domestication (see above) firmly validates monocultures as a strategy for sustainable agriculture. Monocultures – as modern varieties, varietal mixtures, landraces and dual-purpose crops – are the most widely grown type of cropping system on earth. Humanity relies on monocultures for food security, and this is unlikely to change for the foreseeable future, in spite of the protestations of the anti-monoculture lobby. Yields of staple crops such as rice, wheat and maize, grown as monocultures, have increased several-fold in the past 50 years (Spielman and Pandya-Lorch, 2009; see Chapter 5, this volume). History records no increase in food production that was remotely comparable in scale, speed, spread and duration (Lipton and Longhust, 1989). The investments in science and technology, along with complementary investments in irrigation systems, road networks, fertilizer production and food price stabilization policies, have paid off handsomely. Future investment in agricultural science for food security should therefore give highest priority to the sustainable intensification of monocultures – making them even more productive, resource efficient and environmentally stable. They are the foundation of future global food security.

Inter-continental crop introduction has increased food security

The thesis of crop introduction is counter-intuitive: introduced crops perform better than local ones. Yet the value of crop introduction is amply backed by production statistics and by practice going back centuries – not least the 'article of faith' by the British colonial Departments of Agriculture. The reality is that staple food production in many countries is highly reliant on introduced crops (see Chapter 4, this volume). It is therefore of high importance for policy makers to understand the value of and the reasons for crop introduction, as recently there has been a cacophony of support from northern NGOs – with no hands-on experience of agriculture – for funding for 'local adaptation'. This is the idea that local crops and varieties progressively become better and even optimally adapted to local conditions (an idea demolished by Gould, 1997). The claim is always accompanied by calls for 'on-farm conservation', with farmers using their own supposedly better-adapted seed and never replacing it from off-farm sources (see Chapters 4 and 10, this volume).

Crops and varieties may adapt to local abiotic conditions of rainfall, temperature, seasonality and day length and they can be successfully moved to other regions globally with the same conditions. But crops and varieties become progressively 'dis'-adapted to the treadmill of biotic constraints such as pests and diseases. Crop introduction very simply moves crops and varieties across oceans to escape their coevolved biotic constraints (with the hope that quarantine can maintain this status) (see Chapters 4 and 10, this volume).

There are complex policy issues associated with crop introduction. On the one hand, the economic value of introduced crops – of necessity brought from elsewhere – is one of the major justifications for the FAO Seed Treaty (ITPGR): everyone benefits from the free flow of genetic resources needed to capture the benefits of crop introduction. On the other hand, a major funding target for the ITPGR is support for on-farm conservation, based firmly on the premise of local adaptation. Unfortunately, projects for on-farm conservation do not countenance the introduction of new crops and varieties: farmers have to make do with their former suite of crops and varieties, subject to long-term coevolution to local biotic constraints, which may be of more value to some donor countries, but is unethical as it denies farmers access to better varieties (see Chapter 6, this volume). Moreover, some policy makers may be confused by these conflicting approaches.

Most developing countries have already benefited and continue to benefit substantially from food produced from introduced staple food crops, for example maize from Latin America in East and Southern Africa (see Spielman and Pandya-Lorch, 2009). And, there is still great potential for further benefits, especially nutritionally, from intercontinental introduction of vegetables and fruits providing the appropriate policies are in place to support such activities.

Modern plant breeding has substantially increased food security

The most knowledgeable, effective, efficient and successful way to use crop diversity for food production is through modern plant breeding – whether by conventional, biotechnological and/or client-oriented approaches (see Chapters 5, 6 and 7, this volume). The development of modern plant breeding clearly demonstrates the striking impact of investment in scientific research on crop productivity and food security. It has greatly facilitated ever-wider use of a wealth of diversity from many sources and, until recently, has allowed food production to keep pace with population growth. Modern plant breeding therefore greatly increases the potential for broadening the diversity for useful traits in crops locally, regionally and globally and has allowed ongoing use of a wealth of crop diversity by millions of farmers (Wood and Lenné, 1999). The many success stories in feeding millions of poor people in developing countries summarized in Chapter 5, this volume (also see Spielman and Pandya-Lorch, 2009), clearly show that modern varieties have an essential role to play in improving livelihoods and food security. Enhanced support by policy makers and investors for modern plant breeding is essential for future food security.

An analysis of the impact of the introduction of modern varieties on crop diversity through three case studies of client-oriented breeding of rice in Nepal and eastern India in Chapter 6, this volume, showed that the approach can rapidly produce new, highly accepted varieties and in situations where prior breeding efforts have met with less success. Loss of crop diversity can be moderated by using client-oriented breeding and rapid seed delivery methods, providing a greater choice of varieties to farmers. Although this approach appears to be the best way to provide access by small-scale farmers to modern varieties under certain circumstances, it has been neglected by international breeding programmes of the IARCs and investors. To meet future food security needs, policy makers should support all appropriate approaches to modern plant breeding, depending on conditions and circumstances.

The recent biotechnology revolution has shown that transgenics or GM crops can target isolated genes from a much wider range of source organisms than conventional plant breeding (see Chapter 7, this volume). To date, transgenic approaches have been highly successful for insect pest resistance and herbicide tolerance in maize, soybean and cotton. And, in spite of long-term, often ugly, anti-GM campaigns, globally in 2009, 134 million ha of transgenics were cultivated by 14 million farmers in 25 countries (James, 2010). Furthermore, the environmental impact is currently orders of magnitude less than the insecticides used in conventional agriculture and the concomitant savings from far less soil erosion as well as the benefits to soil structure due to less compaction have been incomparable. GM crops are highly environmentally friendly!

Yet, there has been neglect of other crops, critical for food security in developing countries, e.g. sorghum, pulses, oilseed crops etc. (Gressel, 2008). The main reason for neglect is regulation: there are extreme regulatory impediments to enhancing agrobiodiversity through transgenics. Only large multinational companies can afford the costs required for regulatory approval and concentrate on staple crops. Currently, it is impossible to use transgenic techniques to insert needed genes in important food crops in developing countries. If regulatory regimes could be simplified, the public sector, local biotechnology companies and local seed companies could be instrumental in getting more technologies to small-scale farmers (Gressel, 2008). Therefore there is a critical role for policy makers in refining the future focus of transgenic research and simplifying the regulatory systems to ensure that improved, more productive GM crops are available to small-scale farmers.

New molecular methods for transgenics could assist in improving crop productivity, enhancing crop biodiversity and improving food security. The first generation of commercialized transgenic crops is analogous to the first generation of widely purchased automobiles, the Ford Model T. The first transgenic crops are now being replaced by far better models, as excellent as modern vehicles, through timed and tissue-specific gene expression, targeted gene insertions and mini-chromosomes etc. But once the transgenic variety has been successfully produced, conventional breeding remains an essential tool to produce new varieties. This further reinforces increased investment in modern plant breeding for future food security.

Paradoxically, some of the recent advances in food crop science that have fed millions have been made during a period of the ongoing, severe erosion of funding for public sector plant breeding. One must question *what* level of advances might have been achieved in increasing crop productivity if funding had continued at the level of the 1970s? Clearly, there is an urgent need to re-align policy and funding to support crop improvement with increasing human needs for food. Funding support for food security should be firmly anchored in ongoing and increased support for modern crop breeding.

Ex situ conservation has enabled increased food security

Agrobiodiversity conservation is a cost-effective method of maintaining collections of introduced crops for future use: it is cheaper to store than to re-collect and re-introduce. Unlike generic biodiversity conservation – often in protected areas – which has a real problem in demonstrating its practical value to meet future needs, crop genetic resources are of clear present and future demonstrable value in breeding of new varieties for food production. However, this value is best captured if conserved collections are a service to crop breeding institutes with all the facilities for long-term storage and with teams of scientists and technicians and field and laboratories for sample evaluation, selection and breeding. It was this focus that enabled the Green Revolution in wheat and rice production, and is even more necessary today.

The most important facet of agro-biodiversity conservation – crop genetic resources – is of immediate value as an ecosystem service producing our food. This vital service is in contrast to conservationists continually blocking agricultural development of forested areas for the stated reason that these forests produced more nebulous and questionable 'ecosystem services' such as 'clean air'. There is a further parallel with the biodiversity debate: the most important service provided by plant biodiversity is photosynthesis, in turn producing food as a support for all animal life. However, this is a wider character of all green plants, not a feature of rare and endangered species, and not restricted to protected areas, so is never emphasized in conservationist claims.

In summary, *ex situ* conservation was the critical backstop to the advances made in achieving food security through modern plant breeding. However, recently, *ex situ* policy at the international level has compromised the flow of germplasm between countries, which has all but dried up. Here, again, the lessons for policy makers are clear. Despite two international legal instruments for bioconservation – the CBD and the ITPGR – neither works well for conservation-for-use. Furthermore, both fail as mechanisms for development, adopting a regressive, outdoor-museum mindset, rather than a dynamic, diversity-for-development, and highly successful biodiversity to development, approach characteristic of the IARCs. International policies for *ex situ* conservation need comprehensive revision if modern plant breeding for future food production is to continue to benefit as it has in the past.

Biological control has enabled increased food security

Damaging crop-associated pests significantly reduce the stability and sustainability of food security by reducing crop yields, up to 80% for both pre- and postharvest losses, wasting often scarce and costly inputs of energy, water, nutrients and labour. There is little wonder that farmers have laboured for millennia and agricultural science has devoted more than 100 years to developing improved methods for controlling pests (Lenné and Wood, 1999; also see Chapter 8, this volume). The reduction of crop losses by harmful crop-associated biodiversity (C-AB) through management of beneficial C-AB is a key agroecosystem service. The most successful examples are biological control of insect pests and weeds.

The impact of biological control of arthropods in agroecosystems was illustrated by some notable success stories in Chapter 8, this volume. These include: management of cassava mealy bug and cereal stemborers in Africa with parasitoids; pest management in tropical rice systems by generalist predators such as spiders; cassava green mite by predatory mites in Africa; and Green Muscle® for locusts and grasshoppers and SpexNPV for army worm in Africa. The controls are ecologically, economically and environmentally sound and more feasible, efficient, often permanent and lower cost than other methods, especially pesticides (Neuenschwander, 2004; Van Driesche *et al.*, 2008). Huge returns to investment have been documented, for example 200:1 for control of cassava mealybug in Africa.

The impact of biological control of weeds – including natural enemy management by a beetle on prickly pear cactus in Australia, USA and South Africa; a beetle on St John's Wort in the USA; and weevils on water hyacinth in Africa as well as management with pathogens of rush skeleton weed (rust) in Australia and the USA; strangler vine in India and the Pacific; and *Striga* spp. in Africa – was also presented in Chapter 8, this volume. These management strategies have been proven to be cost-effective and environmentally safe, through rigorous host-range testing (Morin *et al.*, 2009).

Biological control programmes can dramatically and successfully reduce food crop losses thus contributing to food security in both developed and developing countries. But policy makers also expect increased yields from successful pest management strategies, achieving *only* reduced losses may be judged as failed technology. It is therefore vitally necessary to make policy makers and investors more aware of the value of the approach, the realistic returns and the environmental benefits from biological control successes.

Adaptation and mitigation for climate change will increase future food security

Global warming and unpredictable rainfall are already affecting agricultural production in many parts of the world, reducing food security and farm incomes. Agrobiodiversity remains the main raw material for agroecosystems to cope with climate change by providing traits for plant breeders to breed climate-adapted crops, as highlighted in Chapter 12, this volume. Furthermore, mitigation through agrobiodiversity management that reduces vulnerability to climate change will also greatly assist in ensuring enough food, feed, fibre and biofuel supply in the future. Mitigation and adaptation are therefore complementary strategies already being used by agricultural scientists to manage agrobiodiversity for climate change.

Sustainable agroecosystem management mitigates climate change through carbon sequestration in soils and biomass, thereby improving soil fertility, as well as reducing emissions through conservation agriculture practices (e.g. minimum tillage) and increased input efficiency (Reynolds and Ortiz, 2010). Although usually ignored by so-called climate change experts, improving the productivity of agriculture has been shown to be a key mitigation strategy as it results in less land being cleared and cultivated for crop production. Green Revolution technologies saved 1.1 billion ha of land from clearing, ploughing and greenhouse gas generation, avoiding emissions of 161 gigatonnes of carbon (Burney *et al.*, 2010). And, since 1996, the soil carbon sequestered, facilitated by transgenic herbicide tolerant crops coupled with other conservation agriculture practices, has been equivalent to 83.2 billion t of CO_2 which would otherwise have been released into the global atmosphere (Brookes and Barfoot, 2009).

Crop breeding, including modern biotechnology such as genomics and transgenics, provides genetically enhanced seed-embedded technologies that adapt crops to stresses. Ongoing adaptation of staple food crops to changing stresses through modern plant breeding has been in progress for the past 50 years. Currently, it is increasingly targeting stress tolerances such as drought, flooding, heat, cold, and changing pest and disease situations, traits most likely to be needed in future. Tackling future food production under changing climatic conditions is basically an extension of existing activities. A key target is modifying photosynthesis, particularly converting C_3 staple crops such as rice to C_4 photosynthesis which could realize yield gains of 30–50%.

The Inter-governmental Panel on Climate Change (IPCC) has not given enough attention to the value of biodiversity for food and agriculture, which will increase with global warming, drought and other stresses. It has largely ignored the critical importance of agrobiodiversity management for climate change and appears not even to be aware of the ongoing efforts of major staple crop IARCs. As a result, policy makers and many governments have largely failed to realize the critical role that agrobiodiversity plays on climate change adaptation and mitigation. Hence, the necessary investments in agrobiodiversity management for climate change are not being made. Global public awareness about the role of agrobiodiversity for dealing with climate change and achieving food security needs to be raised. In this way, policy and investment support for agrobiodiversity management through use can be included as an important building block of adaptation to climate change.

Investments in and support for spillovers will substantially contribute to food security

Not only have the above proven approaches led to notable successes in the managing of agrobiodiversity for food security, but they have also enabled the spillover of science and technologies to other countries and regions with similar needs. Spillovers from agricultural research and development investments have been shown to account for more than half of agricultural productivity growth globally (Alston, 2002). This may or may not require additional adaptation to the particular circumstances and systems in specific countries. Spillovers provide rapid and cost-effective means of applying proven agricultural technologies to solve today's and tomorrow's hunger and malnutrition through increasing the production of, access to, and quality of food (Pardey and Pingali, 2010). A recent analysis has shown that the benefits from transgenic or GM crops, especially from increased yields and increased profits, are greatest for small-scale farmers in developing countries, who have benefited from the spillover of technologies originally targeted at farmers in developed countries (Carpenter, 2010). Greater support from policy makers and increased funding from investors would facilitate more widespread and efficient targeting of spillovers for food security.

Promising Approaches – More Research Needed

Dual-purpose crops have great potential to efficiently increase crop production

Dual-purpose crops can produce food, fuel and fodder very efficiently as several products can be generated from the same inputs of water, fertilizer, labour and land, increasingly important in future for the sustainable intensification of agriculture (see Royal Society, 2009). Policy makers should give priority to policies that support the expansion of dual-purpose crops and donors should increase investments in the science needed to make them more productive and resource efficient.

Alternative cropping systems may contribute more to local food security

Intercrops or polycultures contribute to household food and nutritional security but generally make limited contribution to national food security. But very little new research has been done on the ecology, biology, functionality or productivity of mixed cropping systems in the past 20 years. Hence, the wider application of these cropping systems, except as gardens, requires much further research before it should be recommended to policy makers and investors as a widely applicable strategy for food security.

Home gardens, however, have been proven to make an important contribution to family nutrition, food security and cash income (Landauer and Brazil, 1990; Spielman and Pandya-Lorch, 2009). For example, the homestead food production programme in Bangladesh has reached 5 million poor people and contributed to combating micronutrient deficiencies that can be major causes of diseases among women and children. This model has great potential to spill over to other developing countries where home gardens are an appropriate strategy for improving household nutrition and food security but it cannot replace monocultures for staple food production.

GM crops may be integral to future integrated pest management strategies for food security

Review of extensive scientific knowledge from worldwide research and commercial cultivation over 10 years has not provided any sound scientific evidence that the presently commercialized insect-resistant GM crops have caused any environmental harm at either field or landscape levels (Romeis *et al.*, 2008, 2009; see Chapter 8 this volume). Where *Bt* crops have been deployed with an associated decline in insecticide use, biological control organisms, hence ecosystem services, have benefited significantly. Furthermore, there have been spillover advantages for nearby non-GM varieties as reduced pest populations means reduced pesticide use (Black, 2010). Therefore there is great potential to sustainably and profitably improve food production by fostering insect-resistant GM crop-based integrated pest management systems.

Planned vegetational diversity may enhance pest management for food security

Planned vegetational diversity through agri-environmental schemes is expected to conserve biodiversity and protect natural resources but, if such schemes are to be effectively used to manage crop pests, sound understanding of their ecology is needed for informed management decisions (Marshall, 2002; see Chapter 8, this volume). Planned deployment of crops and varieties on farms such as rotations and crop–livestock systems are common. Their value in terms of nutrient management is widely recognized but their role in managing harmful pests is less well documented. Even for the much researched system of push–pull involving intercrops and grass borders to manage stemborers and the weed striga in East Africa (Khan *et al.*, 2000), there is still doubt as to whether it delivers the benefits claimed (van den Berg *et al.*, 2006).

Planned vegetational diversity based on scientific understanding of crop–pest interactions can make a valuable contribution to improved pest management. However, each agricultural situation must be assessed separately since pest–crop interactions will vary depending on the pest, crop, associated vegetation, associated beneficial biodiversity, location and size of field, climate and cultural practices. The enormity of this challenge helps to explain why very limited new research has been done in the past 20 years on the beneficial role of natural vegetation associated with crops in managing harmful biodiversity (Neueschwander *et al.*, 2003). Further support to this approach will depend on availability of resources.

Sound soil management is likely to improve agroecosystem functioning

Soils contain more species diversity than any other terrestrial habitat but it is difficult to see, to measure, to value accurately, to ascribe critical functions to, and much of it is dormant, unless awakened by disturbance such as tillage as noted in Chapter 9, this volume. These difficulties, however, have not prevented the development of the widely held belief, especially outside science, that greater soil biodiversity contributes to improved agroecosystem functioning and sustainability. And, despite the claims of the anti-agricultural intensification lobby that intensification leads to soil biodiversity loss and impaired sustainability (McIntyre *et al.*, 2009), sound, science-based evidence to support this mantra is very difficult to find. Evidence for a relationship between soil biodiversity and sustainable agroecosystem functioning is at best anecdotal and scattered (Brussaard *et al*., 2007).

Despite a plethora of theory and publications that support positive correlations between species, functional diversity and ecosystem functioning, there is an equally widespread acceptance of the redundancy hypothesis. In fact, soil assemblages show a large degree of redundancy (Giller *et al*., 1997; Wardle, 2006). However, this issue has been down-played by ecologists and biologists because it is a political 'hot potato' – redundancy implies excess biodiversity (Welbaum *et al*., 2004) in agroecosystems!

The main dangers are that: (i) unqualified promotion of the redundancy hypothesis could lead to reduction in the support needed to further understand the relationship between soil biodiversity and agroecosystem functions in the short-term; and/or (ii) over-stating a positive relationship between soil biodiversity and ecosystem functioning could lead to withdrawal of future support to soil biodiversity science in the longer term when such unsupported claims are exposed as flawed. While studies have shown that certain soil management practices can increase soil biodiversity, we should not conclude that higher microbial diversity or biomass drives or controls increases in productivity and food security (see Chapter 9, this volume). Much more research is needed before specific forms of soil agrobiodiversity management can be recommended in preference to or exclusion of proven, judicious soil and crop management practices.

Lessons learned from proven, successful management of agrobiodiversity to achieve food security

The world has already achieved great successes through agrobiodiversity management that has substantially contributed to food security. These successes provide insights that are important for policy makers and investors in agriculture for future research on agrobiodiversity management. A common thread running through many of the proven approaches detailed in this book is the confluence of *science*, *policy* and *leadership* (Neuenschwander, 2004; Spielman and Pandya-Lorch, 2009). The application of *science* and technology to agricultural development is a common determinant of success. Long term and sustained investment in agrobiodiversity management is therefore vital to developing-country agriculture and future food security. The likelihood of success in increasing food production further increases with the right incentives and *policies* that encourage farmers, entrepreneurs and companies to invest in agriculture. Most successes also involve partnerships among diverse actors: research institutes, community-based organizations, private companies, government agencies and international bodies. Community involvement and dedicated *leaders* who will lead initiatives even in the face of serious challenges and mobilize the required political and financial support are also essential ingredients for success. Finally, creating an environment that encourages *leadership* is important to creating success (Spielman and Pandya-Lorch, 2009).

A review of major successes in African agriculture including cassava breeding and pest control in Africa, development of high-yielding maize in East and Southern Africa, and export horticulture in East Africa (also discussed in Chapter 5, this volume), shows two key convergent determinants of or preconditions for successful agricultural performance: (i) sustained productivity enhancing research; and (ii) favourable market incentives for farmers and agribusinesses (Haggblade and Hazell, 2010). But to achieve these requires: both sustained investment in research and development over time; effective extension, input supply and credit systems that enable farmers to access needed inputs such as improved varieties (seed or planting material) and fertilizers; positive market incentives; sufficient infrastructure to facilitate market access; and marketing and pricing policies that encourage trade, storage and processing. Most importantly, all these factors must come together in a coordinated way, a daunting challenge for agricultural policy makers. No matter how good the new technology, it will have no impact if the correct policies are not in place.

The Bad and Ugly: Unproven or Flawed Ideas as Threats to Food Security

Any outstanding success in agricultural research for development such as the Green Revolution will be attacked from two sides: on one side by those saying it was not necessary (the 'counter-revolutionaries'); and on the other side by those saying they could do it better (the 'anti-revolutionaries'). We shall look briefly at examples of counter-revolutionaries and then comment on a classic anti-revolutionary manifesto – the International Assessment of Agricultural Knowledge, Science and Technology for Development (IAASTD), a 5-year, US$15 million, multi-authored attempt to provide a roadmap for future food security, specifically targeted at policy makers (McIntyre *et al.*, 2009; see Chapter 11, this volume). We classify the 'bad' as bad science – very obvious in the IAASTD editorial process. The 'ugly' is overt attempts to prevent efficient agricultural production, often by international conservation corporations and their national franchises. Our perspective is of lessons and cautions for policy makers.

The agricultural counter-revolution

The Green Revolution was a resounding success in addressing global food security. However, some 'counter-revolutionary' entities did not want it to happen and moved mountains to prevent further success. With reference to Africa, Paarlberg (2008) noted 'surprising hostility to scientific advances in farming among some newly influential members of the international NGO community'. For example, the NGO 'Food First' in California had claimed that there was enough food for all: they are named after their book – *Food First: Beyond the Myth of Scarcity* (Lappe and Collins, 1977). As described by Paarlberg (2008), Food First now endorses 'the non-productive, science-starved smallholder farming systems that operate in most of rural Africa today.' The present Food First target, expectedly, has attacks on GM crops for developing countries. But the actions of Food First hide agricultural protectionism. Lappe and Collins (1977: p. 254) specifically notes the 'The Mexican Connection': Mexico exports to the USA asparagus, cucumber, aubergine, squash, tomatoes, strawberries and cantaloupes, damaging US agriculture. Another North American NGO – RAFI (now ETC Group) in Canada – has tried for over 30 years to block agricultural research for development, specifically trying to close down the Consultative Group for International Agricultural Research institutes and move their funding to FAO in Rome (ETC, 2009), a certain way of jeopardizing global food security. The other anti-development strategy of RAFI is to insist on farmers in developing countries using their own farm-saved seed, based on the belief that varieties become 'locally adapted' and therefore better (although, in fact, farm-saved seed deteriorates in many ways and is regularly replaced by farmers). However, RAFI's belief allows it to attack plant breeding, multinational seed companies and, again as expected, GM crops. It should be noticed that both Food First and RAFI are located in crop exporting countries that would benefit from reduced agricultural competition from elsewhere.

Now forewarned, policy makers will no doubt recognize other demands from 'anti-revolutionary' institutes, NGOs and the organic farming industry to turn back the agricultural clock and prevent the results of agricultural research for development reaching poor consumers. When attempted beyond national frontiers, as it typically is, it constitutes 'trans-national Luddism': trying to wreck the economies of competing countries by holding back development.

An anti-revolutionary manifesto: the IAASTD as a threat to food security

We class the IAASTD (see McIntyre *et al.*, 2009) as anti-revolutionary as it promotes new and questionable approaches to replace the Green Revolution. But the IAASTD failed badly as it tried to sell a dubious message. The way it failed offers a clear warning to policy makers of '*caveat emptor*' – or, closer to home: '*Don't buy a pig in a poke*'. The IAASTD is highly biased and presents one side only of a long-running dispute between environmentalists on one side and agricultural scientists (and also most farmers) on the other side. We wish to emphasize strongly that it is part of a pattern: criticism of modern farming is cherry-picked or even invented – with continually repeated and exaggerated 'facts'. The generic approach of the IAASTD is a critique of 'industrial agriculture' as risky and unsustainable and in need of replacement. Note that this name-calling and subjective criticism is to be set against the continued and objective increase in production of modern farming.

In one glaring example of bias, IAASTD (2009) reported that 75% of the genetic base of crops has been lost, with the implication that modern plant breeding has caused this loss. However, this claim is not found elsewhere in the IAASTD reports: why should it be edited into a summary? In any case, there is no factual evidence for this figure whatever – it seems to be a scaremongering device originating from RAFI and the FAO, both with interests in maintaining the deception of massive and damaging genetic erosion. Despite being quite wrong, characteristically it is widely repeated, for example: 'the widespread use of genetically uniform modern crop varieties has caused agricultural crops to lose about 75% of their genetic diversity in the last century.' (WWF, no date). In fact, a recent survey of rice in South and East Asia by agricultural scientists (Ford-Lloyd *et al.*, 2009) showed that 'over many decades, contrary to popular opinion, we have been unable to detect a significant reduction of available genetic diversity in our study material.'

Another error of IAASTD was its promotion of 'agroecology' as an ill-defined alternative to traditional and conventional agriculture and an explicit substitute for Green Revolution agriculture (notably in the IAASTD 'Issues in Brief: Towards Multifunctional Agriculture for Social, Environmental and Economic Sustainability': this also favours 'indigenous' crops – see Chapter 4 for a rebuttal of this). There are many, similar, calls for funding centred in this non-existent space: organic, eco-agriculture, ecosystem agriculture and agroecology. Rather than getting on with the job and adding to food production by legitimate competition with other methods of farming, 'agroecology' seeks to demonize and then replace conventional agriculture – of course, with a suitable transfer of research funds to the proponents of the new and, as yet, untried discipline.

The IAASTD cherry-picks the discipline of ecology to justify its take on 'agroecology'. However, this biased approach fails. Using the IAASTD definition of agroecology as 'The science of applying ecological concepts and principles to the design and management of sustainable agroecosystems', we can readily justify the monocultural farming of introduced soybeans in Brazil (see Chapters 3 and 4, this volume). Both monocultures and also introduced crops are ecologically rational, based on comprehensive knowledge of how components of agrobiodiversity interact through competition and pest and disease management. In contrast, the IAASTD relies on unproven and generic claims that 'agroecology' is based on 'ecological principles': perhaps so, but the few principles chosen from a cast of many are irrelevant or dangerous to food security.

We suspect that the viral nature of agroecology being propagated by NGO websites on the release of the IAASTD report is inversely related to the value of agroecology for food production. This could provide a useful rule for screening requests for funding: the wider the promotion by NGOs, the greater the need for intense technical scrutiny to save people from going to bed hungry. But a serious question for policy makers is this: who is funding these anti-science NGOs and for what purpose?

Unintended (?) consequences

There are growing numbers of agrobiodiversity projects – needing funding and therefore oversight by policy makers – whose consequences may not made explicit, or even guessed by their proposers. These could be a minefield for funding agencies as things go wrong and questions begin to be asked as to just why a project was funded.

We have pointed to some dubious

approaches above: here we list a selection, contrasting their claimed benefits with what could actually happen.

- There are very many environmental NGOs opposing the release of genetically modified crops in Centres of Origin and diversity of crop species. The claimed benefit is that that improved GM crops could 'contaminate' Centres of Origin that continue to provide genetic resources for plant breeding – for example, maize from Mexico or wheat from Ethiopia. Further, there is the belief that somehow (unspecified) this 'contamination' is a bad thing. However, there is no evidence that traces of GM genetic material can in any way harm traditional varieties or reduce their value to plant breeders. What actually happens is that farmers in Centres of Crop Diversity – already poverty hotspots – are denied the latest technology.
- A related problem, with the on-farm conservation of traditional varieties, equally denies farmers access to improved crops (see Chapter 10, this volume). Farmers are expected to go on growing a suite of supposedly locally adapted traditional crops and varieties under the questionable claim that they evolve useful features over time. But there is no actual evidence of this, despite many on-farm conservation projects completed and many more in the pipeline (a major target of the implementation of the FAO Seed Treaty). The downside for farmers near and far is that traditional varieties harbour pests and diseases and these can become more virulent or aggressive and invade other areas over time. On this issue, conservation policy is seriously adrift.
- The Government of Norway has provided a permafrost seed store (on the Arctic island of Svalbard), which has become the largest store of agrobiodiversity in the world (see Chapter 10, this volume). This store is generating problems for national policy makers. As a condition of depositing safety duplicates in Svalbard, original samples in depositing genebanks must be placed under the terms of the FAO Seed Treaty. However, as anyone can deposit any seed sample whatever, the coverage of the Seed Treaty has rapidly expanded well beyond the intentions of the Treaty and its members. Far more seriously, others have deposited over 95,000 samples of Mexican origin, although Mexico is not a member of the Treaty and will not benefit from the use by others of the Mexican samples in the Treaty. Mexico is being penalized by the Treaty for its past benevolence as the largest supplier of samples for global agriculture. As a consequence of the Treaty and Svalbard, inter-country movement of crop genetic resources, the foundation of past, present and future crop breeding, has dried up.
- One of the 'great debates' has been on the need for diversity/complexity in agriculture. We come down firmly on the side of managed – and productive – simplicity. Yet the expected promotion of 'Biodiversity for ecosystem services' will insist to policy makers that biological diversity is vital for clean air, clean water and good soils (and for agriculture). In contrast, we feel that well-managed agriculture based on simple natural models can provide all this and also the bulk of our food. On the vital points of control of water runoff and soil erosion, protected areas can do little to help downstream agriculture that cannot be far better done by agricultural terracing and water catchment (see our cover photograph). For example, in Yemen (once known as '*Arabia felix*'), rivers, even during violent storms, very rarely reach the sea. Every drop of water and suspended silt from soil erosion is caught on a series of terraces, diverted through irrigation canals, or held behind temporary bunds on flooded fields which build up fertile soil year on year. Most of the most important crop on Earth – paddy rice – is grown on silt and water from the Himalayas trapped and managed by the long-term skills of rice farmers.
- International conservation corporations are using concerns over deforestation to block or certify vegetable oil crop production in developing countries (oil palm in Indonesia; soybean in Brazil). This lowers farm income in developing countries,

increases prices to consumers, and increases profits for farmers in developed countries (soybean and canola in North America; rapeseed oil in the EU). There are many other examples of this transboundary interference with commodity production in developing countries where it is impossible to disentangle the conservation claim from the actual impact: reduced global commodity trade from developing countries.

Conclusion

We have come a long way from the origins of agriculture more than 10,000 years ago through an incremental series of agricultural revolutions where the advantages of each stage have been incorporated and further developed to feed growing populations. As we have shown, science-based agrobiodiversity management to increase global food production has led to enormous gains in agricultural productivity, food security and human well-being, especially in the past 40 years. These successes are now adequately feeding more than 5 billion people. But these successes are increasingly under attack from counter-revolutionaries who criticize the massive successes such as the Green Revolution and biotechnology-based modern crop breeding and anti-revolutionaries who promote unproven, flawed pseudo-science-based alternatives to modern intensive agriculture.

The 2008 food crisis has forced policy makers and investors in agriculture to critically reassess future approaches to global food production. Some now agree that modern, intensive agriculture will continue to be the most appropriate approach to meeting rising food demand from burgeoning populations while, at the same time, reducing the rate of conversion of natural ecosystems into agricultural land and dealing with climate change. But there is still a long way to go in convincing *all* policy makers, governments and international donors of the critical need to massively increase support for and investment in sound science for future food production, both nationally and internationally. Unfortunately, too many are listening to the counter-revolutionary and anti-revolutionary rhetoric and supporting unproven alternatives to modern intensive agriculture such as so-called agro-ecological approaches, often because they are technically unable to discriminate between sound scientific approaches and pseudo-science. Scientists therefore have an important role to play in making policy makers and investors in agriculture much more *technically* aware of sound scientific approaches and should seize opportunities to participate in policy debates to influence investment decisions on the science that underpins food production:

> By hesitating to enter the debate, we can only accede the field to the biologically naive and find ourselves able to serve only as peripherally significant technicians in the pursuit of the objectives of the uninformed.
>
> Namkoong (1991)

Future growth in agricultural productivity and global food security can continue to be achieved through concerted application of science-based agrobiodiversity management, *provided there is enhanced policy support, and significant increases in government and international donor investment.* Substantial increases in commitment and support for research – sooner rather than later – are essential to enable the food system to cope with both known and unknown challenges in the coming decades to achieve global food security.

We think that the long and successful tradition of agricultural research for development is the best resource for future food security. But this needs continued, secured and increased funding. Precious development funding should not be spent on biodiversity conservation in protected areas. Indeed, unless increased crop productivity is possible, protected areas will need to be surveyed for agricultural land-use and re-integration of human populations.

References

Alston, J.M. (2002) Spillovers. *Australian Journal of Agricultural and Resource Economics* 46, 315–346.

Bailey, R. (2009) Norman Borlaug: the man who saved more human lives than any other has died. Available at: http://reason.com/blog/show/136043.html (accessed 20 March 2010).

Black, R. (2010) GM crops bring cash harvest to non-GM varieties. BBC News: Science and Environment, 8 October 2010. Available at: www.bbc.co.uk/news/science-environment-11496710 (accessed 18 October 2010).

Brookes, G. and Barfoot, P. (2009) *GM Crops: Global Socio-economic and Environmental Impacts 1996-2007.* PG Economics Limited, Dorchester, Dorset, UK, 128 pp. Available at: www.pgeconomics.co.uk/pdf/2009globalimpactstudy.pdf (accessed 5 October 2010).

Brussaard, L., de Ruier, P.C. and Brown, G.G. (2007) Soil biodiversity for agricultural sustainability. *Agriculture, Ecosystems and Environment* 121, 233–244.

Burney, J.A., Davis, S.J. and Lobell, D.B. (2010) Greenhouse gas mitigation by agricultural intensification. *Proceedings of the National Academy of Sciences* doi:10.1073/pnas.0914216107.

Byerlee, D., de Janvry, A. and Sadoulet, E. (2009) Agriculture for development: toward a new paradigm. *Annual Review of Resource Economics* 1, 15–31.

Carpenter, J.E. (2010) Peer-reviewed surveys indicate positive impact of commercialised GM crops. *Nature Biotechnology* 28, 319–321.

ETC (2009) H(a)lf a Loaf. Available at: www.etcgroup.org/upload/publication/pdf_file2/ecom101_romanforum090123sm.pdf (accessed 18 September 2010).

Evans, L.T. (1998) *Feeding the Ten Billion.* Cambridge University Press, Cambridge.

Evans, L.T. (2003) Agricultural intensification and sustainability. *Outlook on Agriculture* 32, 83–89.

Evenson, R.E. and Gollin, D. (2003) Assessing the impact of the Green Revolution, 1960 to 2000. *Science* 300, 758–762.

Fan, S. (2010) Halving hunger: meeting the first Millennium Goal through business unusual. International Food Policy Research Institute Report. Available at: http://dx.doi.org/10.2499/0896295389 (accessed 9 September 2010).

Ford-Lloyd, B., Brar, D., Khush, G., Jackson, M. and Virk, P. (2009) Genetic erosion over time of rice landrace agrobiodiversity. *Plant Genetic Resources* 7, 163–168.

Giller, K.E., Beare, M.H., Lavelle, P., Izac, A.-M.N. and Swift, M.J. (1997) Agricultural intensification, soil biodiversity and ecosystem function. *Applied Soil Ecology* 6, 3–16.

Gould, S.J. (1997) An evolutionary perspective on strengths, fallacies, and confusions in the concept of native plants. In: Wolschke-Bulmahn, J. (ed.) *Nature and Ideology: Natural Garden Design in the Twentieth Century.* Dumbarton Oaks, Washington, DC, pp. 11–19.

Gressel, J. (2008) *Genetic Glass Ceilings: Transgenics for Crop Biodiversity.* Johns Hopkins University Press, Baltimore, Maryland.

Grime, J.P. (2002) Declining plant diversity: empty niches or functional shifts? *Journal of Vegetation Science* 13, 457–460.

Haggblade, S. and Hazell, P.B.R. (2010) Successes in African agriculture: lessons for the future. *International Food Policy Institute Issue Brief* 63.

IAASTD (2009) *Summary for Decision Makers of the Global Report.* Island Press, Washington, DC.

James, C. (2010) Global status of commercialized biotech/GM crops in 2009: ISAAA Brief 41. Available at: www.isaaa.org/resources/publications/briefs/41/executivesummary (accessed 26 January 2010).

Khan, Z.R., Pickett, J.A., Van Den Berg, J., Wadhams, L. and Woodcock, C.M. (2000) Exploiting chemical ecology and species diversity: stem borer and striga control for maize and sorghum in Africa. *Pest Management Science* 56, 957–962.

Landauer, K. and Brazil, M. (eds) (1990) *Tropical Home Gardens.* United Nations, University Press, Tokyo.

Lappe, F.M. and Collins, J. (1977) *Food First: Beyond the Myth of Scarcity.* Houghton Mifflin, Boston, Massachusetts.

Lenné, J.M. and Wood, D. (1999) Vegetational diversity in agroecosystems: a mixed blessing for successful pest management? In: Terry, P.J. (ed.) *International Crop Protection: Achievements and Ambitions.* 1999 British Crop Protection Council Proceedings No. 73, pp. 75–98.

Lipton, M. and Longhurst, R. (1989) *New Seeds and Poor People.* Unwin Hyman, London.

Marshall, E.J.P. (2002) Introducing field margin ecology in Europe. *Agriculture, Ecosystems and Environment* 89, 1–4.

McIntyre, B.D., Herren, H.R., Wakhungu, J. and Watson, R.T. (eds) (2009) *Agriculture at the Crossroads. The global report of the International Assessment of Agricultural Knowledge, Science and Technology.* Island Press, Washington, DC.

Morin, L., Reid, A.M., Sims-Chilton, N.M., Buckley, Y.M., Dhileepan, K., Hastwell, G.T., Nordblom, T.L. and Raghu, S. (2009) Review of approaches to evaluate the effectiveness of weed biological control agents. *Biological Control* 51, 1–15.

Namkoong, G. (1991) Biodiversity issues in genetics, forestry and ethics. *The Forestry Chronicle* 68, 438–443.

Neuenschwander, P. (2004) Harnessing nature in Africa: biological pest control can benefit the pocket, health and the environment. *Nature* 432, 801–802.

Neuenschwander, P., Borgemeister, C. and Langewald, J. (2003) *Biological Control in IPM Systems in Africa.* CAB International, Wallingford, UK.

Paarlberg, R.L. (2008) *Starved for Science: How Biotechnology is Being Kept Out of Africa.* Harvard University Press, Massachusetts.

Pardey, P.G. and Pingali, P.L. (2010) *Reassessing International Agricultural Research for Food and Agriculture.* Global Conference on Agricultural Research for Development 2010, Background Paper. Available at: www.gcard2010.net (accessed 30 March 2010).

Pardey, P.G., Alston, J.M. and Piggott, R.R. (eds) (2006) *Agricultural R&D in the Developing World: Too Little, Too Late?* International Food Policy Research Institute, Washington, DC.

Raudsepp-Hearne, C., Peterson, G.D., Tengo, M., Bennett, E.M., Holland, T., Benessaiah, K., MacDonald, G.K. and Pfeifer, L. (2010) Untangling the environmentalist's paradox: why is human well-being increasing as ecosystems degrade? *BioScience* 60, 576–589.

Reynolds, M.P. and Ortiz, R. (2010) Adapting crops to climate change: a summary. In: Reynolds, M.P. (ed.) *Climate Change and Crop Production.* CAB International, Wallingford, UK. In press.

Romeis, J., Van Driesche, R.G., Barratt, B.I.P. and Bigler, F. (2008) Insect-resistant transgenic crops and biological control. In: Romeis, J., Shelton, A.M. and Kennedy, G.G. (eds) *Integration of Insect-Resistant Genetically Modified Crops within IPM Programs.* Springer Science + Business Media B.V., pp. 87–117.

Romeis, J., Meissle, M., Raybould, A. and Hellmich, R.L. (2009) Impact of insect-resistant transgenic crops on above-ground non-target arthropods. In: Ferry, N. and Gatehouse, A.M.R. (eds) *Environmental Impact of Genetically Modified Crops.* CAB International, Wallingford, UK, pp. 165–198.

Royal Society (2009) *Reaping the Benefits: Science and the Sustainable Intensification of Global Agriculture.* RS Policy Document 11/09, Royal Society, London.

Sarewitz, D. (2010) Advancing the science of science and innovation policy. Testimony before the US House of Representatives Committee on Science and Technology. Available at: http://democrats.science.house.gov/Media/file/Commdocs/hearings/2010/Research/23sep/Sarewitz_Testimony.pdf (accessed 15 September 2010).

Spielman, D.J. and Pandya-Lorch, R. (2009) *Millions Fed: Proven Successes in Agricultural Development.* International Food Policy Research Institute, Washington, DC.

van den Berg, J., de Bruyn, A.J.M. and van Hamburg, H. (2006) Oviposition preference and survival of the maize stem borer, *Busseola fusca* (Lepidoptera: Noctuidae), on Napier grasses, and maize. *African Entomology* 14, 211–218.

Van Driesche, R., Hoddle, M. and Center, T. (2008) *Control of Pests and Weeds by Natural Enemies: An Introduction to Biological Control.* Blackwell Publishing, Oxford.

Wardle, D.A. (2006) The influence of biotic interactions on soil biodiversity. *Ecology Letters* 9, 870–886.

Welbaum, G.E., Sturtz, A.V., Dong, Z. and Nowak, J. (2004) Managing soil microorganisms to improve productivity of agro-ecosystems. *Critical Reviews in Plant Sciences* 23, 175–193.

Wood, D. and Lenné, J.M. (eds) (1999) *Agrobiodiversity: Characterization, Utilization and Management.* CAB International, Wallingford, UK.

Wood, D. and Lenné, J. (2001) Nature's Fields: a neglected model for increasing food production. *Outlook on Agriculture* 30, 165–174.

WWF (no date) Arguments for protection: food. Available at: wwf.panda.org/what_we_do/how_we_work/protected_areas/arguments_for_protection/goods_services/food (accessed 7 September 2010).

Index

Page numbers in **bold** refer to information in tables, figures or boxes.